Vauxh
Viva H
Owners Workshop Manual

by J H Haynes
Member of the Guild of Motoring Writers

and D H Stead

Models covered:

Viva two-door saloon
Viva De-Luxe two-door saloon
Viva De-Luxe two-door estate car
Viva De-Luxe four-door saloon
Viva SL two-door saloon
Viva SL two-door estate car
Viva SL four-door saloon
Viva 1300
Viva 1300L
Viva 1300 SL
Viva E (saloon only) and GLS
October 1970 onwards, fitted with 1159 cc
1159 cc extra performance and 1256 cc ohv engines

ISBN 90055 047 3

2510/047

Printed in England

Haynes

HAYNES PUBLISHING GROUP
SPARKFORD YEOVIL SOMERSET ENGLAND
distributed in the USA by
HAYNES PUBLICATIONS INC
861 LAWRENCE DRIVE
NEWBURY PARK
CALIFORNIA, 91320
USA

Acknowledgements

Thanks are due to General Motors for the assistance given in the supply of technical material and illustrations. To Castrol Limited for advice on lubrication. The Champion Sparking Plug Company for the illustrations showing the various spark plug conditions. The bodywork repair photographs used in this manual were provided by Lloyds Industries Limited who supply 'Turtle Wax', 'Dupli-color Holts', and other Holts range products.

Lastly, thanks are due to all of those people at Sparkford who helped in the production of this manual. Particularly, Peter Ward, who originated Chapter 13, Ted Frenchum and Stanley Randolph for designing the layout of each page and John Rose for the editorial work.

Whilst every care is taken to ensure that the information in this manual is correct no liability can be accepted by the authors or publishers for loss, damage or injury caused by any errors in, or omissions from, the information given.

Photographic captions and cross references

The book is divided into thirteen chapters. Each chapter is divided into numbered sections which are headed in **bold type** between horizontal lines. Each section consists of serially numbered paragraphs.

There are two types of illustration: (1) Figures which are numbered according to Chapter and sequence of occurrence in that chapter and having an individual caption to each figure. (2) Photographs which have a reference number in the bottom left hand corner. All photographs apply to the chapter in which they occur so that the reference figures pinpoint the pertinent section and paragraph numbers.

Procedures, once described in the text, are not normally repeated. If it is necessary to refer to another chapter the reference will be given in chapter number and section number thus: Chapter 1/6.

If it is considered necessary to refer to a particular paragraph in another chapter the reference is eg 'Chapter 1/6:5'. Cross references given without use of the word 'Chapter' apply to sections and/or paragraphs in the same chapter, eg, 'see Section 8' means also 'in this Chapter'.

When the left or right hand side of a car is mentioned it is as if one were looking in the forward direction of travel.

Chapter 13 is in the form of a Supplement, which has been added to bring this manual up-to-date by describing all modifications and changes applicable to the 1256 cc models built after September 1973. Wherever possible, modifications have also been incorporated into the existing Chapter.

1256 VIVA SL

VIVA SL ESTATE

Introduction

This owners workshop manual is intended for the owner of an HC series Vauxhall Viva, who, having acquired his car either new or secondhand, does not wish to spend a fortune in keeping it in a safe, economical and good value condition.

No one, however, particularly with all round commitments and interests, wishes to spend all his leisure time in looking after his transport. This manual is designed, therefore, to enable the practically minded owner to maintain his car and at the same time show him the way to carry out most of the repairs that a car normally requires during its life, as economically and quickly as possible.

Modern trends in motor car design and production are towards the fitting of components which are renewed completely if they fail - the economics being that the cost of the time required to repair them is little less than the combined cost of labour and material needed to produce them in the first place. However, this trend has not yet moved things completely out of the owner's control and many items such as starter motors, fuel pumps, distributors, brake cylinders, and so on can be repaired for an expenditure of pence on parts where replacement would cost pounds. Where a component is only obtainable as a complete unit, an owner should be quite able to install it correctly with the aid of this manual.

The HC Viva, the latest model in the range which started production in 1963, shows developments in assembly and design which make access to, and removal of, many parts much simpler than it was on the earlier HA and HB series. For the do-it-yourself owner, careful study of the electrical system, for example, (and all the 'fiddly' bits that one tends to associate with it) will show how it has been simplified in a most sensible manner. The same goes for other parts as well. Modifications and changes applicable only to 1256 cc models built after September 1973 are contained in the supplementary Chapter 13.

Safety is of permanent concern to the responsible owner who appreciates the hazards of increased traffic volume and motorway speeds. All the checks for safety are fully covered in this manual and the conscientious do-it-yourself maintenance man will not just regard these checks as something to pay attention to only when the DOE test becomes a regular necessity. A three year old car that has been totally neglected can be a far more lethal article than a ten year old one that has been properly looked after.

For the more ambitious owner, this manual gives step by step details of all the other repairs and overhauls which we consider are within the capabilities of a practically minded person who is in possession of a reasonably comprehensive set of tools. With regard to the latter, this is one area where only good quality equipment will do. Borrowing of tools is discouraged except for certain special items which may only be used once in a blue moon. Certainly we would say that you should not be annoyed if someone should refuse to lend them. Appreciate how much they cost if lost or damaged - apart from the fact that keen owners regard their tools as particularly personal belongings.

Much of the work involved in looking after a car and carrying out repairs depends on accurate diagnosis in the first place. Where possible, therefore, a methodical and progressive way of diagnosis is presented. The time that can be wasted in hopping from one possible source of trouble to another, suggested at random quite often by self styled 'experts' must have been experienced by many people. It is best to say at the start therefore, 'This could be one of several things - let's get the book out'.

Contents

Routine maintenance

The manufacturers base their own servicing operations on a time rather than mileage factor. They take 12000 miles per annum as an average to base this service plan. This system is very satisfactory as it enables both owner and service station to plan servicing in advance on a regular basis and confirms that deterioration of a vehicle's performance and safety is not necessarily connected with the number of miles covered. Where mileage is consistently and significantly in excess of the average the time intervals between services may be reduced in proportion.

By implication the servicing cycle recommended by the manufacturers gives a 6000 mile interval between engine oil changes. Many owners prefer to change the oil more frequently particularly where much of the driving is in short runs, or stop/start situations, where the engine either does not get many opportunities to warm up completely or operates constantly in heavy traffic. These conditions take far more out of an engine than steady runs along motorways in top gear.

The maintenance information given is not detailed in this Section as the full information is given in the appropriate Chapters of the book.

Weekly

COOLANT LEVEL IN RADIATOR
1 inch (25 mm) below bottom edge of filler neck

ENGINE OIL LEVEL WITH THE DIPSTICK
Level must be above 'Add oil' mark. Quantity required from 'Add oil' to 'Full' is 2 pints.

BATTERY ELECTROLYTE LEVEL
Should just cover the plate separators. Wipe away any moisture or dirt.

TYRE PRESSURES
Also examine tread depth and for signs of damage.

Safety Check Service 'S'

BRAKES
- Master cylinder fluid level
- Hydraulic pipes and hoses inspection
- Wheel cylinder inspection
- Shoes - adjustment
- Handbrake lever setting

STEERING
- Tyre condition
- Front wheel hub bearings adjustment
- Track rods and ball joints damage and/or wear

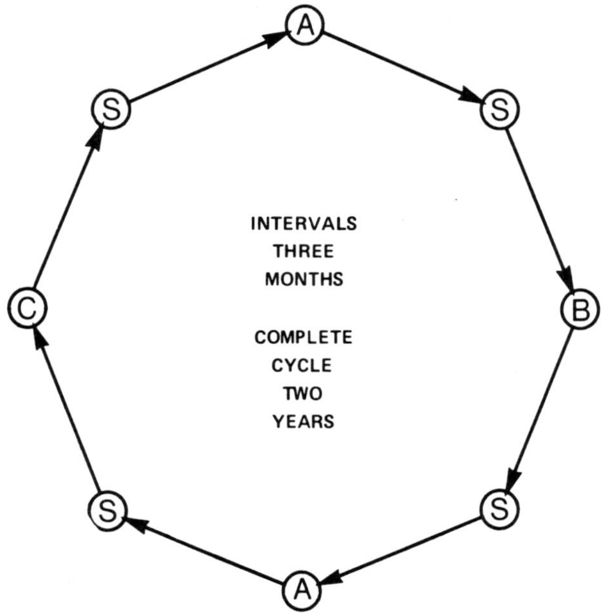

Illustration of the servicing cycle which includes checks under the 'S' service covering all safety requirements for the DOE Test

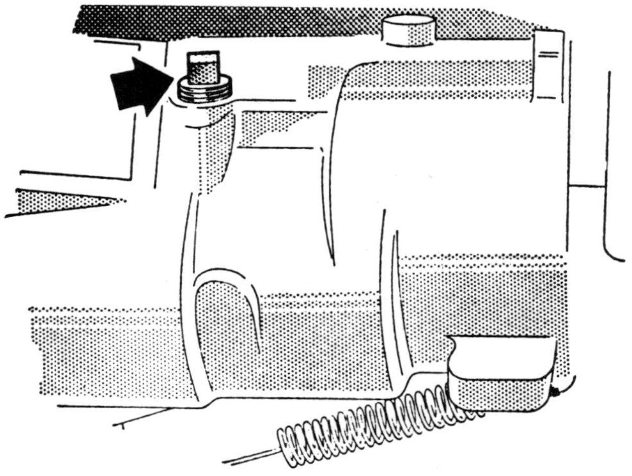

Gearbox filler/level plug

SUSPENSION
Suspension arm upper and lower ball joints - wear
Springs - level and unbroken

AUTOMATIC TRANSMISSION
Fluid level correct

GENERAL
Lights in order
Exhaust system intact
Windscreen wipers - blades serviceable
Seat belts and anchorage points in order

Service 'A'

BRAKES
Examine pads and linings for wear - renew if necessary and adjust.

CLUTCH AND TRANSMISSION
Check and adjust clutch pedal free play
Gearbox oil level - check and top up
Rear axle oil level - check and top up

ENGINE
Renew oil and oil filter element
Clean crankcase ventilator air cleaner
Carburettor air cleaner - clean filter element
Carburettor damper dashpot (Stromberg carburettors) - check oil level
Fuel pump - clean filter
Spark plugs - remove, clean and reset
Distributor contact breaker points - adjust gap. Clean or renew if necessary
Valve clearances - check and adjust
Crankcase ventilation valve - clean and check
Fan belt - adjust tension if required.
Engine idling speed - adjust carburettor if required.

Service 'B'

Service 'A' to be carried out with the following additons:
BRAKES
Disc brake servo filter - renew

SUSPENSION
Grease upper and lower arm ball joints

AUTOMATIC TRANSMISSION
Clean all ventilation holes and slots around the torque converter cover. Scrape all accumulations of dirt from surrounding areas also.

ENGINE
Renew spark plugs
Renew paper element in air cleaner

Service 'C'

Service 'B' to be carried out with the following additions:

STEERING
Remove front wheel bearings - clean and repack with grease.

BRAKES
Renew hydraulic fluid

AUTOMATIC TRANSMISSION
Renew fluid

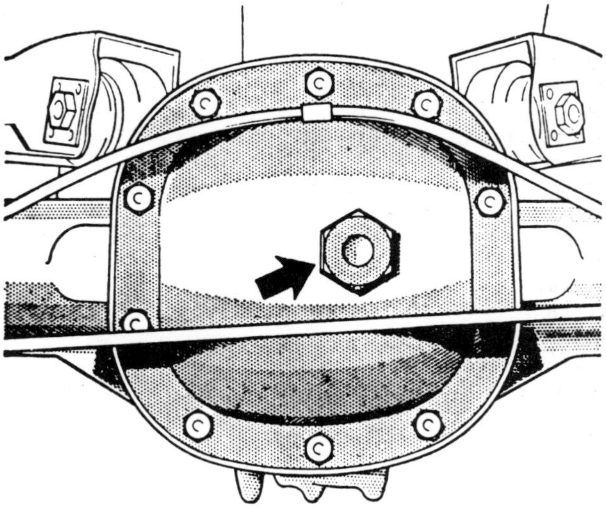

Rear axle filler/level plug

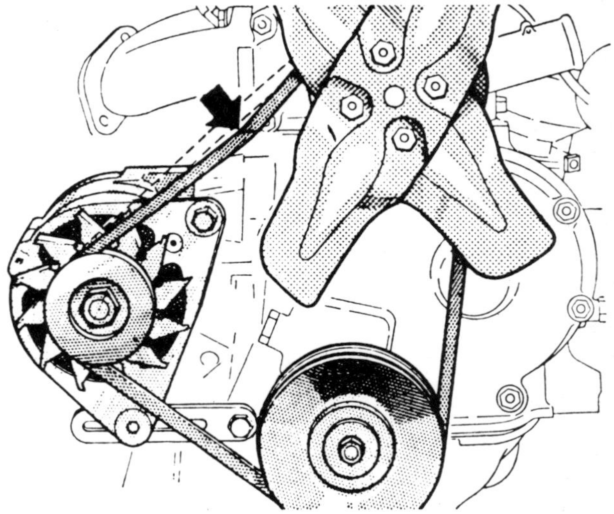

Fan belt adjustment

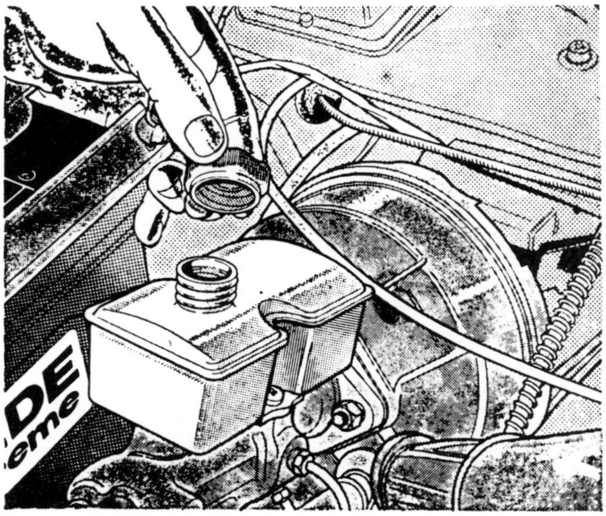

Hydraulic fluid reservoir

Recommended lubricants

COMPONENT	TYPE OF LUBRICANT OR FLUID	CORRECT CASTROL PRODUCTS
ENGINE	Multi-grade engine oil 20W/50	Castrol GTX
GEARBOX/FINAL DRIVE	Gear oil of SAE 90EP standard	Castrol Hypoy
FRONT HUBS	Medium Grade multi-purpose grease	Castrol LM Grease
BRAKE CABLES	Medium Grade multi-purpose grease	Castrol LM Grease
DISTRIBUTOR AND GENERATOR BEARINGS	Engine or light oil	Castrol GTX
CONTACT BREAKER CAM	Medium Grade multi-purpose grease	Castrol LM Grease
BATTERY TERMINALS	Petroleum jelly	
CARBURETTOR DASHPOT	Engine oil 10W/30	Castrol 'Everyman'
UPPER CYLINDER LUBRICANT		Castrollo
BRAKE AND CLUTCH MASTER CYLINDERS	Hydraulic fluid SAE 70R3	Castrol Girling Universal Brake and Clutch Fluid
ANTIFREEZE	BSI 3151/3152	Castrol Antifreeze
FRONT SUSPENSION BALLJOINTS	Graphite or Moly grease	Castrol MS3 grease
STEERING GEAR (NOT ROUTINE)	Gear oil to SAE 90EP standard	Castrol Hypoy
AUTOMATIC TRANSMISSION (NOT ROUTINE)	Dexron type	Castrol TQ

Additionally, Castrol GTX can be used to lubricate door locks, strikers and hinges; bonnet and engine compartment locks, hinges and catches; rear screen locks and hinges.

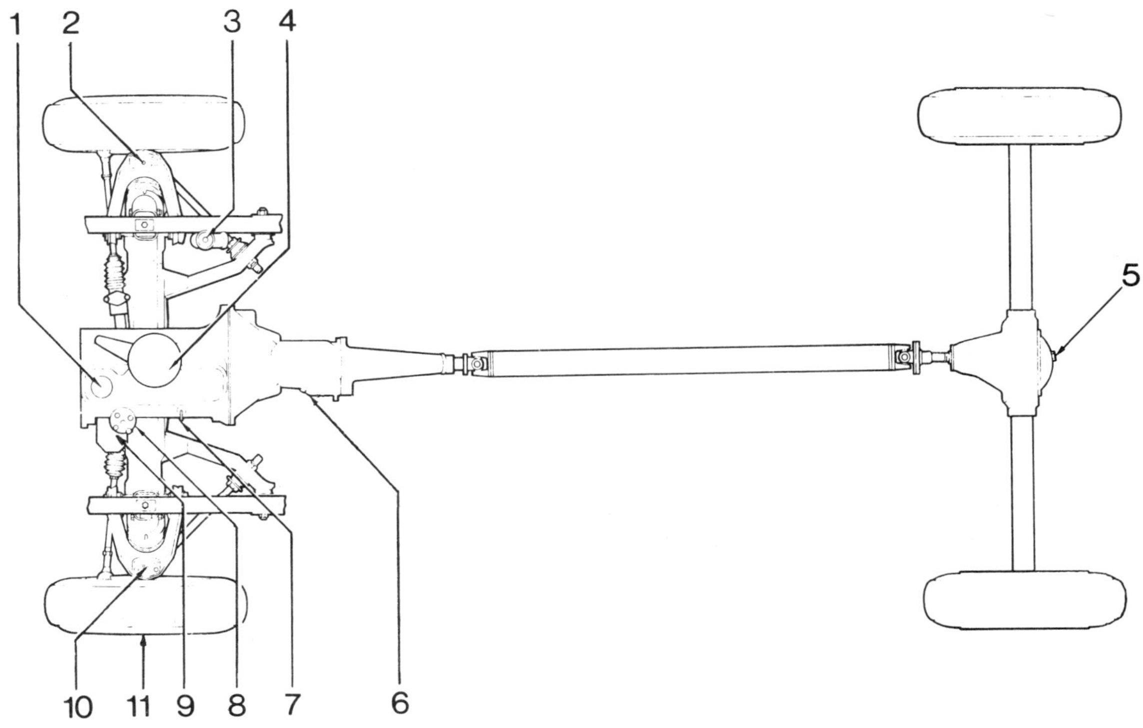

1 Engine Oil
Weekly
Check the oil level using the dipstick (Item 7) and, if necessary, add sufficient Castrol GTX to bring the level up to the 'Full' mark. Two pints are needed to raise the level from 'Add oil' to 'Full'.
Service 'A'
When warm undo the sump drain plug and drain off the engine oil. Renew the oil filter element also. Replace the drain plug and refill with fresh Castrol GTX. Under the adverse conditions encountered in town area operation with frequent stops and starts, particularly in cold weather or where much driving is done over dusty roads, it is recommended that the oil is changed more frequently.
Capacity:— 5½ pints, dry
5 pints, filter element change

2,10 Front suspension arm ball joints - 2 nipples each side
Service 'B'
Apply Castrol MS3 Grease with the grease gun to the top and bottom nipples.

3 Hydraulic brake master cylinder reservoir
Service 'S'
Check fluid level and, if necessary, top up with the recommended fluid to within ¼ inch below the filler neck.

4 Air cleaner
Service 'B'
With a wire gauze type filter element, remove, wash in paraffin, oil with Castrol GTX, drain and replace.

5 Rear axle oil
Service 'A'
Remove filler plug and, if necessary, replenish to the bottom of the filler plug orifice with Castrol Hypoy 90 Gear Oil (Castrol Hypoy is available in 1 pint 'Handipacks' fitted with a filling tube which greatly facilitates this operation). Examine casing for signs of leakage.
Capacity:— 1¼ pints.

6 Gearbox oil
Service 'A'
Remove the filler plug and, if necessary, replenish to the bottom of the filler plug orifice with Castrol Hypoy 90. If for any reason the gearbox has been drained, or a new gearbox fitted, refill with fresh Castrol Hypoy 90 (Castrol Hypoy is available in 1 pint 'Handipacks' fitted with a filling tube which greatly facilitates this operation). Examine casing for signs of leakage.
Capacity:— 0.9 pint

7 Engine oil level dipstick - see Item 1.

8 Distributor
Service 'B'
Remove distributor cap and rotor arm. Impregnate felt pad in top of spindle with a few drops of Castrol 'Everyman' or GTX oil and put a few drops also through the indicated hole. Smear the spindle cam faces sparingly with petroleum jelly.

9 Engine oil filter - see Item 1.

10 As for 2.

11 Front wheel bearings
Service 'C'
Remove bearings, flush with paraffin and repack with Castrol LM Grease.

Buying spare parts and vehicle identification numbers

Buying spare parts

Spare parts are available from many sources, for example Vauxhall garages, other garages and accessory shops, and motor factors. Our advice regarding spare part sources is as follows:

Officially appointed Vauxhall garages - This is the best source of parts which are peculiar to your car and are otherwise not generally available (eg; complete cylinder heads, internal gearbox components, badges, interior trim etc). It is also the only place at which you should buy parts if your car is still under warranty; non-Vauxhall components may invalidate the warranty. To be sure of obtaining the correct parts, it will always be necessary to give the storeman your car's engine and chassis number, and if possible, to take the 'old' part along for positive identification. Remember that many parts are available on a factory exchange scheme - any parts returned should always be clean! It obviously makes good sense to go straight to the specialists on your car for this type of part for they are best equipped to supply you.

Other garages and accessory shops - These are often very good places to buy materials and components needed for the maintenance of your car (eg spark plugs, bulbs, fanbelts, oils and greases, touch-up paint, filler paste, etc). They also sell general accessories, usually have convenient opening hours, charge lower prices and can often be found not far from home.

Motor factors - Good factors will stock all of the more important components which wear out relatively quickly (eg cylinders/pipes/hoses/seals/shoes and pads etc). Motor factors will often provide new or reconditioned components on a part exchange basis - this can save a considerable amount of money.

Vehicle identification numbers

The car identification number is attached to the top of the instrument panel on the left hand side and can be read through the windscreen.

A further identification plate, giving the model designation and service numbers with other code numbers, is fitted to the right hand wheel arch panel in front of the battery.

The engine number is stamped on the block just underneath the front two spark plugs.

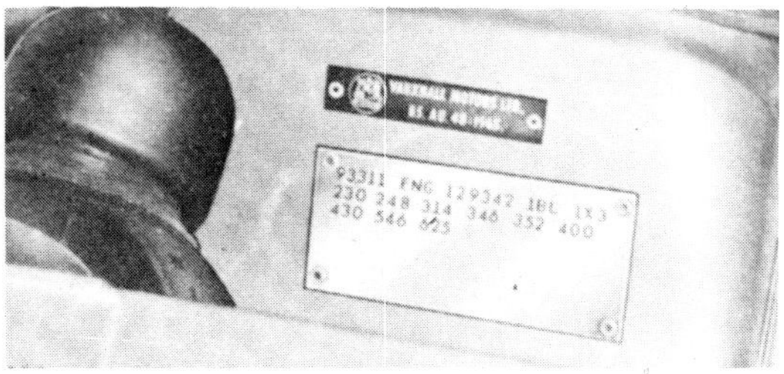

Model designation, serial and code numbers

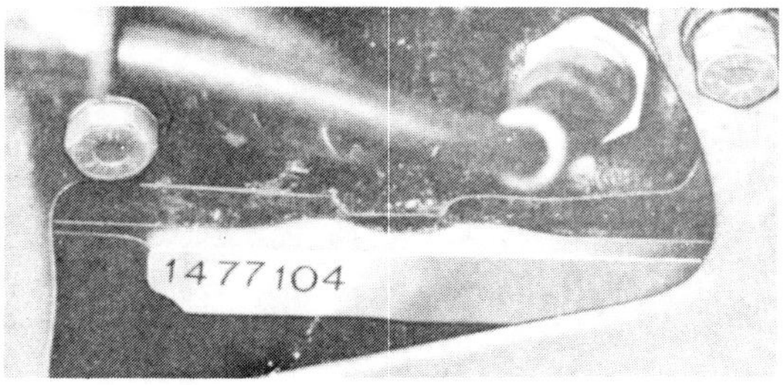

Engine serial number

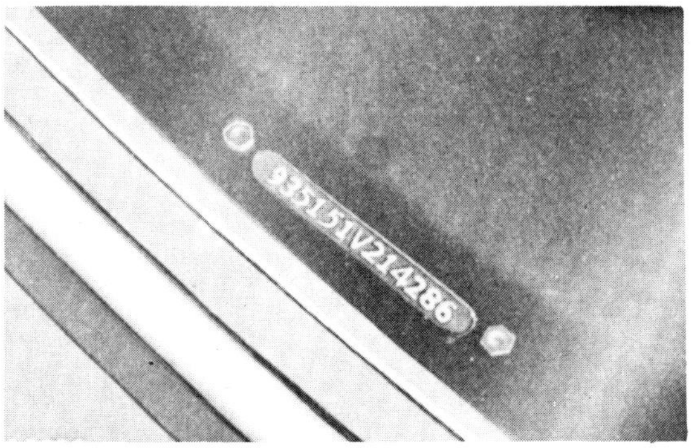

Car Identification Number

Chapter 1 Engine

Contents

Specifications

Engine - General	1159 cc (70.7 cu in)	1256 cc (76.6 cu in)
Type	4 cylinder in line OHV pushrod operated	
Bore	77.7 mm (3.062 in)	81 mm (3.188 in)
Stroke	61.0 mm (2.400 in)	61 mm (2.400 in)
Weight	227 lbs (270 lbs with gearbox) approximately	

Compression ratios

1159 cc Standard	8.5:1
1159 cc Low compression	7.3:1
1159 cc High performance	9.0:1
1256 cc Standard	8.5:1
1256 cc Low compression	7.3:1

Firing order ... 1 3 4 2
(No 1 cylinder at front of engine)

Compression pressure

1159 cc and 1256 cc Standard	125 lb/sq in
1159 cc Low compression	110 lb/sq in
1159 cc High performance	135 lb/sq in

Brake horse power		
1159 cc Standard	56 bhp @ 5400 rpm	
1159 cc Low compression	52 bhp @ 5400 rpm	
1159 cc High performance	69 bhp @ 5800 rpm	
1256 cc	61 bhp @ 5600 rpm	
Engine details - 1159 cc engines		
Camshaft and bearings		
Camshaft drive	Single row endless chain	
Bearings	3 renewable shell type	
Camshaft journal diameter - Front	1.6127 - 1.6132 in	
- Rear	1.5733 - 1.5738 in	
Camshaft clearance in bearing	.0010 to .0025 in	
Camshaft end float	.006 to .013 in	
Cam dimension (minimum) - peak to base up to 1973	Intake	Exhaust
Standard	1.3185 in	1.3125 in
Extra performance	1.322 in	1.3222 in
Cam dimension (minimum) - after 1973		
Standard	1.2980 in	1.2920 in
Extra performance	1.3017 in	1.3017 in
Camshaft thrust plate thickness	.119 into .122 in	
Connecting rods and big end bearings		
Bearing type	Shell	
Bearing material	White metal/lead indium Aluminium/tin (extra performance engines)	
Connecting rod end float on crank pin	.004 in to .010 in	
Bearing housing bore	1.8960 in to 1.8965 in	
Crank pin diameter - Standard	1.7705 in to 1.7712 in	
- Grade 'P'	1.7605 in to 1.7612 in	
Crank pin to bearing clearance	.0010 in to .0029 in	
Crank throw	1.197 in to 1.202 in	
Crankshaft and main bearings		
Bearing type	Shell	
Journal diameter - Standard	2.1255 in to 2.1260 in	
- Grade 'J'	2.1155 in to 2.1160 in	
Journal to bearing clearance	.0010 in to .0025 in	
Crankshaft flange diameter	2.998 in to 3.002 in	
Crankshaft end float	.002 in to .008 in	
Crankshaft run out.	.0015 in maximum	
Main bearing housing bore	2.2835 in to 2.2840 in	
Main bearing centre upper half width	1.287 in to 1.289 in	
Thrust washers	Incorporated in the upper half of centre main bearing shell	
Cylinder block and crankcase		
Type	Cast iron - cylinders cast integrally with upper half of crankcase	
Water jackets	Full length	
Oversize bores	.005 in, .020 in, .040 in	
Permissible distortion on top face - Longitudinal	.005 in maximum	
- Transverse	.003 in maximum	
Minimum block depth after re-facing (top face to bearing cap face)	7.508 in	
Cylinder head and valves		
Permissible distortion on mating face - Longitudinal	.005 in maximum	
- Transverse	.003 in maximum	
- Manifold faces	.002 in maximum	
Permissible head depth - Standard	3.185 in minimum	
- Extra performance	3.157 in minimum	
Porting	Inlet ports on top of head, exhaust ports at side	
Valve stem bore	.2765 in to .2773 in	
Valve seating angle	45°	
Valve seating width - Inlet	.05 in to .06 in	
- Exhaust	.06 in to .08 in	
Rocker stud diameter - Standard	.3535 in to .3543 in	
Rocker stud height above cylinder head top face	1.08 in to 1.12 in	
Rocker ball clearance on stud	.006 in to .0028 in	
Valve seat angle	44°	

Valve stem diameter - Intake	.2748 in to .2755 in
- Exhaust	.2745 in to .2752 in
Valve head thickness - Inlet	.03 in minimum
- Exhaust	.04 in minimum
Valve springs - assembled height	1.34 in maximum
- nominal free length	Early 1.48 in maximum - late 1.50 in
- load at 1.31 in	Early 46 to 54 lbs - late 44 to 52 lbs at 1.32 in
Tappet diameter	.4712 in to .4718 in
Valve timing - Inlet valve open maximum	107° ATDC
Valve clearance - hot	
Standard - Inlet	.006 in
- Exhaust	.010 in
High performance - Inlet and Exhaust	.008 in

Pistons and Piston rings (including gudgeon pins)

Piston clearance in cylinder bore	.0009–.0014 in	
Ring gap in cylinder bore	.009–.014 in	
Ring thickness - Top	.077–.078 in	
- Centre	.077–.078 in	
- Scraper	.1865–.1875 in	
Ring clearance in piston groove - Top	.0019–.0039 in	
- Centre	.0016–.0026 in	
- Scraper	.0015–.0035 in	
Piston size	Grade	Dimensions
Standard	5	3.06085 to 3.06110 in
	6	3.06110 to 3.06135 in
	7	3.06135 to 3.06160 in
	8	3.06160 to 3.06185 in
.005 in oversize	5	3.06585 to 3.06635 in
	8	3.06635 to 3.06685 in
.020 in oversize	5	3.08085 to 3.08135 in
	8	3.08135 to 3.08185 in
.040 in oversize	5	3.10085 to 3.10135 in
	8	3.10135 to 3.10185 in
Gudgeon pin	Semi-floating interference fit in connecting rod	

Lubrication System

Oil pump	
Driving impeller spindle diameter	.4327–.4331 in
Spindle to bush clearance	.0006–.0017 in
Spindle end float	.007–.010 in
Driven impeller spindle diameter	.4303–.4307 in
Spindle fit in body	.0015–.0026 in interference
Impellers	
Teeth backlash	.004 to .008 in
Driving impeller to spindle fit	.0009 to .0021 in interference
Driven impeller to spindle fit	.0003 to .0015 in clearance
End float in body	.002 to .005 in
Radial clearance in body	.002 to .005 in
Drive gear - bore diameter	.3221–.4325
- spindle fit	.0002–.0010 interference
Oil pressure relief valve - plunger diameter	.4353–.4358
- plunger fit in cover	.0012–.0027 clearance
- spring free length	1.92 in
- spring load at 1.66 in	6 lbs 10 oz to 6 lb 14 oz
Oil pressure - hot	35–45 lbs/in^2 @ 3000 rpm
- pressure switch opens at	3–5 lbs/in^2
Oil capacity - total	5½ pints Imp
Oil change capacity	4½ pints Imp
Oil change capacity with new filter element	5 pints Imp

Torque Wrench Settings

Big end bearing cap bolts	25 lb/ft oiled threads
Main bearing cap bolts	58 lb/ft oiled threads
Flywheel bolts	25 lb/ft sealed threads
Cylinder head bolts	49lb/ft (55 lb/ft where Hylomar jointing compound is used)
Valve rocker stud adjusting nuts	3 lb/ft min oiled threads
Oil filter housing bolt	14 lb/ft dry threads

Engine details - 1256 cc engines (up to engine no. 1400000 Sepember 1973). For later models see Chapter 13.

All details for the 1256 cc engine (not included under 'Engine - General') are the same as for 1159 cc high performance versions with the following exceptions.

Cylinder head depth	3.235 minimum
Pistons	Low compression versions marked 'L'

Piston sizes and gradings for the 81 mm bore are not given but are available in the same oversizes for reboring.

Compression pressure at cranking speed - hot

High compression	140 lb/sq in minimum
Low compression	120 lb/sq in minimum

1. General Description

1 Both 1159 cc and 1256 cc engines are over square, four cylinder, overhead valve pushrod operated types, the difference between the two being mainly in the increased cylinder bore size of the later version.

Both have optional low compression ratios and on the 1159 cc engines the low compression engines have a thicker copper asbestos cylinder head gasket fitted. The low compression version of the 1256 cc engine is achieved by using different pistons with a lower crown. The cylinder head gasket is the same.

Two valves per cylinder are mounted at an inclination in line, in a cast iron cylinder head. They are operated by tappets, short pushrods and rocker arms from a camshaft located to the right side of the cylinder bores. Adjustment of valve to rocker clearances is effected by a ball joint pivot on which each rocker arm bears and is adjustable on the mounting stud.

The cylinder block/crankcase casting is distinctive for its width relative to depth, even though the centre line of the crankshaft is at the level of the lower edge of the casting. This is due to the over-square nature of the design and the unusual inlet porting in the head which has the manifold mounted on top, with the exhaust ports coming from the side.

The crankshaft is mounted in three bearings and the top half of the centre bearing shell is flanged to control crankshaft end float.

The camshaft is driven by a single timing chain from the forward end of the crankshaft and a mechanical tension adjuster is fitted.

Pistons have a solid cut-away skirt and have three rings, two compression and one oil control. The gudgeon pin floats in the piston and is an interference fit in the connecting rod.

The centrifugal water pump and cooling fan are V-belt driven from a crankshaft pulley wheel. The distributor is mounted at the left side of the engine and is advanced by centrifugal and vacuum means. There is no vernier control for the static ignition setting.

The oil pump is of the gear type and is driven from the camshaft skew gear.

The clutch is a single dry plate diaphragm type operated mechanically by Bowden cable.

The engine and transmission unit is supported at three points: on each side of the engine between crankcase and chassis frame, and underneath the gearbox casing to a crossmember bolted to the bodyshell.

2 Major operations with engine in place

The following work may be conveniently carried out with the engine in place:

1 Removal and replacement of the cylinder head assembly.
2 Removal and replacement of the clutch assembly.
3 Removal and replacement of the engine front mountings.

The following work can be carried out with the engine in place, but is inadvisable unless there are very special reasons:

4 Removal and replacement of the sump (the front suspension assembly must be removed first).
5 Removal and replacement of big end bearings (after sump removal).
6 Removal and replacement of pistons and connecting rods (after removing cylinder head and sump).
7 Removal and replacement of the timing chain and sprockets (after removal of the sump).
8 Removal and replacement of the oil pump (after removal of the sump).

3 Major operations for which the engine must be removed

1 Removal and replacement of crankshaft and crankshaft main bearings.
2 Removal and replacement of flywheel.
3 Removal and replacement of rear crankshaft oil seal.

4 Methods of engine removal

1 The engine complete with gearbox can be lifted as a unit from the engine compartment. Alternatively, the engine and gearbox can be split at the front of the bellhousing, the gearbox supported and left in position, and the engine removed. Whether or not components like the carburettor, manifolds, generator and starter are removed first depends to some extent on what work is to be done.

5 Engine removal - with gearbox

1 The average do-it-yourself owner should be able to remove the engine fairly easily in about 3½ hours. It is essential to have a good hoist, and two strong axle stands if an inspection pit is not available. Engine removal will be much easier if you have someone to help you. Before beginning work it is worthwhile to get all the accumulated debris cleaned off the engine unit at a service station which is equipped with steam or high pressure air and water cleaning equipment. It helps to make the job quicker, easier and, of course, much cleaner. Decide whether you are going to jack up the car and support it on axle stands or raise the front end on wheel ramps. If the latter, run the car up now (and chock the rear wheels) whilst you still have engine power available. Remember that with the front wheels supported on ramps the working height and engine lifting height is going to be increased.

2 Open the bonnet and prop it up to expose the engine and ancillary components. Disconnect the battery leads and lift the battery out of the car. This prevents accidental short circuits while working on the engine.

3 Remove the windscreen washer pipe from the rear underside of the bonnet (photo).

4 Undo the two nuts and bolts from the bonnet side of the hinges and lift the bonnet off. Place it somewhere safe where it will not be knocked over or bumped into (photos).

5 Drain the cooling system.

6 Remove the sump drain plug and drain the oil out of the engine

5.3

5.4a

5.4b

Remove the windscreen washer pipe, bonnet hinge bolts and bonnet

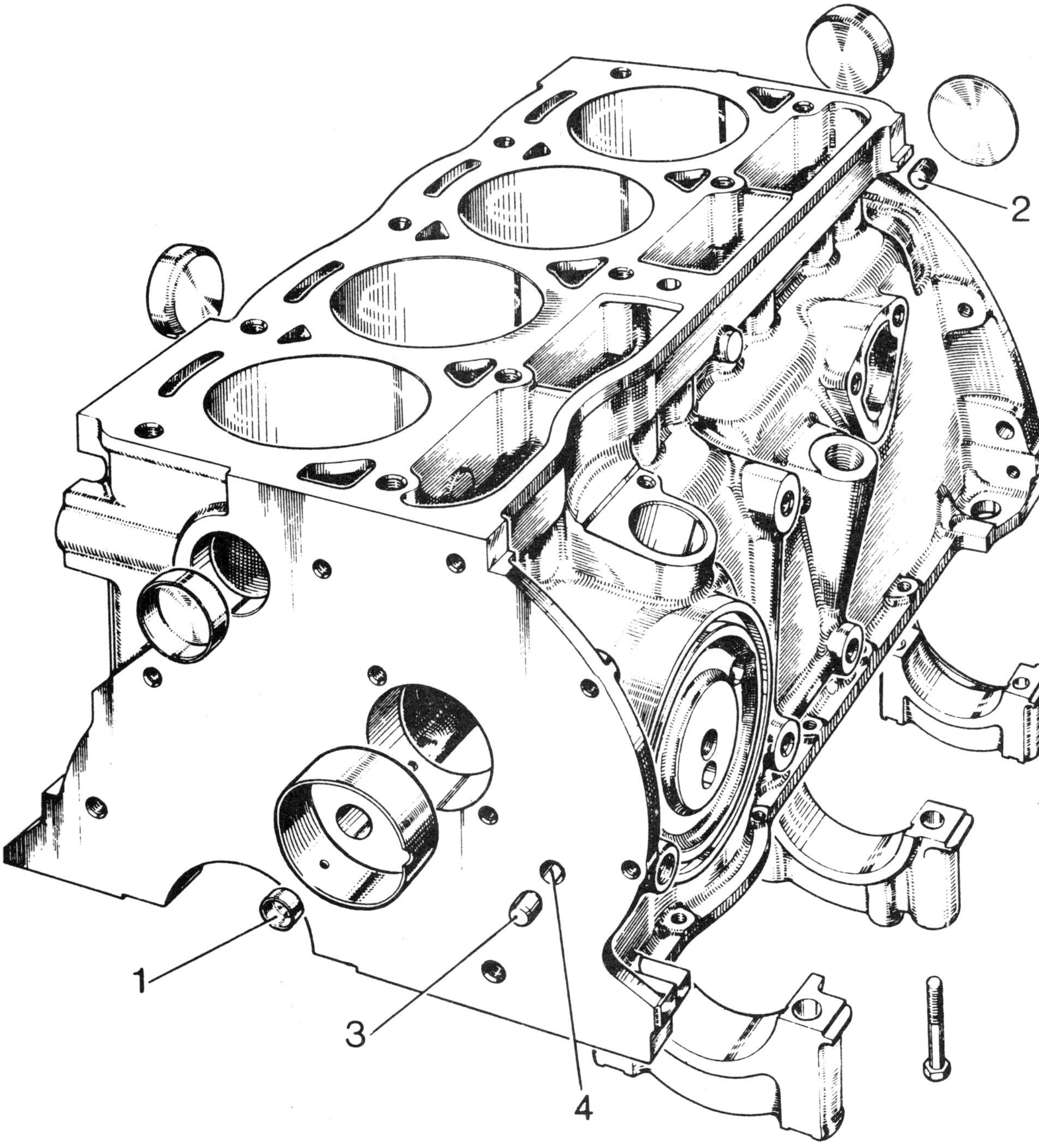

Fig. 1.1. VIVA CYLINDER BLOCK

1 & 2 Oil gallery plugs 3 & 4 Oilway and plug

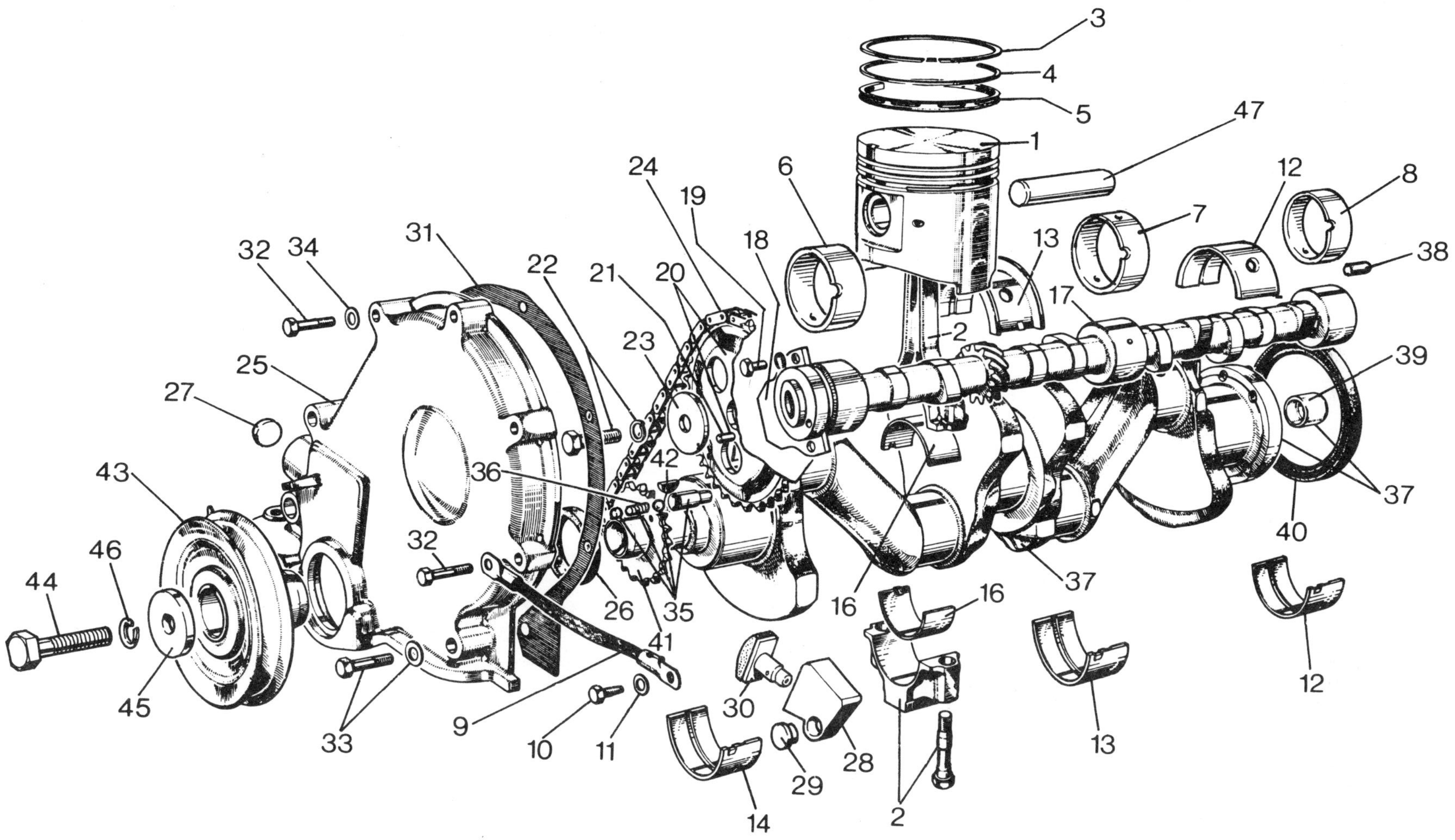

Fig. 1.2. ENGINE INTERNAL COMPONENTS

1	Piston	11	Washer	21	Dowel peg	30	Plunger and slipper head	39	Pilot bush
2	Connecting rod assembly	12	Rear crankshaft bearing	22	Retaining screw and washer	31	Gasket	40	Oil seal
3	Top compression ring	13	Centre crankshaft bearing	23	Washer	32	Bolt	41	Timing wheel
4	Lower compression ring	14	Front crankshaft bearing	24	Timing chain	33	Screw	42	Woodruff key
5	Oil scraper ring	16	Connecting rod bearing	25	Timing case	34	Washer	43	Crankshaft pulley
6	Front camshaft bearing	17	Camshaft	26	Oil seal	35	Oil nozzle assembly	44	Pulley bolt
7	Centre camshaft bearing	18	Camshaft thrust plate	27	Cap plug	36	Nozzle spring	45	Washer
8	Rear camshaft bearing	19	Bolt	28	Chain tensioner body	37	Crankshaft	46	Lockwasher
9	Earth cable	20	Camshaft sprocket	29	Sealing cap	38	Flywheel dowel	47	Gudgeon pin
10	Screw								

into a container of a minimum capacity of six pints.

7 There is no gearbox drain plug. As there will certainly be some small spillage from the gearbox rear cover extension later when the propeller shaft comes out, provide another tray-like receptacle with shallow sides to catch any oil which may drip out before the hole can be plugged. (Makeshift oil receptacles can be made from one-gallon oil tins from which one of the large sides has been cut out).

8 Remove the air cleaner assembly from the carburettor as described in Chapter 3.

9 Remove the radiator and the water hoses which connect it to the engine. If a heater is fitted in the car remove also the hose which connects it to the cylinder head (photo).

10 On the right hand side of the engine compartment the windscreen washer reservoir hangs from two hooks. Remove the cap together with the pipe and then lift out the reservoir. On models with two speed wipers there is a washer pump mounted on the top of a tank.

11 Remove the electrical lead from the terminal marked 'SW' (or +) on the ignition coil.

12 Remove the cap from the distributor and take the leads from the spark plugs. Remove also the HT lead from the centre terminal of the coil. Two of the plug leads should be detached from the clip secured to the inlet manifold.

13 Disconnect the oil pressure switch sender lead from the sender unit on the engine next to the oil filter. Disconnect also the lead to the water temperature gauge fitted at the forward end of the cylinder head.

14 Disconnect the braided earthing wire which connects the engine to the bodyframe by undoing the nut where it is attached to the timing cover (photo).

15 Disconnect the choke and accelerator cables from the carburettor as described in Chapter 3.

16 Remove the starter motor cable from the terminal on the solenoid switch (Chapter 10). It is easier to remove the lead at this terminal than at the starter terminal.

17 Disconnect the exhaust pipe from the exhaust manifold. This is done by removing the two brass nuts. The nuts may be quite easy to move, in which case an open ended spanner is adequate. If tight, however, a ring, socket or box spanner should be used to avoid the possibility of burring over the flats in the soft metal. Do not attempt to force the pipe flange off the manifold studs at this stage (photo).

18 Remove the fuel line from the suction side of the fuel pump (Chapter 3).

19 From inside the car remove the knob from the gear change lever by slackening the locknut underneath it and screwing it off. Lift off the rubber grommet from around the base of the lever.

20 Through a hole in the floor tunnel another rubber cover can be seen round the lever at its base. This covers the screw cap which holds the gear lever in position in the gearbox extension tube. Grip the rubber cover with the cap underneath it and unscrew it.

21 When the cap is unscrewed the gear lever complete with cap, spring and retaining plate may be lifted out.

22 Work may now start on disconnecting the various necessary items under the car. If it has not already been raised on wheel ramps jack the car up at the front so as to give sufficient working clearance underneath and support it on proper stands or blocks. Position the stands under the front crossmember braces where they run back to the bodyframe. Under no circumstances use odd tins, the vehicle jack or other makeshift devices to support the car - it is foolishly dangerous.

23 Undo the knurled screw on the end of the speedometer drive cable where it goes into the gearbox casing at the rear on the right hand side. Take care not to lose the oil seal disc after the cable is removed (photo).

24 Remove the clutch operating cable from the clutch operating lever and bellhousing as described in Chapter 5.

25 Unhook the clutch lever return spring.

26 Remove the nut and large washer from the centre of the crossmember, supporting the gearbox by the rear cover (Chapter 6).

27 The gearbox should next be supported just forward of the crossmember by a jack or blocks.

28 The engine/gearbox unit is now free except for the forward engine mounting brackets which should be detached from the flexible mounting studs by undoing the single nut at each side (photo).

29 Because the engine/gearbox are being taken out of the car together they will have to be tilted to a very steep angle. It is easier, therefore, if the sling for lifting is only a single loop to facilitate tilting the unit. The sling should pass BEHIND (ie the gearbox side) the engine mounting brackets. These, being still attached to the engine, will provide the main lifting lugs.

30 When the mounting bolts have been removed, set the sling so that the lifting hook of the hoist is as close to the engine as possible. A lift of at least 3 feet will be necessary to enable the unit to come out clear of the body (photo).

31 When the weight has been taken by the sling, draw the engine forward and up at the front.

32 Prepare for the disengagement of the propeller shaft which will come off the splines at the rear of the gearbox. If you are short handed put something soft on the floor to absorb the shock as it drops; and provide a tray to catch the oil which will drain from the gearbox as soon as the prop shaft comes out.

33 Once the rear of the gearbox is clear it is then simply a question of raising and tilting the unit until it is finally completely clear of the car.

34 At all times ensure that the sling is secure and not straining against any ancillary parts on the engine which could be damaged.

6 Engine removal - without gearbox

1 Begin by following the instructions in Section 5 from paragraphs 1-25 inclusive.

2 Remove the exhaust manifold from the engine by undoing the six mounting bolts. One of these bolts holds a bracing strip attached to the inlet manifold. Remove the other bolt on this brace and take it off (photo).

3 Remove the bolts securing the engine to the gearbox bellhousing. Some of these are accessible from above and others from below.

4 Two of the bolts also locate the starter motor which should be removed when free (photo).

5 Undo the two forward engine mounting bolts (one each side) which hold the brackets to the flexible mountings.

6 Now support the engine in a sling. As it does not come out at such an acute angle as with the gearbox attached, it requires support. Make sure the hoist hook is as close to the engine as possible when slung, in order to provide maximum lift.

7 Support the gearbox forward of the crossmember by a jack or block. Otherwise, when the engine is drawn away, the full weight of the gearbox will try and pivot forward, imposing a severe strain on the rear mounting.

8 Lift the engine a little and draw it forward so that the clutch is drawn off the gearbox input shaft splines. It is important not to raise or tilt the engine until it is clear, otherwise serious damage could be caused to either the shaft, clutch mechanism or both.

9 Once clear the engine can be lifted up and away.

7 Engine dismantling - general

1 Really keen owners who dismantle a lot of engines will probably have a stand on which to mount them, but most will make do with a work bench. This should be large enough on which to spread the inevitable bits and pieces and tools, and strong enough to support the engine weight. If the floor is the only possible place, try and ensure that the engine rests on a hardwood platform or similar rather than concrete (or beaten earth!!).

2 Spend some time on cleaning the unit. If you have been wise this will have been done before the engine was removed, at a service bay. Good solvents such as 'Gunk' will help to 'float' off caked dirt/grease under a water jet. Once the exterior is clean, dismantling may begin. As parts are removed clean them in petrol or paraffin (do not

5.9 Disconnecting heater hose connection from inlet manifold

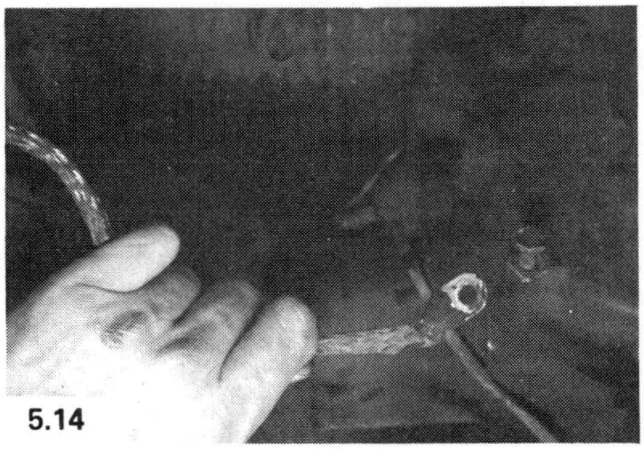

5.14 Detaching the engine to frame earth cable

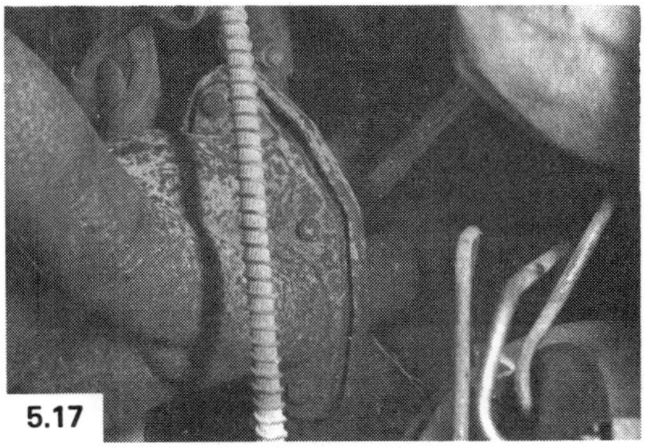

5.17 Removing the exhaust manifold flange bolts

5.23 Detaching the speedometer cable from the gearbox

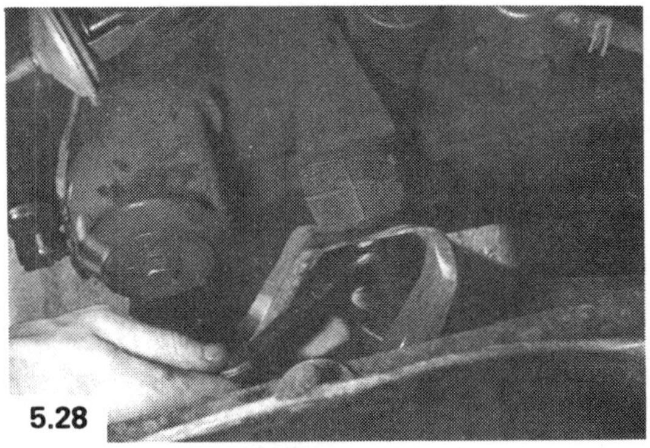

5.28 Unscrewing the front mounting nuts

5.30 Preparing to lift out the engine

6.2 Removing exhaust manifold brace

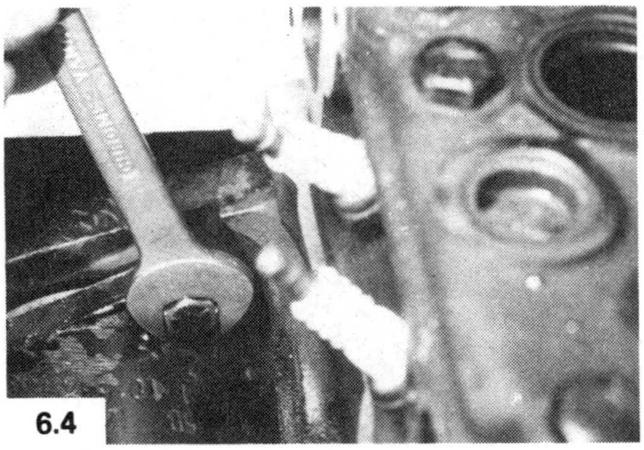

6.4 Undoing top starter motor bolt

8.4a

Unscrewing oil filter centre bolt

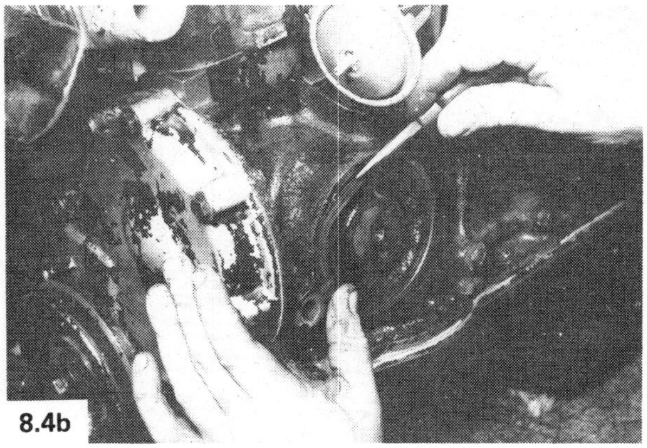

8.4b

Removing oil filter housing sealing ring

8.5

Removing inlet manifold centre bolt

immerse parts with oilways in paraffin - clean them with a petrol soaked cloth and clear oilways with wire. If an air line is available so much the better for final cleaning off. Paraffin, which could possibly remain in oilways would dilute the oil for initial lubrication after reassembly).

3 Where components are fitted with seals and gaskets it is always best to fit new ones - but do NOT throw the old ones away until you have the new ones to hand. A pattern is then available if they have to be specially made. Hang them on a convenient hook.

4 In general it is best to work from the top of the engine downwards. In any case support the engine firmly so that it does not topple over when you are undoing stubborn nuts and bolts.

5 Always place nuts and bolts back together in their components or place of attachment if possible - it saves so much confusion later. Otherwise put them in small, separate pots or jars so that their groups are easily identified.

8 Engine dismantling - ancillary components

1 If you are obtaining a factory replacement reconditioned engine all ancillaries must come off first - just as they will if you are doing a thorough engine inspection/overhaul yourself. These are:

Dynamo or alternator	(Chapter 10)
Distributor	(Chapter 4)
Thermostat and cover	(Chapter 2)
Oil filter and cover	**(Section 8 this Chapter)**
Carburettor	(Chapter 3)
Inlet manifold	**(Section 8 this Chapter)**
Exhaust manifold	**(Section 6 this Chapter)**
Water pump	(Chapter 2)
Fuel pump	(Chapter 3)
Engine mounting brackets	**(Section 8 this Chapter)**
Spark plugs	(Chapter 4)

2 If you are obtaining what is called a 'short engine' (or sometimes 'half-engine') comprising cylinder block, crankcase, crankshaft, pistons, and connecting rods all assembled, then the cylinder head, flywheel, sump and oil pump will need removal also.

3 Remove all the ancillaries according to the removal instructions for them described in the chapters and sections as indicated in paragraph 1.

4 Oil filter and cover - removal

a) Undo the centre bolt holding the cover to the side of the crankcase and pull off the cover. Discard the oil filter element (photo).

b) Remove the sealing ring from the groove in the crankcase (photo).

5 Inlet manifold - removal

Unscrew the two holding bolts, one at each end of the manifold, and with a socket wrench unscrew the third bolt located inside the intake orifice (photo).

6 Exhaust manifold - removal

See Section 6.

7 Engine front mounting brackets - removal

Each bracket is held to the block by three bolts. These bolts have been sealed in, so will require considerable torque to turn them. Do NOT use anything other than a ring or socket spanner to remove them. Note which side of the engine each bracket comes from as they are not interchangeable. The brackets may be removed if necessary, with the engine still in place. The engine should be supported underneath and the bolts securing them to the engine and frame removed. In this way the flexible mountings may also be renewed.

9 Cylinder head removal - engine out of car

1 Position the engine on a bench (or floor) with the cylinder head uppermost.

2 Remove the four screws holding the rocker cover and lift it off together with the cork sealing gasket which may be re-used if it is not over compressed or damaged.

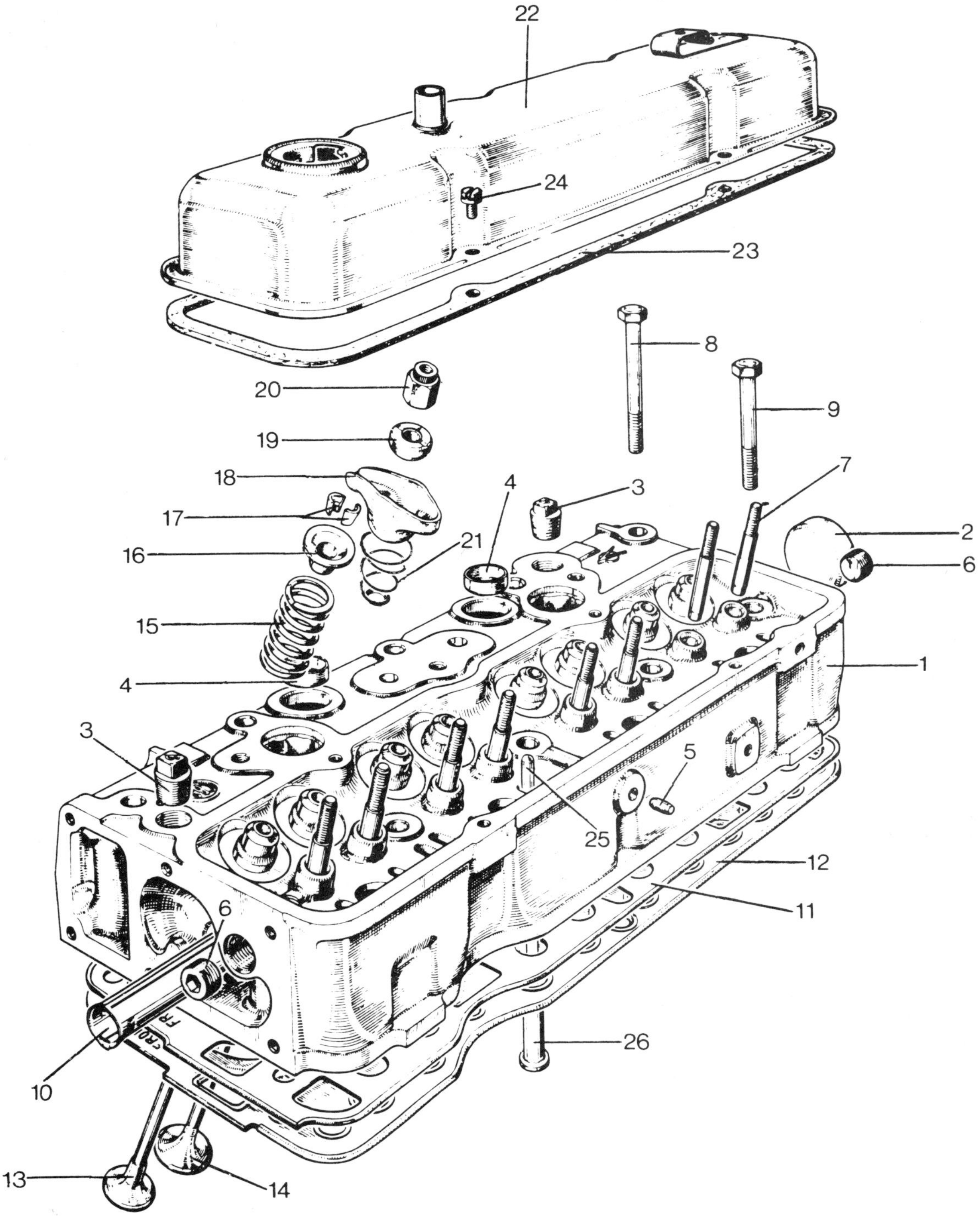

Fig 1:3 CYLINDER HEAD ASSEMBLY – EXPLODED VIEW

1 Cylinder head
2 Core plug-water jacket
3 Blanking plug - water jacket
4 Core plug - water jacket
5 Plug - oilway
6 Plug - oilway
7 Rocker arm stud
8 Cylinder head bolt (long)
9 Cylinder head bolt (short)
10 Water distribution tube
11 Gasket - high compression
12 Gasket - low compression
13 Exhaust valve
14 Inlet valve
15 Valve spring
16 Valve collar
17 Collets
18 Rocker arm
19 Rocker arm pivot
20 Rocker adjusting nut
21 Rocker support spring
22 Rocker cover
23 Rocker cover gasket
24 Rocker cover screw
25 Pushrod
26 Tappet

3 Remove the carburettor (Chapter 3). The inlet and exhaust manifolds may be removed but it is not essential.
4 Slacken off the rocker clearance adjusting nuts (in the centre of each rocker arm) just enough (2-3 turns) to permit each rocker to be swung aside when the valve is closed so that the pushrods may be lifted out. The crankshaft will need rotating to do this. If it is intended to remove the valves anyway in due course, then it will save time if the rocker arms are removed completely at this stage.
5 Put the pushrods into a piece of pierced cardboard so that each can be identified to its relative valve.
6 Slacken off the ten cylinder head holding down bolts in the reverse order of tightening sequence (see Fig 1:18).
7 The cylinder head should now lift off easily. If not, try turning the engine over by the flywheel (with the spark plugs in position) so that compression in the cylinders can force it upwards. A few smart taps with a soft headed mallet or wood cushioned hammer may also be needed. Under no circumstances whatsoever try to prise the head off by forcing a lever of any sort into the joint. This can cause damage to the machined surfaces of the block and cylinder head.

10 Cylinder head removal - engine in car

1 Before proceeding as described for 'Cylinder head removal - engine out of car' it is first of all necessary to carry out the following, including the removal of the parts as stated:
2 Disconnect both battery leads.
3 Drain the cooling system
4 Remove the hoses from the water pump.
5 Remove the carburettor air cleaner unit
6 Disconnect the vacuum advance suction pipe from both the carburettor and distributor, unclip it from the fuel feed pipe and take it off.
7 Disconnect the fuel feed pipe by unscrewing the union at the carburettor. To move the pipe out of the way slacken one of the two clips on the flexible connection at the fuel pump and turn the pipe away. Try and avoid bending the pipe.
8 Disconnect the accelerator linkage.
9 Disconnect the 'Lucar' connector for the lead from the water temperature gauge sender unit in the cylinder head.
10 Remove the distributor cap complete with the plug leads and coil which should be disconnected at the plugs and coil respectively.
11 Proceed as for removing the head with the engine out of the car (Section 9). If the engine needs turning to assist in breaking the joint, reconnect the battery leads and give the engine a quick turn with the starter motor.

11 Valve rocker arms - removal

1 The rocker arms can be removed as soon as the rocker cover is off. Each arm is located on a stud and is lightly supported on the stud by an inverted cone shaped coil spring. The vertical location of each rocker arm is controlled by a hemispherical ball on which the rocker pivots, and which is held in position by a self-locking nut.
2 When the nut is removed the rocker can be lifted off, followed by the spring and washer at the base of the stud.

12 Valves - removal

1 Remove the cylinder head (Sections 9 and 10).
2 Removal all the rocker arms.
3 The valves are located by a collar on a compressed spring which grips two colletts (or a split collar) into a groove in the stem of the valve. The spring must be compressed with a special G clamp in order to release the colletts and then the valve. Place the specially shaped end of the clamp over the spring collar with the end of the screw squarely on the face of the valve. Screw up the clamp to compress the spring and expose the colletts on the valve stem. Sometimes the spring collar sticks and the clamp screw cannot be turned. In such instances, with the clamp pressure still on, give the head of the clamp (over the spring) a tap with a hammer, at the same time gripping the clamp frame firmly to prevent it slipping off the valve.

Take off the two colletts, release the clamp, and the collar and spring can be lifted off. The valve can then be pushed out through its guide and removed. Pull off the seal and clip. If the end of the valve sticks at the guide when removing it, it is due to burring. Carefully grind off the corner of the stem to permit it to pass through the guide. If you force it through it will score the guide way. Make sure that each valve is kept in a way that its position is known for replacement. Unless new valves are to be fitted each valve MUST go back where it came from. The springs, collars and colletts should also be kept with their respective valves. A piece of card with eight holes punched in is a good way to keep the valves in order.

13 Valve guides - reconditioning

If the valves are a slack fit in the guides, ie if there is noticeable movement when side to side pressure is exerted on the stem whilst in the guide, then the procedure is to ream out the valve guide bores and fit new valves with oversize stems.

Reaming is a skilled operation and the non-qualified owner would be well advised to have this done by a fitter.

14 Sump - removal

The engine must be out of the car in order to remove the sump, unless you wish to remove the front suspension and axle assembly which we do not recommend. With the engine out, it is better to wait until the cylinder head is removed. Then invert the engine and undo the set screws holding the sump to the crankcase and lift it off. If the cylinder head is not being removed (for example if the oil pump only is being removed) the engine should be placed on its side and the sump removed as described.

15 Crankshaft pulley, timing cover, sprockets and chain - removal

1 The timing gear is accessible with the engine in the car but unless the sump, and, therefore, the front axle are removed first, the sump gasket will have to be broken and a section replaced. This is not ideal but can be done with careful attention to the re-sealing of the timing case lower end to the sump, on replacement. It will be necessary to remove the radiator and the fan belt as described in Chapter 2. Then undo the four set screws at the front end of the sump which locate into the lower edge of the timing case. Then proceed as described hereafter.
2 First remove the large bolt which holds the fan belt driving pulley to the crankshaft. It will be necessary to prevent the crankshaft from turning, by locking the flywheel with a bar in the starter ring teeth against one of the dowel pegs in the end of the crankcase.
3 Another way is to block one of the crankshaft journals with a piece of wood against the side of the crankcase. With the engine in the car, the pulley may also be removed. Put the car in gear while undoing the bolt.
4 Remove the spring washer and pulley washer. The pulley is keyed onto a straight shaft and should pull off easily. If not, lever it off with two screwdrivers at 180° to each other
5 Take care not to damage either the pulley flange or the timing case cover which is made only of light alloy. Do not lose the Woodruff key from the shaft - check that it fits tightly, and if not get another one, or make one from a piece of mild steel, that is a tight fit in the shaft and the keyway of the pulley boss.
6 Remove the bolts holding the timing case cover in place and draw it off.
7 Pull the chain tensioning device off the locating peg.
8 Remove the bolt, lockwasher and plain washer from the camshaft

timing sprocket. The sprocket may be prevented from turning by blocking one of the crankshaft journals against the crankcase with a piece of wood.
9 Both wheels and the chain may now be drawn off together - the sprockets are a push fit onto their respective shafts. Do not lose the Woodruff keys, and if they are loose, check for fit.

16 Pistons, connecting rods and big end bearings - removal

1 As it is necessary to remove the cylinder head and the sump from the engine in order to remove pistons and connecting rods, the removal of the engine is the logical thing to do first. With the engine on the bench and the cylinder head and sump removed, stand the block inverted (with crankshaft uppermost).
2 Each connecting rod and its bearing cap is matched, and held by two high tensile steel bolts. Before anything else, mark each connecting rod and cap with its cylinder number and relationship - preferably with the appropriate number of dabs of paint. Using punch or file marks may be satisfactory, but it has been known for tools to slip - or the marks even to cause metal fatigue in the connecting rod. Once marked, undo the bearing cap bolts using a good quality socket spanner. Lift off each bearing cap and put it in a safe place. Carefully turn the engine on its side. Each piston can now be pushed out from the block by its connecting rod. Note that if the pistons are standard, there is a small notch in each which indicates the side of the piston towards the front of the engine. If there is no such notch, clean a small area on the front of each piston crown and place an indicative dab of paint. Do not use a punch, or file marks, on the pistons under any circumstances. The shell bearings in the connecting rods and caps can be removed simply by pressing the edge of the end opposite the notch in the shell and they will slide round to be lifted out. Note from where each comes.

17 Gudgeon pins

The gudgeon pins float in the piston and are an interference fit in the connecting rods. This 'interference fit' between gudgeon pin and connecting rod, means that heat is required (230-260°C) before a pin can be satisfactorily replaced in the connecting rod. If it is, therefore, necessary to replace either the piston or connecting rod, we strongly recommend that the assembly of the two be entrusted to someone with experience. Misapplied heat can ruin one, or all, of the components very easily (Vauxhall now supply pistons and connecting rods separately).

18 Piston rings - removal

Unless new rings are to be fitted for certain, care has to be taken that rings are not broken on removal. Starting with the top ring (all rings are to be removed from the top of the piston) ease one end out of its groove and place a piece of steel band (shim, old feeler gauge blade, strip of cocoa tin!) behind it.

Then move the metal strip carefully round behind the ring, at the same time nudging the ring upwards so that it rests on the surface of the piston above, until the whole ring is clear and can be slid off. With the second and third rings which must also come off the top, arrange the strip of metal to carry them over the other grooves.

Note where each ring has come from (pierce a piece of paper with each ring showing 'top 1', 'middle 1' etc).

19 Flywheel - removal

1 **The flywheel can be removed with the engine in the car if the gearbox is first withdrawn but it is not recommended, and the following procedures prevail when the engine has been lifted out:**
2 Remove the clutch assembly (Chapter 5).
3 Remove the four bolts from the centre of the flywheel. There are no washers as the bolts are locked with a sealing compound.
4 Using a soft headed mallet, tap the periphery of the flywheel progressively all round, gradually drawing it off the crankshaft flange and locating dowel. Do not allow it to assume a skew angle as the fit on the flange and dowel are at very close tolerances to maintain proper balance and concentricity with the crankshaft. When the flywheel is nearly off make sure it is well supported so that it does not drop. It is heavy.

20 Oil pump - removal

1 Remove the engine from the car and detach the sump.
2 Assuming that no dismantling of the camshaft and timing gear has taken place, set the crankshaft pulley pointer to the TDC marker on the timing cover, (see Chapter 4) and ensure that the distributor rotor arm corresponds with the No 1 piston firing position. This will facilitate reassembly as the oil pump spindle also drives the distributor.
3 Remove the distributor (Chapter 4).
4 Disconnect the pump suction pipe support bracket by undoing the bolt holding it to the centre main bearing cap. Remove the two large bolts attaching the pump to the crankcase. Lift out the pump and gasket.

21 Camshaft - removal

1 Remove the engine from the car.
2 Remove the sump, timing gear cover, timing chain and sprockets, oil pump and distributor.
3 Slacken all the rocker arm nuts sufficiently to allow withdrawal of the pushrods.
4 Remove the fuel pump.
5 If the cylinder head has been removed stand the engine inverted - if not lie it on its right hand side and rotate the camshaft several times to push the tappets out of the way.
6 Undo the two bolts retaining the thrust plate and slide the thrust plate out.
7 The camshaft can now be drawn out, and care must be taken that it is manoeuvred past the tappets without damage to either the cams or the tappets. This will be easy if the engine is completely inverted. If, however, it is lying on its side with the tappets in such a position that they could fall out under their own weight, more care is necessary. In the event it would be advisable to prop the engine so that the tappets cannot fall out of their bores. Take care also not to damage the camshaft bearings with the edges of the cams as the shaft is withdrawn.

22 Tappets - removal

1 With the camshaft removed, (Section 21) simply lift out the tappets and place them so that they can be returned similarly. A papier maché egg box is a useful container for this. (Note the pushrods through the cardboard in the background of the photo).

23 Crankshaft and main bearings - removal

1 With the engine removed from the car, remove the sump, oil pump, timing chain and sprockets, and flywheel. If the cylinder head is also removed so much the better as the engine can be stood firmly in an inverted position.
2 Remove the connecting rod bearing caps. This will already have been done if the pistons are removed.
3 Using a good quality socket wrench, remove the two cap bolts from each of the three main bearing caps.
4 Lift off each cap carefully noting its position. Each one is different in shape.

Keeping tappets in order using an egg box

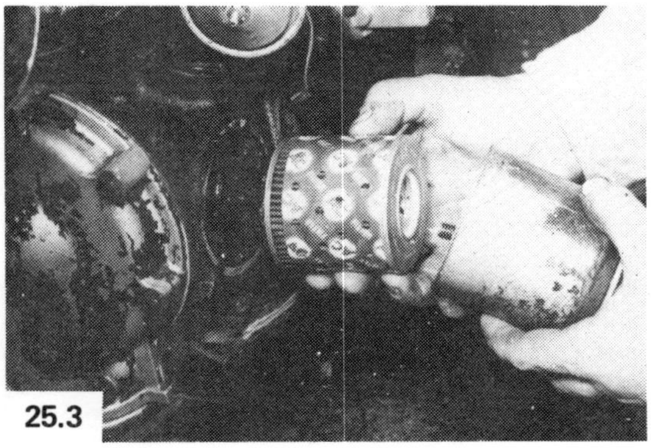

Fitting a new oil filter element

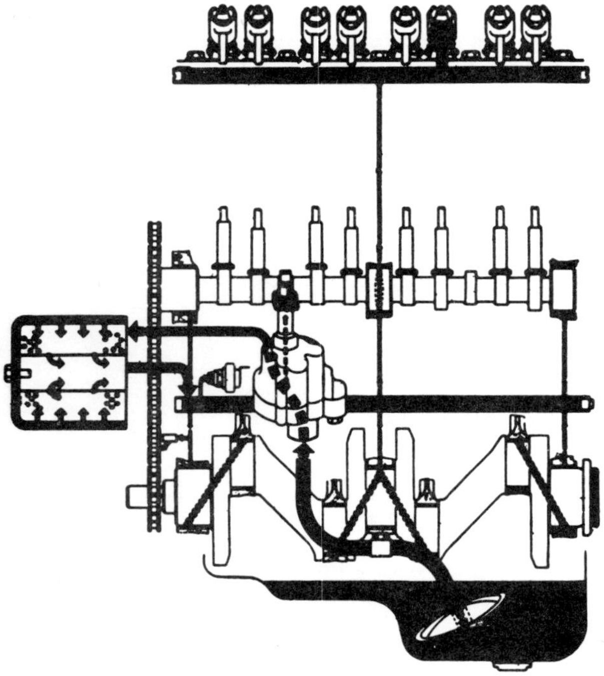
Fig 1:4 Engine lubrication system - schematic diagram

5 The bearing cap shells will probably come off with the caps, in which case they can be removed by pushing them round from the end opposite the notch and lifting them out.
6 Grip the crankshaft firmly at each end and lift it out. Put it somewhere safe where it cannot fall. Remove the shell bearings from the **crankcase, noting that the centre one has a flange on each side.**

24 Engine lubrication system - description

A forced feed system of lubrication is used, with oil circulated to all the engine bearing surfaces under pressure by a pump which draws oil from the sump under the crankcase. The oil is first pumped through a full flow oil filter (which means that all oil is passed through the filter. A bypass oil filter is one through which only part of the oil in circulation passes).

From the filter, oil flows into a main oil gallery - which is cast integrally into the cylinder block. From this gallery, oil is fed via oilways in the block to the crankshaft main bearings, and then from the main bearings along oilways in the crankshaft to the connecting rod bearings. From the same gallery, oilways carry the oil to the camshaft bearings.

From the centre camshaft bearing a further oilway passes oil to a gallery in the cylinder head. This gallery delivers oil through the hollow rocker mounting studs to lubricate the rocker pivots. The tappets are lubricated by oil returning from the rocker gear via the pushrods and is not under pressure. Once oil has passed through the bearings and out, it finds its own way by gravity back to the sump.

If the filter gets blocked, oil will continue to flow because a pressure relief valve will open, permitting oil to circulate past the filter element. Similarly, a blockage in oilways (resulting in greatly increased pressure) will cause the oil pressure relief valve in the oil pump to operate, returning oil direct to the sump.

Oil pressure is measured after oil has passed through the filter. As the oil pressure warning light only comes on when the pressure is as low as 3-5 lbs/sq in it is most important that the filter element is regularly changed, and the oil changed at the recommended intervals, in order that the lubrication system remains clean.

Should the warning light ever come on when the engine is running at any speed above idling, stop at once and investigate - serious bearing and cylinder damage may otherwise result.

The crankcase is ventilated to prevent pressure building up from the action of the pistons and also to cause oil, and sometimes fuel, vapour to be carried away. Air enters via an oil wetted gauze filter at the dipstick hole, and crankcase fumes are extracted via a pipe from the rocker cover into the air cleaner. From here it passes into the combustion chambers with the fuel/air mixture.

25 Oil filter - renewal of element

1 Remove the filter casing as described in Section 8.
2 Clean out the interior of the bowl with paraffin ensuring that any sludge deposits are wiped out.
3 A new sealing ring will be supplied with the new element, and this should be carefully fitted into the groove in the block. Do not stretch the ring by forcing one part into the groove and pressing the rest in afterwards, otherwise it will not seat properly - it will remain stretched. Get the whole ring into the groove at once and carefully bed it down with a small screwdriver, ensuring that it does not twist and jam across the groove. Put the new element in the bowl, place the bowl into the groove and screw up the centre bolt (this bolt is not detachable from the cover) (photo).
4 Before the cover is quite tight, revolve it to ensure it is bedded into the sealing ring. Check the seating for leaks at the first opportunity, by running the engine.

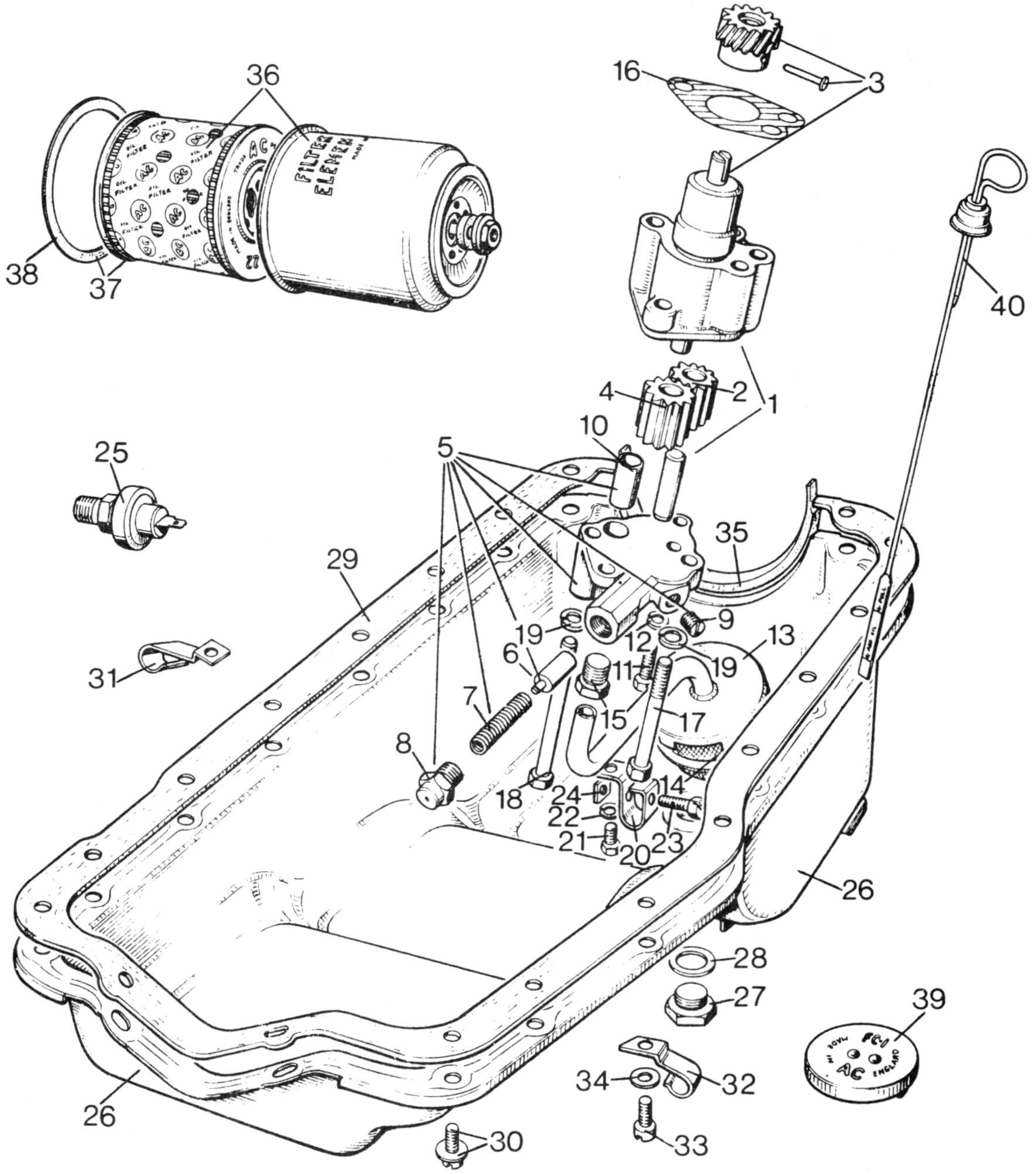

Fig 1:5 SUMP, OIL PUMP AND FILTER ASSEMBLIES

1 Oil pump body
2 Pump gears - driven
3 Pump drive gear and spindle
4 Pump gear - driving
5 Pump cover and relief valve assembly
6 Pressure relief valve
7 Spring
8 Retaining plug
9 Plug
10 Reservoir tube
11 Bolt - cover to body
12 Washer
13 Strainer
14 Strainer gauze
15 Strainer pipe union
16 Pump to crankcase gasket
17 Pump mounting bolt
18 Pump mounting bolt
19 Washer
20 Clip
21 Bolt
22 Washer
23 Screw
24 Nut
25 Oil pressure switch
26 Sump
27 Drain plug
28 Drain plug washer
29 Sump gasket
30 Sump screw washer
31 Speedo cable clip
32 Clutch cable clip
33 Screw
34 Washer
35 Rear bearing cap to sump gasket
36 Filter element and bowl
37 Element and washer
38 Washer
39 Oil filler cap
40 Dipstick

26 Engine examination and renovation - general

With the engine stripped down as described in the preceding sections, all parts should be thoroughly cleaned in preparation for examination. Details of what to look for are described in the following sections, together with instructions regarding the necessary repairs or renewals.

27 Crankshaft - examination and renovation

1 Examine all the crankpins and main bearing journals for signs of scoring or scratches. If all surfaces are undamaged, check next that all the bearing journals are round. This can be done with a micrometer or caliper gauge, taking readings across the diameter at six or seven points for each journal. If you do not own, or know how to use, a micrometer, take the crankshaft to your local engineering works and ask them to 'mike it up' for you (photo).
2 If the crankshaft is ridged or scored, it must be reground. If the ovality exceeds .002 in. on measurement, but there are no signs of scoring or scratching on the surfaces, regrinding may be necessary. It would be advisable to ask the advice of the engineering works to whom you would entrust the work of regrinding in such instances.

28 Big end (connecting rod) bearings and main bearings - examination and renovation

1 Big end bearing failure is normally indicated by a pronounced knocking from the crankcase and a slight drop in oil pressure. Main bearing failure is normally accompanied by vibration, which can be quite severe at high engine speeds, and a more significant drop in oil pressure. Oil pressure drop can only be verified, of course, if you have fitted a gauge.
2 The shell bearing surfaces should be matt grey in colour with no sign of pitting or scoring.
3 Replacement shell bearings are supplied in a series of thicknesses dependent on the degree of regrinding that the crankshaft requires, which is done in multiples of .010 in. So depending on how much it is necessary to grind off, bearing shells are supplied as '.010 in undersize' and so on. The engineering works regrinding the crankshaft will normally supply the correct shells with the reground crank.
4 If an engine is removed for overhaul regularly, it is worthwhile renewing big end bearings every 30000 miles as a matter of course, and main bearings every 50000 miles. This will add many thousands of miles to the life of the engine before any regrinding of crankshafts is necessary. Make sure that bearing shells renewed are to standard dimensions if the crankshaft has not been reground.

29 Cylinder bores - examination and renovation

1 The bores must be checked for ovality, scoring, scratching and pitting. Starting from the top, look for a ridge where the top piston ring reaches the limit of its upward travel. The depth of this ridge will give a good indication of the degree of wear. It can be checked with the engine in the car and the cylinder head removed. Other indications are excessive oil consumption and a smoky exhaust.
2 Measure the bore diameter across the block and just below any ridge. This can be done with an internal micrometer or a Mercer gauge. Compare this with the diameter of the bottom of the bore, which is not subject to wear. If no micrometer measuring instruments are available, use a piston from which the rings have been removed, and measure the gap between it and the cylinder wall with a feeler gauge. Another way of measuring the bore to determine the need for rebore, is to measure the difference in the gap of a piston ring below and above the ridge. The difference divided by three will give an indication of the bore wear.
3 If the difference in bore diameters at top and bottom is .010 in. or more, then the cylinders need reboring. If less than .010 in, then the fitting of new and special rings to the pistons can cure the trouble.
4 If the cylinders have already been bored out to their maximum, it is possible to have liners fitted. This situation will not often be encountered.

30 Pistons and rings - examination and renovation

1 Examine the pistons (with the rings removed as described in Section 18) for signs of damage on the crown and around the top edge. If any of the piston rings have broken, there could be quite noticeable damage to the grooves, in which case the piston must be renewed. Deep scores in the piston walls also call for renewal. If the cylinders are being rebored, new oversize pistons and rings will be needed anyway. If the cylinders do not need reboring and the pistons are in good condition, only the rings need to be checked.

Pistons should be measured across the diameter at right angles to the gudgeon pin - 3 inches up from the bottom of the skirt.
2 To check the existing rings, place them in the cylinder bore and press each one down in turn to the bottom of the stroke. In this case a distance of 2½ inches from the top of the cylinder will be satisfactory. Use an inverted piston to press them down square. With a feeler gauge, measure the gap for each ring which should be as given in the 'Specifications' at the beginning of this Chapter. If the gap is too large, the rings will need renewal.
3 Check also that each ring gives a clearance in the piston groove according to 'Specifications'. If the gap is too great, new pistons and rings will be required. Independent specialist producers of pistons and rings can normally provide the rings required separately. If new Vauxhall pistons and rings are being obtained it will be necessary to have the ridge ground away from the top of each cylinder bore. If specialist oil control rings are being obtained from an independent supplier, the ridge removal may not be necessary as the top rings will be stepped to provide the necessary clearance. If the top ring of a new set is not stepped it will hit the ridge and break.
4 If new pistons are obtained the rings will be included, so it must be emphasised that the top ring be stepped if fitted to an un-reground bore (or un-deridged bore). It is worth noting that standard size pistons supplied are, in fact, .005 in. larger in diameter than those fitted on original engine assembly.
5 The new rings should be placed in the bores as described in paragraph 2, and the gap checked. Any gaps which are too small should be increased by filing one end of the ring with a fine file. Be careful not to break the ring as they are brittle (and expensive). On no account make the gap less than specification. If the gap should close when under normal operating temperatures, the ring will break.
6 The groove clearance of new rings in old pistons should be within the specified tolerances. If it is not enough, the rings could stick in the piston grooves causing loss of compression. The piston grooves in this case will need machining out to accept the new rings.

31 Camshaft and camshaft bearings - examination and renovation

1 With the camshaft removed, examine the bearings for signs of obvious wear and pitting. If there are signs, then the three bearings will need renewal. This is not a common requirement, and to have to do so is indicative of severe engine neglect at some time. As special removal and replacement tools are necessary to do this work properly, it is recommended that it is done by a 'specialist'. Check that the bearings are located properly so that the oilways from the bearing housings are not obstructed. Each camshaft bearing shell has a notch in the front edge on the side away from the crankshaft, as a position indicator.
2 The camshaft itself should show no marks on either the bearing journals or the profiles. If it does, it should be renewed. Check that the overall height of each cam from base to peak is within specification. If not, the camshaft should be renewed.

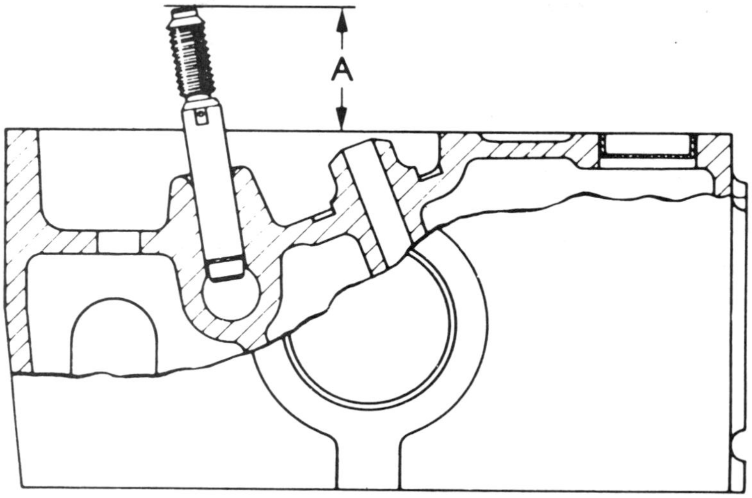

Fig 1:6 Valve rocker stud. Dimension 'A' should be equal for all studs and within specifications

3 Examine the skew gear for signs of wear or damage. If this is badly worn it will mean renewing the camshaft.
4 The thrust plate (which also acts as the locating plate) should not be ridged or worn in any way. If it is, renew it.

32 Timing chain, sprockets and tensioner - examination and renovation

1 Examine the teeth of both sprockets for wear. Each tooth is the shape of an inverted 'V' and if the driving (or driven) side is concave in shape, the tooth is worn and the sprocket should be replaced. The chain should also be replaced if the sprocket teeth are worn or if the tensioner adjustment is fully taken up. It is sensible practice to replace the chain anyway.
2 The tensioner device is mounted on a stud (which also feeds oil to the chain). It is held on the stud by the pressure of the timing chain cover against the rubber plug in the body of the unit. It is prevented from rotating by butting up to two extensions of the camshaft thrust plate which is also modified to suit. The principal is that the tensioner is spring loaded against the chain and, as it moves out to take up slack, is locked by means of a piston with a spiral notched groove holding against a lug in the bore of the tensioner sleeve. If the pressure shoe is badly worn or ridged it should be renewed.

33 Valve rocker arms and pushrods - examination and renovation

1 Each rocker arm has three wearing surfaces, namely, the pushrod recess, the valve stem contact, and the centre pivot recess. If any of these surfaces appears severely grooved or worn the arm should be replaced. If only the valve stem contact area is worn, it is possible to clean it up with a fine stone.
2 If the rocker ball is pitted, or has flats worn in it, this should also be replaced.
3 The nut is a self-locking type on the stud. If it has been removed or adjusted many times, the self-locking ring may have become ineffective and the nut may be slack enough to turn involuntarily and alter the tappet clearance. If the tightening torque is less than the specified 3 lb/ft minimum, new nuts should be fitted.
4 The rocker studs should be examined to ensure that the threads are undamaged and that the oil delivery hole in the side of the stud at the base of the thread is clear. Place a straight edge along the top of all the studs to ensure that none is standing higher than the rest. If any are, it means that they have come out of the head some distance (Fig 1:6). They should be removed and replaced with an oversize stud. As this involves reaming out the stud hole in the head, you should seek professional advice and assistance to ensure that the new oversize stud is securely fitted at the correct angle.
5 Any pushrods which are bent should be renewed. On no account attempt to straighten them.

34 Tappets - examination and renovation

Examine the bearing surfaces of the tappets which lie on the camshaft. Any indentation in these surfaces, or any cracks, indicate serious wear and the tappets should be renewed. Thoroughly clean them out, removing all traces of sludge. It is most unlikely that the sides of the tappets will prove worn, but, if they are a very loose fit in their bores and can readily be rocked, they should be exchanged for new units. It is very unusual to find any wear in the tappets, and any present is likely to occur only at very high mileages, or in cases of neglect. If tappets are worn, examine the camshaft carefully as well.

35 Connecting rods - examination and renovation

1 Examine the mating faces of the big end caps to see if they have ever been filed in a mistaken attempt to take up wear. If so, the

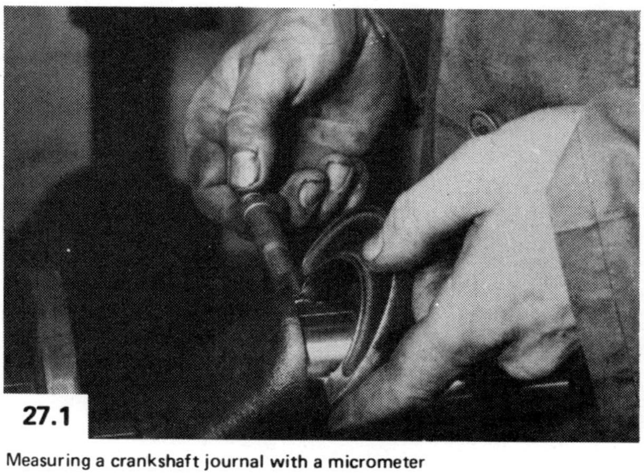

27.1

Measuring a crankshaft journal with a micrometer

offending rods must be renewed.

2 Check the alignment of the rods visually, and if all is not well, take the rods to your local Vauxhall agent for checking on a special jig.

37.3a

37.3b

37.3c

Measuring the oil pump clearances with a feeler blade

36 Flywheel starter ring - examination and renovation

1 If the teeth on the flywheel starter ring are badly worn, or if some are missing, then it will be necessary to remove the ring and fit a new one.

2 Either split the ring with a cold chisel after making a cut with a hacksaw blade between two teeth, or use a soft headed hammer (not steel) to knock the ring off, striking it evenly and alternately, at equally spaced points. Take great care not to damage the flywheel during this process.

3 Clean and polish with emery cloth, four evenly spaced areas on the outside face of the new starter ring.

4 Heat the ring evenly with a flame until the polished portions turn dark blue. Alternatively, heat the ring in a bath of oil to a temperature of 200°C. (If a naked flame is used, take careful fire precautions). Keep the ring at this temperature for five minutes and then quickly fit it to the flywheel so the chamfered portion of the teeth faces the gearbox side of the flywheel. Wipe all oil off the ring before fitting it.

5 The ring should be tapped gently down onto its register and left to cool naturally, when the contraction of the metal on cooling will ensure that it is a secure and permanent fit. Great care must be taken not to overheat the ring, indicated by it turning light metallic blue, as if this happens the temper of the ring will be lost.

37 Oil pump - examination and renovation

1 If the oil pump is worn, it is best to purchase an exchange reconditioned unit, as to rebuild the oil pump is a job that calls for engineering shop facilities.

2 To check if the pump is still serviceable, first check if there is any slackness in the spindle bushes, and then remove the bottom cover held by two bolts.

3 Then check the two gears (the impellers) and the inside of the pump body for wear with the aid of a feeler gauge. Measure (a) the backlash between the gearwheel (blade inserted between the sides of teeth that are meshed together), (b) the gearwheels radial clearance (blade inserted between the end of the gearwheel teeth and the inside of the body), (c) the gearwheel end clearance (place a straight edge across the bottom flange of the pump body and measure with the feeler blades the gap between the straight edge and the sides of the gearwheel). The correct clearances are listed in the 'Specifications' (photos).

4 Fit a replacement pump if the clearances are incorrect.

38 Decarbonisation

1 This can be carried out with the engine either in or out of the car. With the cylinder head off, carefully remove with a wire brush and blunt scraper all traces of carbon deposits from the combustion spaces and the ports. The valve head stems and valve guides should also be freed from any carbon deposits. Wash the combustion spaces and ports down with petrol and scrape the cylinder head surface free of any foreign matter with the side of a steel blade, or a similar article.

2 Clean the pistons and top of the cylinder bores. If the pistons are still in the block, then it is essential that great care is taken to ensure that no carbon gets into the cylinder bores as this could scratch the cylinder walls or cause damage to the piston and rings. To ensure this does not happen, first turn the crankshaft so that two of the pistons are at the top of their bores. Stuff rag into the other two bores or seal them off with paper and masking tape. The waterways should also be covered with small pieces of masking tape to prevent particles of carbon entering the cooling system and damaging the water pump.

3 Remove all traces of carbon. The old idea of leaving a ring of

carbon as a seal round the periphery of the piston crown does not apply to modern engines.
4 Press a little grease into the gap between the cylinder walls and the two pistons which are to be worked on. With a blunt scraper carefully scrape away the carbon from the piston crown, taking great care not to scratch the aluminium. Also scrape away the carbon from the surrounding lip of the cylinder wall. When all carbon has been removed, scrape away the grease which will now be contaminated with carbon particles, taking care not to press any into the bores. To assist prevention of carbon build-up, the piston crown can be polished with a metal polish. Remove the rags or masking tape from the other two cylinders and turn the crankshaft so that the two pistons which were at the bottom are now at the top. Place rag or masking tape in the cylinders which have been decarbonised and proceed as just described.

39 Valves, valve seats and valve springs - examination and renovation

1 Examine the heads of the valves for pitting and burning, especially the heads of the exhaust valves. The valve seatings should be examined at the same time. If the pitting on valve and seat is very slight, the marks can be removed by grinding the seats and valves together with coarse, and then fine, valve grinding paste.
2 Where bad pitting has occurred to the valve seats it will be necessary to recut them and fit new valves. This latter job should be entrusted to the local Vauxhall agent or engineering works. In practice it is very seldom that the seats are so badly worn. Normally, it is the valve that is too badly worn for replacement, and the owner can easily purchase a new set of valves and match them to the seats by valve grinding.
3 Valve grinding is carried out as follows: Smear a trace of coarse carborundum paste on the seat face and apply a suction grinder tool to the valve head. With a semi-rotary motion, grind the valve head to its seat, lifting the valve occasionally to redistribute the grinding paste. When a dull matt even surface finish is produced on both the valve seat and the valve, then wipe off the paste and repeat the process with fine carborundum paste, lifting and turning the valve to redistribute the paste as before. A light spring placed under the valve head will greatly ease this operation. When a smooth unbroken ring of light grey matt finish is produced, on both valve and valve seat faces, the grinding operation is complete. It is important though that the thickness of the valve head is correct and that the seat widths of the grinding do not exceed specification (Fig 1:7). If they do the valve will need renewal and the seat recutting by a specialist.
4 Scrape away all carbon from the valve head and the valve stem. Carefully clean away every trace of grinding compound, taking great care to leave none in the ports or in the valve guides. Clean the valves and valve seats with a paraffin soaked rag, then with a clean rag, and finally, if an air line is available, blow the valves, valve guides and valve ports clean.
5 Check that all valve springs are intact. If any one is broken, all should be replaced. Check that the free height of the springs is within specifications. If some springs are not within specification, replace them all. Springs suffer from fatigue and it is a good idea to replace them even if they look all right.

40 Cylinder head - examination

1 With the valves removed and all carbon deposits cleaned away, the valve seats must be examined for signs of cracking or pitting. Mild pitting can be cured by grinding in the valves with carborundum paste, but any hair line cracks or severe ridging and pitting mean that at least the seats will need recutting or renewing. This is a specialist task. Cracks visible anywhere else in the head, mean that it must be scrapped.
2 The head must be perfectly flat where it joins the cylinder block. Use a metal straight edge at various positions along and across the head to see if it is warped in any way. The least one can expect from a warped head, is persistent blowing of gaskets and loss of coolant.
3 Check that the bores for the valves (there are no removable guides) are not worn. If they are they will need reaming out for the fitting of oversize valves.

Examine also the cooling water distributor tube inside the head. It may be removed for cleaning if necessary. Renew it if it is badly pitted or corroded (see Chapter 2 for details).

41 Engine reassembly - general

1 To ensure maximum life with minimum trouble from a rebuilt engine, not only must everything be correctly assembled, but everything must be spotlessly clean, all the oilways must be clear, locking washers and spring washers must always be fitted where indicated and all bearing and other working surfaces must be thoroughly lubricated during assembly.
2 Before assembly begins renew any bolts or studs, the threads of which are in any way damaged, and whenever possible use new spring washers.
3 Apart from your normal tools, a supply of clean rag, an oil can filled with engine oil (an empty plastic detergent bottle thoroughly cleaned and washed out, will do), a new supply of assorted spring washers, a set of new gaskets, and a torque spanner, should be collected together.

42 Crankshaft - replacement

1 Ensure that the crankcase is thoroughly clean and that all oilways are clear. If possible blow the drillings out with compressed air (photo).
2 It is best to take out the plug at each end of the main oil gallery and so clean out the oilways to the crankshaft bearing housings, and camshaft bearings. Replace the plugs using jointing compound to make an oil tight seal.
3 Treat the crankshaft in the same fashion and then inject engine oil into the crankshaft oilways.
4 If the old main bearing shells are to be replaced, (not to do so is false economy unless they are virtually as new), fit the three upper halves of the main bearings shells to their locations in the crankcase, after wiping the locations clean (photo).
5 The centre upper shell bearing is flanged when installed on cars with manual gearbox but both the upper and lower halves are flanged when the car is equipped with automatic transmission. New bearings have overthick flanges which must be reduced in order to permit the crankshaft to be replaced and fitted with the correct amount of end float, which is from .002 in to .008 in. End float, which is the amount a crankshaft can move endways, is measured between the centre bearing upper shell flange and the bearing surface on the web of the crankshaft, with the crankshaft moved to one extreme of its end float travel (Fig. 1:8) (photo).
6 It will be necessary to reduce the shell bearing flange thickness by rubbing it down evenly on an engineers flat bed covered with fine emery cloth. This is done progressively until a feeler blade of .002 in thickness can be placed between the flange and the crankshaft web (photo).
7 NOTE that at the back of each bearing is a tab which engages in locating grooves in either the crankcase or the main bearing cap housings.
8 If new bearings are being fitted, carefully clean away all traces of any protective grease or coating with which they may have been treated. There should be a leaflet, indicating the need for this, with the bearings.
9 With the three upper bearing shells securely in place, wipe the lower bearing cap housings and fit the three lower shell bearings to their caps (ensuring that the right shell goes into the right cap, if the old bearings are being refitted) (photo).
10 Next install the new rear bearing oil seal. First lubricate the flange and oil seal lip with anti-scuffing paste (molybdenum disulphide) and place the seal on the flange with the lip facing the centre of the crankshaft. Make sure it is fitted squarely, that the lip is not turned

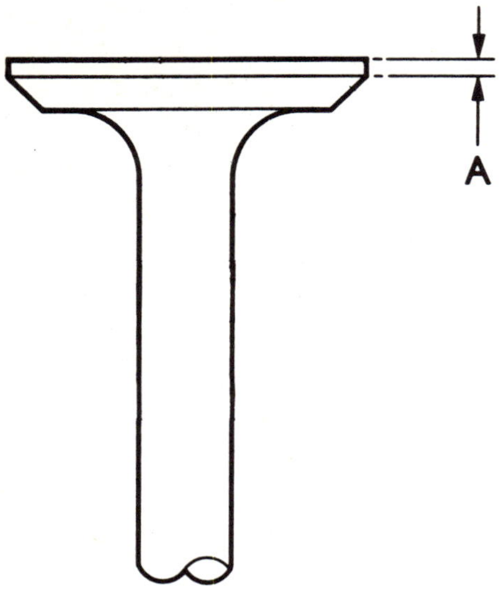

Fig 1:7 Valve head thickness 'A' must not be less than specification

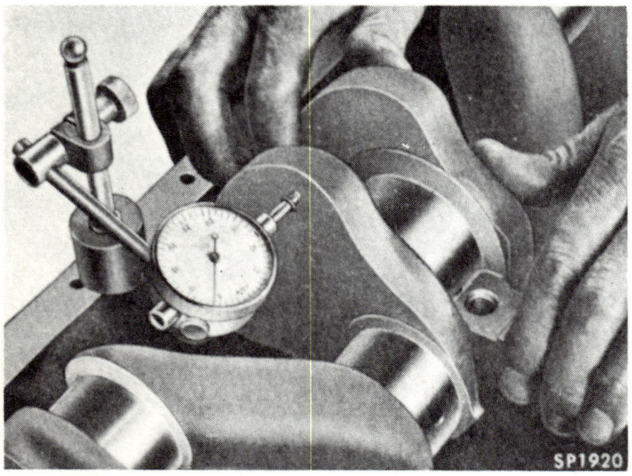

Fig. 1:8 Using a dial gauge to measure crankshaft end float

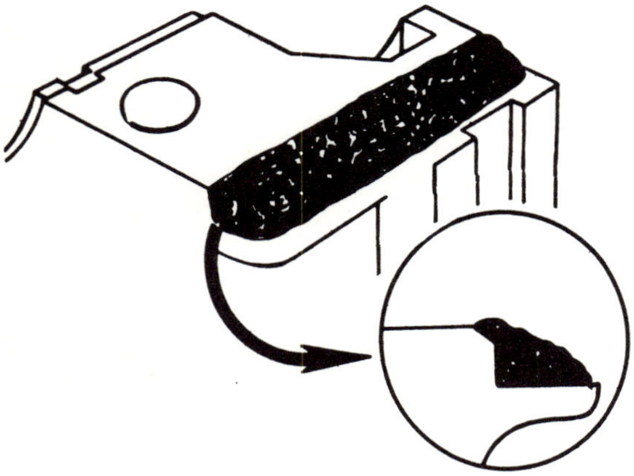

Fig 1:9 Showing application of sealing compound - along rear bearing cap land and chamfer

back in any way and that the flange is completely free of burrs, scores or scratches, which will damage the seal and make it useless (photo).

11 Check that the oil seal groove in the crankcase is completely free of old jointing compound and that the bearing cap faces throughout are similarly clean. Remove all traces of oil.When quite clean and dry apply 'Hylomar SQ32/M' jointing compound sparingly into the seal groove in the crankcase.

12 Thoroughly lubricate the main bearing shells with engine oil and with the oil seal fitted, lower the crankshaft carefully into position. The weight of the crankshaft must be supported to ensure that the seal goes into the crankcase groove without damage or disturbance (photos).

13 The main bearing caps with new shell bearings are next fitted in their respective positions, ensuring that the mating faces are perfectly clean to ensure perfect fitting. The rear bearing cap must have a bead of approved sealing compound applied along the register and chamfer on each side as shown in Fig. 1:9. Jointing compound should also be applied sparingly to the seal groove of this cap as was done in the seal groove of the crankcase (Hylomar).

14 The centre bearing cap should have the tapped hole (for the oil pump pipe support bracket bolt) on the camshaft side.

15 Oil all the threads and replace and tighten the bolts to the specified torque of 58 lb/ft (photo).

16 Any sealing compound which exudes from around the rear main bearing cap should be left where it is.

43 Pistons and connecting rods - reassembly

As mentioned in Section 17 the connecting rods and gudgeon pins are an interference fit requiring heat to enable them to be correctly assembled. This work should be entrusted to someone with the necessary experience and equipment.

44 Piston ring - replacement

1 Ensure that the piston and piston rings have been inspected and renewed in accordance with the procedures described in Section 30.

2 Check that the ring grooves are completely clean.

3 Fit the rings over the top of the piston, starting with the bottom oil control ring.

4 The ring may be spread with the fingers sufficiently to go around the piston, but it could be difficult getting the first ring past the other grooves. It is well worth spending a little time cutting a strip of thin tin plate from any handy can, say 1 in wide and slightly shorter in length than the piston circumference. Place the ring round this and then slide the strip with the ring on it over the piston, until the ring can be conveniently slipped off into its groove.

5 Follow in the same way with the other two rings - remembering that the ring with the cut-out step goes in the top groove with the step towards the top of the piston.

6 The words 'TOP' or 'BOTTOM' which may be marked on the rings indicate which way up the ring goes in its groove in the piston, ie the side marked 'TOP' should face the top of the piston, and does not mean that the ring concerned should necessarily go into the top groove.

45 Piston - replacement in cylinder

1 The pistons, complete with connecting rods and new shell bearings, can be fitted to the cylinder bores in the following sequence:

2 With a wad of clean rag wipe the cylinder bores clean. If new rings are being fitted, any surface oil 'glaze' on the walls should be removed by rubbing with a very fine abrasive. This can be a very fine emery cloth, or a fine cutting paste as used for rubbing down paintwork. This enables new rings to bed into the cylinders properly which would otherwise be prevented, or at least delayed for a long time. Make sure that all traces of abrasive are confined to the cylinder bores and are

39.3

Grinding in valves

42.1

Cleaning the crankshaft journals

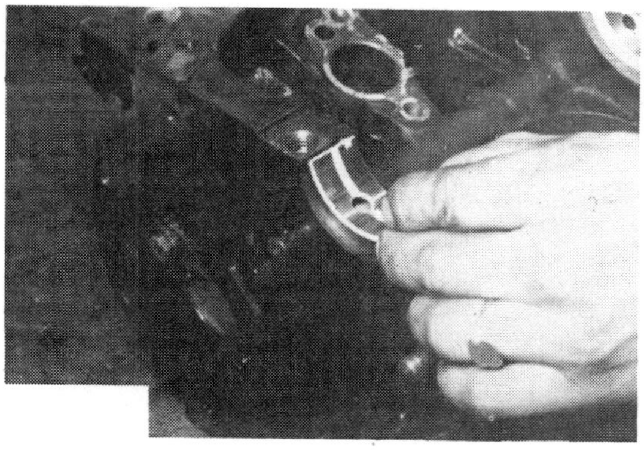

Fitting main bearing shells into crankcase

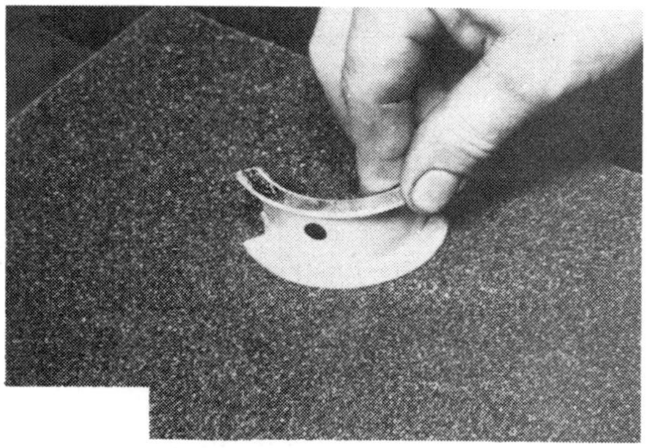

Reducing centre bearing flange thickness on abrasive paper

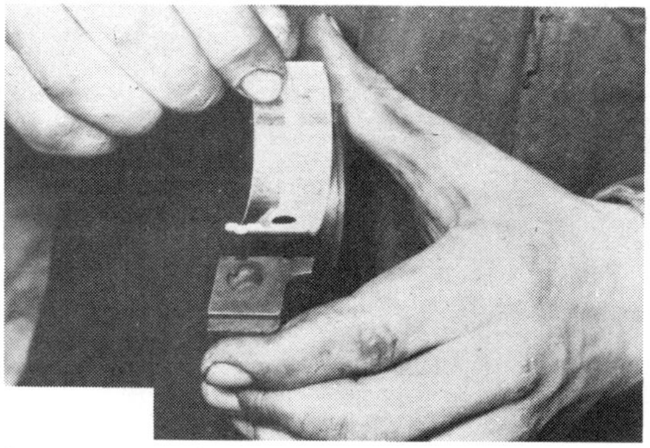

Fitting main bearing shell into cap

Fitting crankshaft rear oil seal

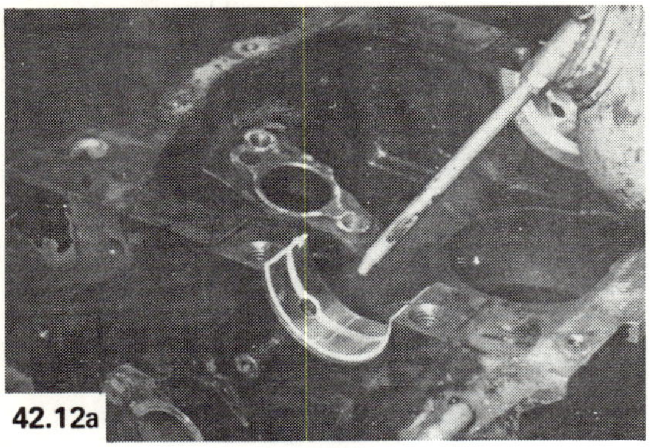

42.12a

42.12b

42.12c

Lubricating main bearings and lowering crankshaft into position ensuring seal locates properly in groove

42.15

Tightening centre main bearing cap bolt (arrow indicates oil pump bracket bolt hole)

completely cleaned off before assembling the pistons into the cylinders. Then oil the pistons, rings, and cylinder bores generously with engine oil. Space the piston ring gaps equally around the piston.

3 The pistons, complete with connecting rods, are fitted to their bores from above.

4 As each piston is inserted into its bore, ensure that it is the correct piston/connecting rod assembly for that particular bore, that the connecting rod is the right way round, and that the front of the piston (which is marked with a notch) is towards the front of the engine (photo).

5 The piston will only slide into the bore as far as the oil control ring. It is then necessary to compress the piston rings into a clamp and to gently tap the piston into the cylinder bore with a wooden or plastic hammer. If a proper piston ring clamp is not available, then a suitable hose clip may be tried (photo).

6 If new pistons and rings are being fitted to a rebored block, the clearances are very small and care has to be taken to make sure that no part of a piston ring catches the edge of the bore before being pressed down. They are very brittle and easily broken. For this reason it is acceptable practice to chamfer the lip of the cylinder very slightly to provide a lead for the rings into the cylinder. The chamfer should be at an angle of 45° and should not be cut back more than .010 in. If some form of hose clip is being used to compress the piston rings, it may be found that the screw housing prevents the clip from lying exactly flush with the cylinder head. Here again watch carefully to ensure that no part of the ring slips from under the control of the clamp.

46 Connecting rod to crankshaft - reassembly

1 Wipe the connecting rod half of the big end bearing and the underside of the shell bearing, clean, and fit the shell bearing in position with its locating tongue engaged with the corresponding groove in the connecting rod (photo).

2 If the old bearings are nearly new and are being refitted, then ensure they are replaced in their correct locations on the correct rods.

3 Generously lubricate the crankpin journals with engine oil, and turn the crankshaft so that the crankpin is in the most advantageous position for the connecting rod to be drawn on to it.

4 If not already done, wipe the connecting rod bearing cap and back of the shell bearing clean, and fit the shell bearing in position ensuring that the locating tongue at the back of the bearing engages with the locating groove in the connecting rod cap.

5 Make sure the cap fits the correct rod by checking the matching marks already made (photo).

6 Generously lubricate the shell bearing and offer up the connecting rod bearing cap to the connecting rod. It is advisable to use new cap bolts (photo).

7 Fit the connecting rod bolts on oiled threads and tighten them with a torque spanner to 25 lb/ft (photo).

8 Oil the cylinder bores well for initial lubrication.

47 Tappet/Camshaft - replacement

1 IMPORTANT. Replace the tappets in their respective bores before replacing the camshaft. They cannot be replaced afterwards (photo).

2 Wipe the camshaft bearing journals clean and lubricate them generously with engine oil. Ensure the small oil hole in the centre of the camshaft is clear.

3 Insert the camshaft into the crankcase gently, taking care not to damage the camshaft bearings with the cams (photo).

4 Replace the camshaft locating plate with the extension legs pointing towards the crankshaft, and tighten down the two retaining bolts and washers (Fig 1:10).

45.4

45.5

Inserting piston into cylinder and clamping rings with a ring clamp. Arrow indicates notch at front edge of piston

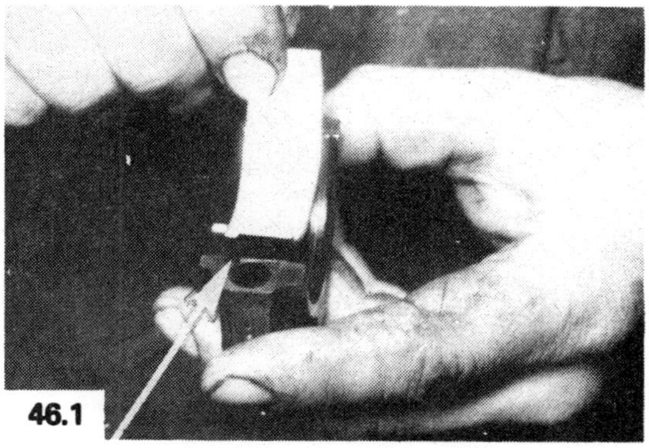
46.1

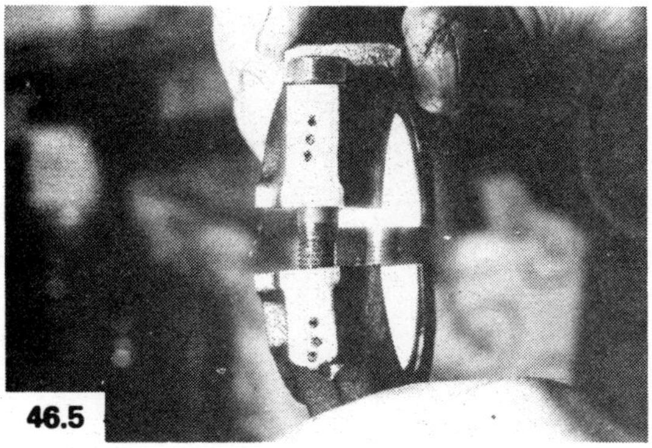
46.5

Showing the matching notches for the shell bearing and the alignment marks for the connecting rod and cap

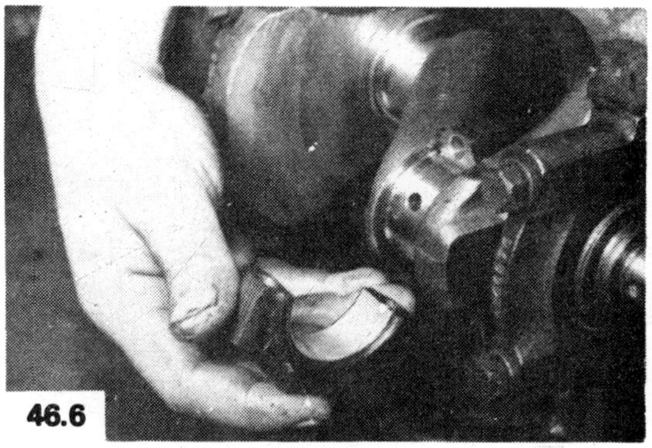
46.6

Offering up the connecting rod bearing cap

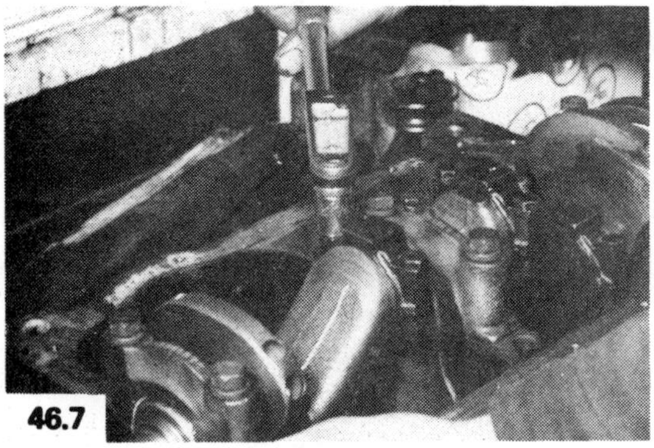
46.7

Tightening the connecting rod cap bolts

47.1

Replacing tappets

47.3

Replacing the camshaft

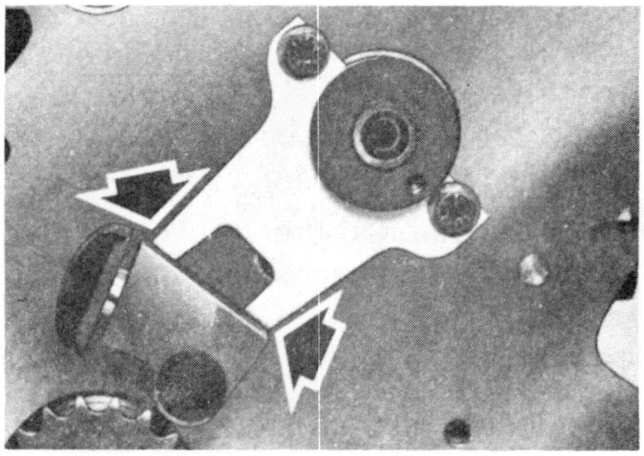
Fig 1:10 Camshaft locating plate showing correct positioning of two extension legs for positioning the timing chain tensioner unit

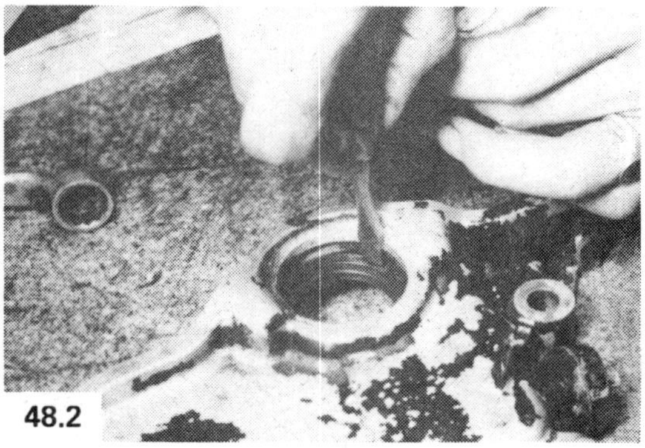
48.2

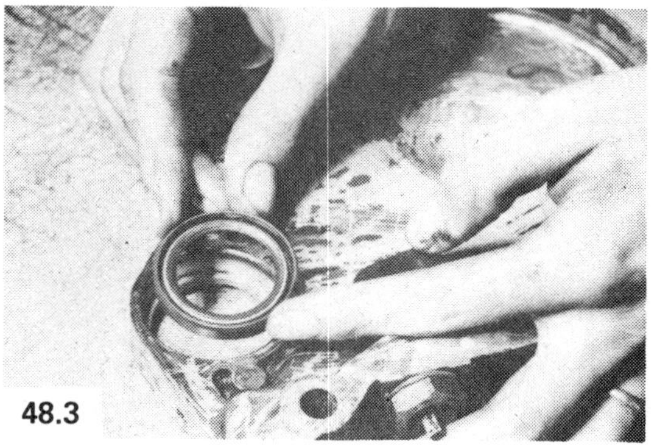
48.3

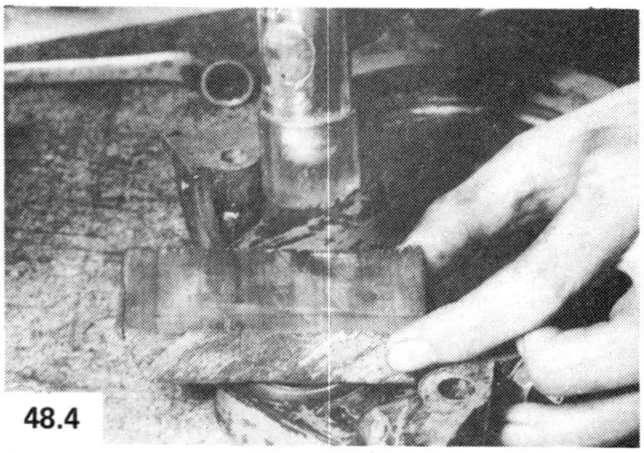
48.4

Removing and refitting the timing cover crankshaft oil seal

48 Timing sprockets, timing chain, cover and tensioner - replacement

1 This section describes the replacement procedure as part of the general overhaul of the engine, and assumes that the engine is removed from the car. If, however, the timing gear has been removed with the engine in the car, the following additional points should be noted when refitting the timing case as described in paragraph 7 following: The front edge of the sump where the section of gasket between it and the timing case is fitted, must be thoroughly cleaned of all remaining traces of gasket and sealing compound. A new piece of gasket, cut from either a whole new sump gasket or material of identical composition and thickness, must be put into position using a sealing compound such as Hermetite or Wellseal. It must fit exactly, particularly where it joins in the angle between the sump and the front face of the engine block. The lower edge of the timing case must be similarly covered with sealing compound. When the timing case is refitted, the gap between the bottom of the timing case and the sump will be minimal and great care will be required to keep the piece of sump gasket in position. Replace the four set screws finger tight only and finally tighten them when all the other procedures for refitting the timing case have been completed.

2 It is advisable to fit a new oil seal into the timing case, so first of all drive out the old one (photo).

3 Place the new seal in position with the lip facing the inside of the cover (photo).

4 Drive the seal home with a block of wood and a mallet and then, using a suitably shaped piece of wood, recess it ¼ in.below the face of the cover. It is possible to do this work with the cover fitted to the engine and the engine in the car (photo).

5 If a new chain has not been obtained as a complete unit, it may have been necessary to buy a length of chain and a connecting link. Having ensured that the total number of links, including the connector, is the same as on the old chain, first join the two ends to the connector pin link. Place the link bar over the pins and clip the spring so that both ends engage in the pin grooves fully.

6 IMPORTANT: When assembling the chain to the sprockets ensure that the closed end of the chain link clip leads in the direction of travel of the chain.

7 Note that the camshaft sprocket has a locating peg on the inner face which engages with a hole in the end of the camshaft (photo).

8 Next assemble the chain to the two timing sprockets so that a straight line will pass through the centre of both wheels and the timing marks, with the timing marks facing each other (photo).

9 Next position the camshaft and the crankshaft so that the locating hole and key are at 10 o'clock and 6 o'clock respectively (with the engine inverted). This position will mean the minimum of fiddling when the sprockets and chain assembly is replaced in its correct position (photo).

10 Holding the assembled timing sprockets and chain so that they cannot separate, place them on to the camshaft and crankshaft together so that the lug and keyway fit in their respective places. Care is necessary to keep the sprockets in their relative positions in the chain as assembled (Fig 1:11) (photo).

11 Replace the bolt and washer holding the camshaft sprocket in position and, locking the sprocket with a screwdriver, tighten the bolt. DO NOT REVOLVE THE CRANKSHAFT DURING OR AFTER THIS – THE OIL PUMP INSTALLATION WILL THUS BE SIMPLIFIED (photo).

12 Take the automatic tensioning device and compress the shoe into the body until it reaches the last notch on the piston where it will lock in position. The whole unit may then be placed into position on the projecting dowel between the two timing sprockets. The top edge of the unit should abut the two extensions of the camshaft thrust plate. When the chain and sprockets are reassembled, the pad can be pressed in to release it from the notch in the piston. Spring pressure will then be applied to the chain.

13 With a new gasket fitted to the face of the block, lift the timing case into position.

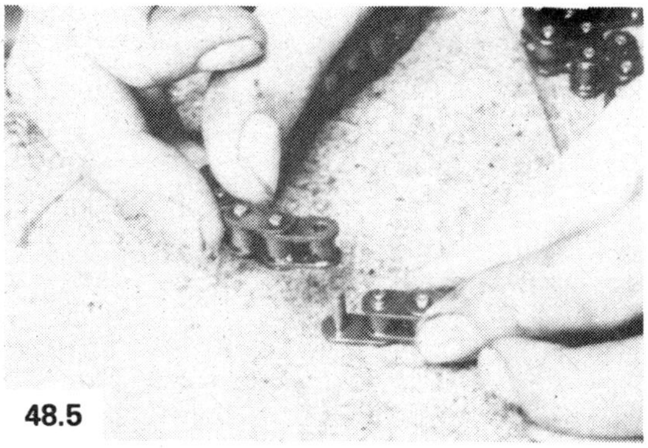

48.5 Joining up the timing chain

48.7 Camshaft sprocket locating peg

48.8 Sprocket alignment marks

48.9 The positioning of camshaft peg and crankshaft key for ease of assembly

48.10

48.11

Replacing the timing chain and sprockets assembly and tightening the camshaft sprocket bolt

48.15 Replacing the crankshaft pulley to line up the oil seal and cover

Fig 1:11 Timing sprockets, chain and tensioner installed and aligned

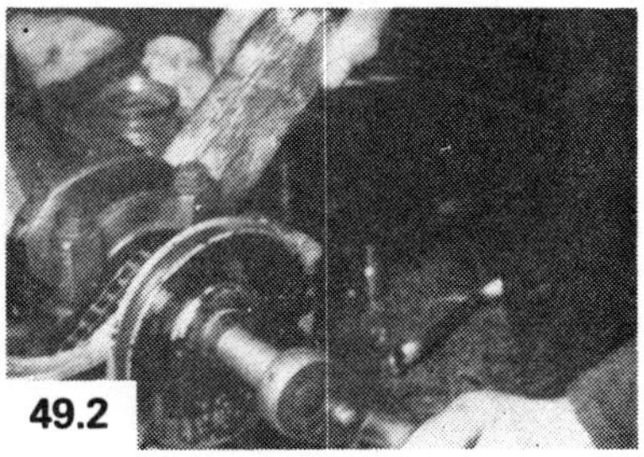

49.2

Tightening crankshaft pulley bolt

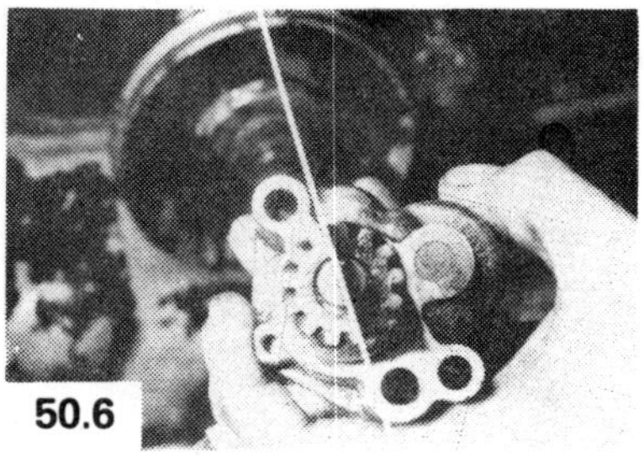

50.6

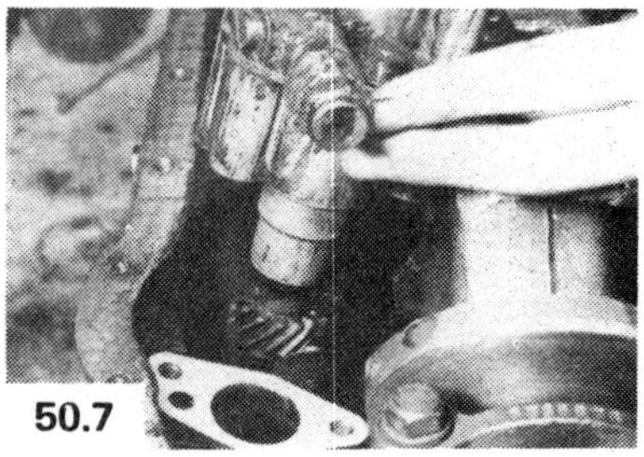

50.7

Aligning the oil pump spindle slot prior to offering up the pump to the crankcase

50.10

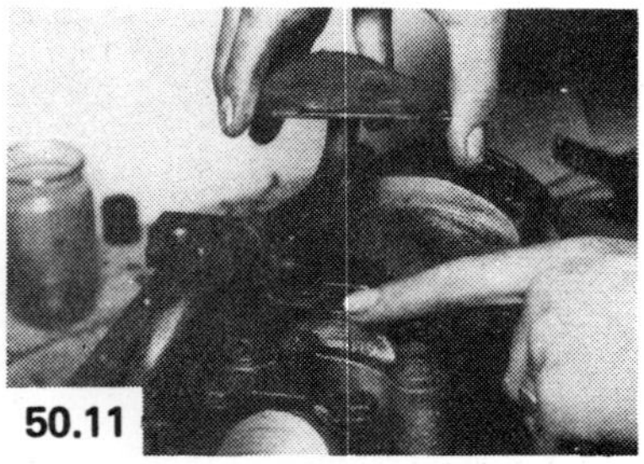

50.11

Tightening the pump securing bolts and fitting the suction pipe and strainer

14 Replace the mounting bolts, finger tight, and put the alternator bracket on the stud before replacing the nut.
15 Temporarily replace the crankshaft pulley wheel on to the end of the crankshaft, lining up the key in the shaft with the keyway in the pulley. This centralises the oil seal before the timing case bolts have been tightened up (photo).
16 Tighten down the cover bolts.

49 Crankshaft pulley wheel - replacement

1 With the timing cover located and bolted up as described in Section 48, fit the pulley on the crankshaft so that the key in the crankshaft fits in the pulley keyway. If there is any slackness of fit of the key in the shaft or the keyway, replace the key and clean up the slot or keyway as necessary.
2 Replace the large pulley washer followed by the lockwasher and bolt. Hold the crankshaft steady with a wooden block placed between the crankshaft web and the crankcase, and tighten up the bolt. If the engine is in the car, engage a gear to hold the crankshaft while the bolt is being tightened (photo).

50 Oil pump - replacement

1 The oil pump spindle has an offset slot in the end of the impeller shaft which drives the distributor. It is, therefore, important that the drive gear is correctly meshed to the camshaft skew gear, otherwise the ignition timing will be incorrect
2 If for any reason the drive gear has been removed from the oil pump spindle, make sure that the slot in the shaft is lined up with a space between two gear teeth.
3 Next examine the pump, be it a new one or original, and ensure that the oilway drilling at the end of the body in the upper side of the spindle bush is clear. This oilway delivers oil for lubrication of the drive gear. Also check that the vent plug hole is clear.
4 Provided that the crankshaft has not been revolved since it was set at TDC on No 1 compression (for pump removal), or since the valve timing gear was reassembled as described in Section 48 the next paragraph may be ignored.
5 If the crankshaft has been turned, it is necessary to ensure that the piston TDC for No 1 is set on the compression stroke and not the exhaust stroke. To do this set No 1 piston at TDC and examine the first two cams on the camshaft. If the lowest points on the cams are uppermost (meaning both valves would be closed), then TDC No 1 is on compression which is what is required. If this is not the case turn the crankshaft through one revolution.
6 With the crankshaft (and therefore the camshaft) in the correct position, set the spindle of the oil pump so that the long side of the offset slot lines up with the holes in the body of the pump (Fig 1:13) (photo).
7 Place a new gasket on the crankcase flange, put the mounting bolts through the pump body and line it up above the crankcase so that when installed it will not need any movement to line up the bolt holes (photo).
8 When replaced and viewed from the other end the slot offset is to the rear and the slot is angled 14° anticlockwise from a line at right angles to the centre line of the crankshaft (Fig 1:14).
9 If the gear should be mistakenly meshed, even one tooth out of position, it will be quite obvious. In such cases remove the pump, re-align the spindle as described in paragraph 6, and replace and check it again.
10 Tighten down the holding bolts when the pump is correctly positioned (photo).
11 Replace the suction pipe and strainer and, before finally tightening the pipe union into the pump body, fix the mounting clip to the centre main bearing cap (photo).

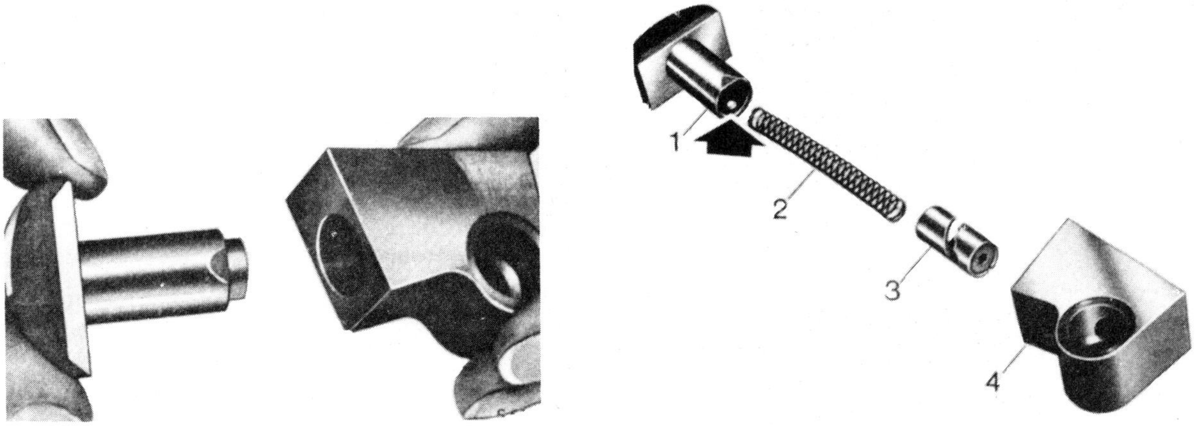

Fig 1:12 EXPLODED AND ASSEMBLED VIEW OF TIMING CHAIN TENSIONER UNIT
1 Sleeve 2 Spring 3 Plunger 4 Body
Arrow: Locking pip

Fig 1:13 Oil pump. Correct position of spindle slot before inserting the pump into the crankcase

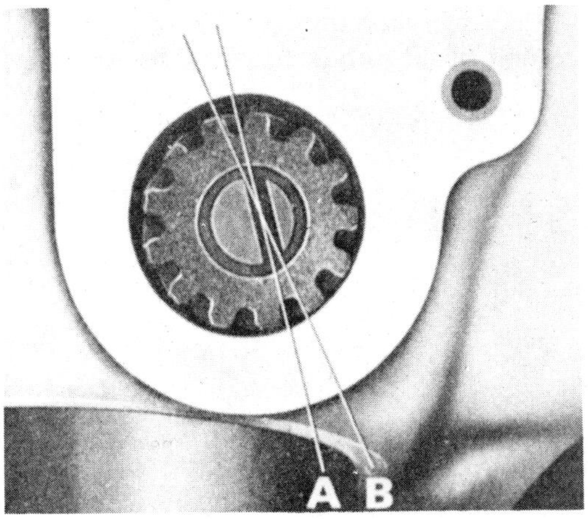

Fig 1:14 Oil pump. Correct position of spindle slot after pump installed. Angle A—B is tolerance

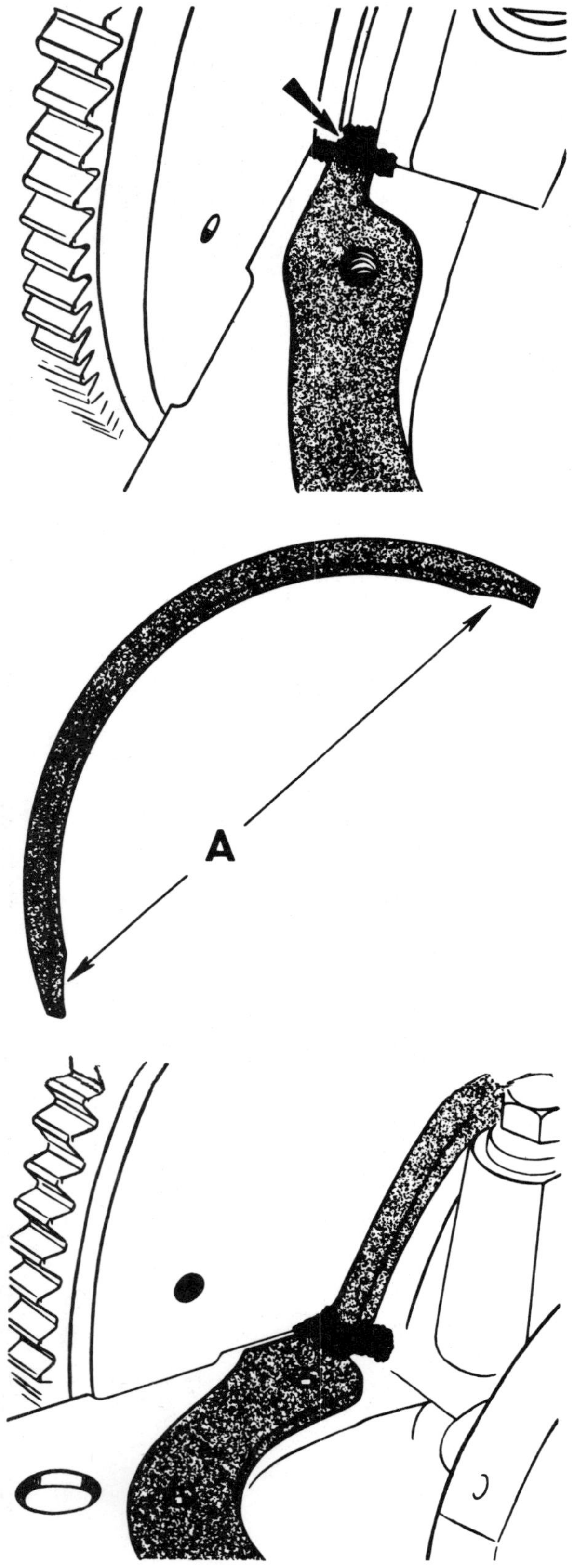

Fig 1:15 SUMP GASKET REPLACEMENT
Diagram to show fitting at rear main bearing cap. Note chamfers on cork strip 'A'

51 Sump - replacement

1 Before replacing the sump, the timing case must be fitted. If the engine has undergone an overhaul to big end and main bearings, check that all bearing caps have been properly tightened down, the oil pump replaced and that nothing that does not belong there has been left inside the crankcase.
2 Clean up, if not already done, the mating surfaces of the crankcase and the sump, and ensure that the outer groove in the rear main bearing cap is clean. Do NOT remove any of the sealer which may have exuded from the recent replacement of the bearing cap, unless it has been left so long that it has gone completely hard.
3 Apply sealer to the ends of the groove in the bearing cap and apply a proprietary non-hardening jointing compound to the face of the crankcase (photo).
4 Place the sump gasket in position ensuring the rear ends are firmly engaged in the groove and bedded in the sealing compound (photo).
5 Apply a little more sealer and then place the cork strip gasket into the groove, making sure that the chamfered ends face inwards overlapping the ends of the main sump gasket. Make sure that an equal amount at each end overlaps the sump gasket ends. It is a good idea to warm and shape the strip before fitting it (photo).
6 Replace the sump and, when fitting, ensure that the gaskets are not displaced. Some care is needed at the front where the gasket curves over the timing case (photo).
7 Replace all the set screws and tighten the sump down evenly.

52 Engine mounting brackets - replacement

1 If the engine mounting brackets have been removed, replace them now, ensuring that the bolts are treated with a suitable locking compound such as 'Loctite' (photo).

53 Flywheel - replacement

1 If the starter ring gear needs renewal proceed as described in Section 36.
2 Usually the flywheel is such a close fit on the crankshaft flange that it is not possible to simply put it on and replace the bolts. Help will be needed.
3 Note that there is a dowel peg on the crankshaft which locates in a hole in the flywheel (photo).
4 Offer up the flywheel to the crankshaft, hold it square and pick up the threads with the four mounting bolts. Then turn each bolt no more than ½ turn at a time and steadily draw the flywheel on.
5 When the flywheel is safely located on the shaft, remove the bolts **and apply a small quantity of sealer to the centre only of each bolt thread (This seals the holes in the crankshaft flange through which oil could otherwise seep, and also locks the bolts. The sealer is kept away from the end of the bolt to avoid the possibility of any falling inside the crankcase).**
6 Replace the bolts and tighten them up evenly in rotation so that the flywheel is held square. If this is not done carefully, there is a possibility of damaging the flange and corresponding bore in the flywheel with resulting disturbance of the finely set balance.
7 Tighten up the bolts to the specified torque of 25 lb/ft (overtightening can distort the mounting flange with serious consequences such as imbalance or the flywheel running out of true) (photo).
8 See that the bush for the gearbox input shaft spigot is in good condition and in position in the end of the crankshaft (photo).

54 Valves and valve springs - reassembly to cylinder head

1 Gather together all the new or reground valves and ensure that if the old valves are being replaced they will return into their original positions.

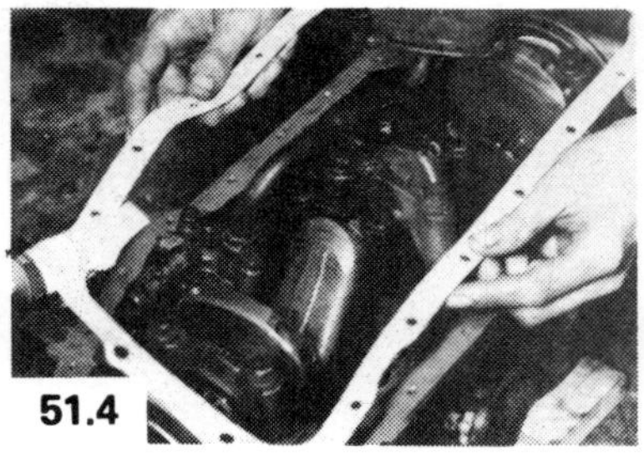

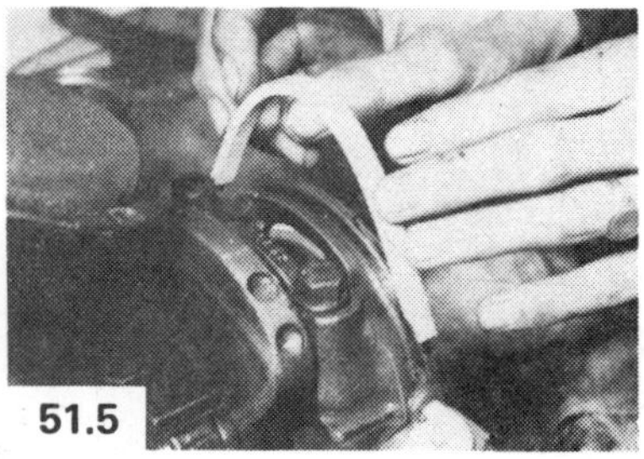

Replacing the sump gaskets ...

... and sump

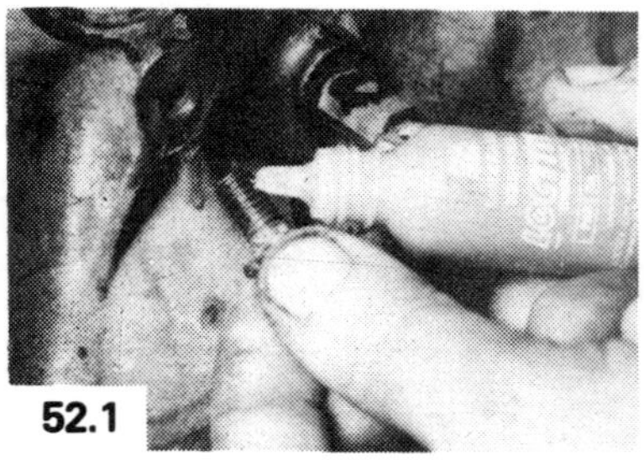

Applying 'Loctite' to the engine mounting bracket bolts

Matching the flywheel dowel peg and hole and ...

... tightening the securing bolts to the correct torque

Checking the condition of the bush for the gearbox input shaft location

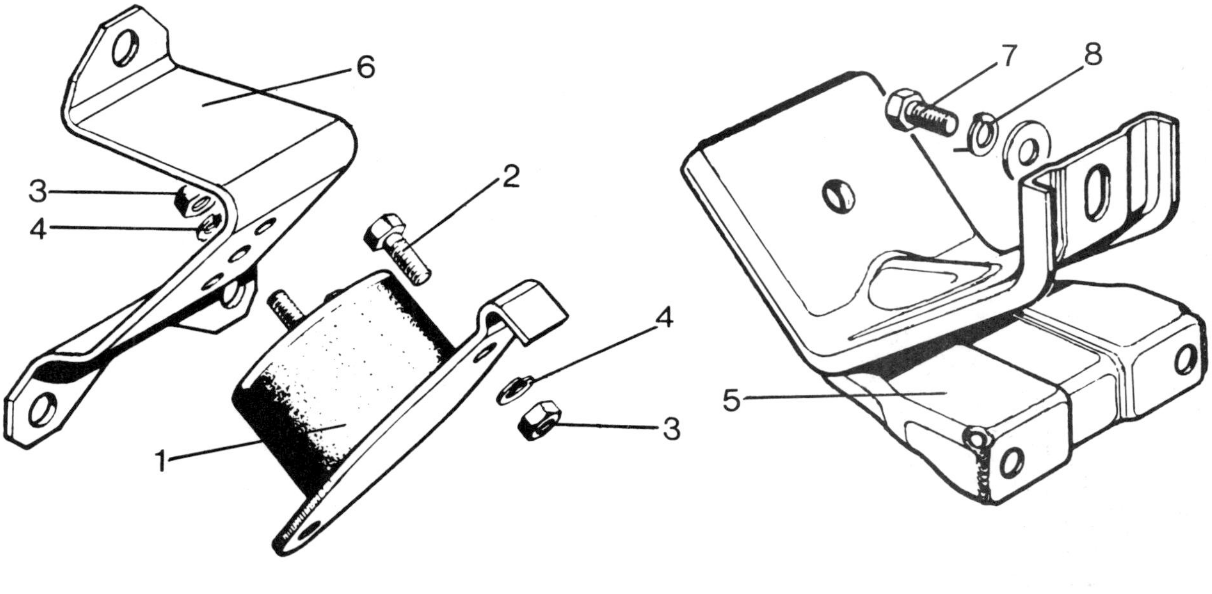

Fig 1:16 ENGINE MOUNTINGS AND BRACKETS

1 Flexible mounting
2 Bolt
3 Nut
4 Lockwasher
5 Bracket - right hand
6 Bracket - left hand
7 Bolt
8 Lockwasher

54.3

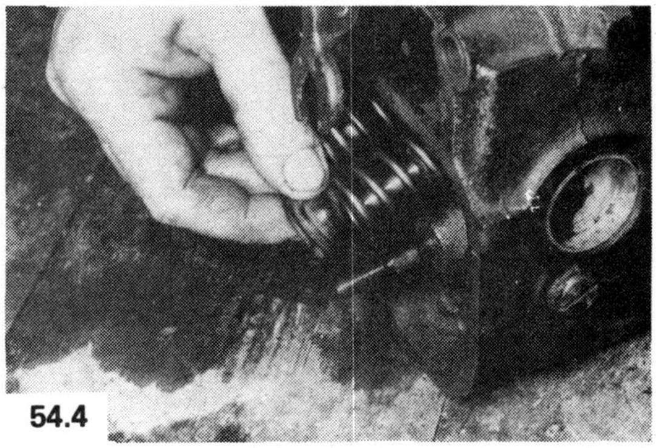

54.4

54.5

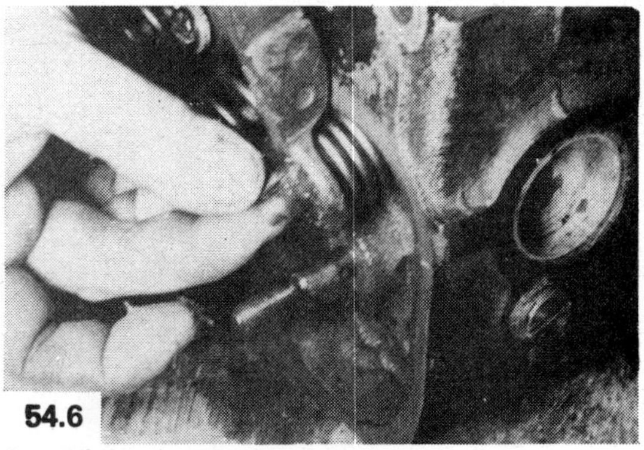

54.6

Assembling the valves and springs to the cylinder head

2 Ensure that all valves and springs are clean and free from carbon deposits and that the ports and valve guides in the cylinder head have no carbon dust or valve grinding paste left in them. Then fit a new sealing collar and securing clip over each inlet valve guide hole shoulder (Fig. 1:17).

3 Starting at one end of the cylinder head, take the appropriate valve, oil the stem and put it in the guide. Then put the screw head of the valve spring compressor over the valve head and place the valve spring over the other end of the valve stem (photo).

4 Then place the cap over the spring with the recessed part inside the coil of the spring (photo).

5 Place the end of the spring compressor over the cap and valve stem and screw it up until the spring is compressed past the groove in the valve stem. Then put a little grease round the groove (photo).

6 Place the two halves of the split collar (collets) into the groove with the narrow ends pointing towards the head. The grease will hold them in the groove (photo).

7 Release the compressor slowly and carefully, making sure that the collets are not dislodged from the groove. When the compressor is fully released, the top edges of the split collets should be in line with each other. It is quite good practice to give the top of each spring a smart tap with a soft mallet when assembly is complete to ensure that the collets are properly settled.

55 Cylinder head - replacement

1 With the valves and springs reassembled, examine the head to make sure that the mating face is perfectly clean and smooth and that no traces of gasket or other compounds are left. Any scores, grooves or burrs should be carefully cleaned up with a fine file.

2 Examine the face of the cylinder block in the same way as the head. Make sure that the concave cups of the tappets are clean and free from sludge.

3 Prior to December 1975, Vauxhall recommended that a replacement head gasket should be coated on each side with a jointing compound such as 'Wellseal'. After this date it was recommended that a 1 to 2 mm (0.04 to 0.08 in) bead of Holymar PL32/M jointing compound was placed along the pushrod rail of the cylinder block face and on the top face of the cylinder block gasket.

4 Most head gaskets indicate which side is the top, but on the Viva there can be no confusion as it is not symmetrical, there being two distinctive shaped water jacket holes at the front end.

5 Place the gasket in position on the block and lower the head onto it. Replace all the cylinder head bolts (it will be obvious where the longer ones go, but in fact they are Nos 7,4,2,5 and 9 in Fig 1:18 and lightly tighten them (photo).

6 Proceed with a torque wrench to tighten down the bolts ¼-½ turn at a time in the progressive order as indicated in Fig 1:18. This tightening sequence should continue until each bolt is down to a torque of 43 lb/ft. If, in the early stages, any one bolt is obviously slacker than the rest, it should be tightened equal to the others even if it may require a turn or so out of sequence. The whole point of the procedure is to keep the tightening stresses even over the whole head, so that it goes down level and undistorted (photo).

56 Valve rocker arms and pushrods - replacement

1 The rocker gear can be replaced with the head either on or off the engine. The only part of the procedure to watch is that the rocker nuts must not be screwed down too far or it will not be possible to replace the pushrods.

2 First place the washer over the rocker stud and then the light spring, with the narrower end down (photo).

3 Next put the rocker arm over the stud followed by the pivot ball. Make sure that the spring fits snugly round the rocker arm centre

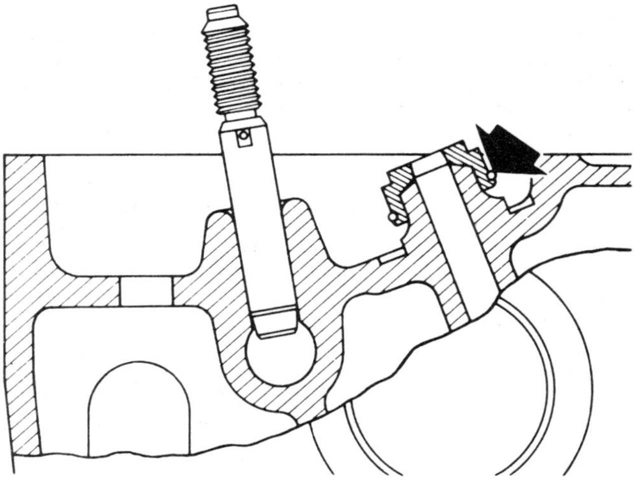

Fig. 1:17 Inlet valve stem sealing collar and securing clip (arrowed)

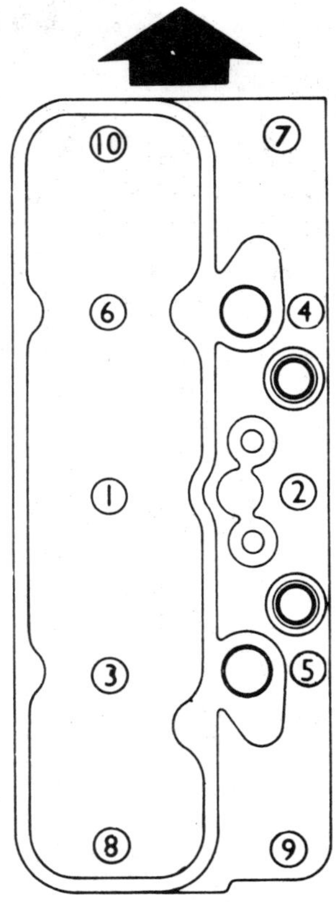

Fig 1:18 Cylinder head bolt tightening sequence

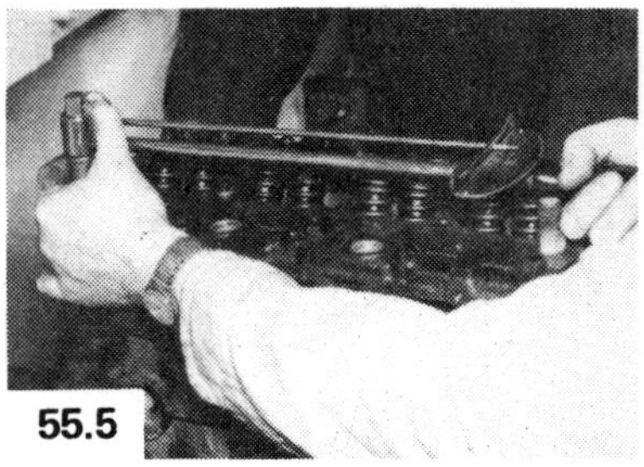

Replacing the cylinder head and tightening the bolts

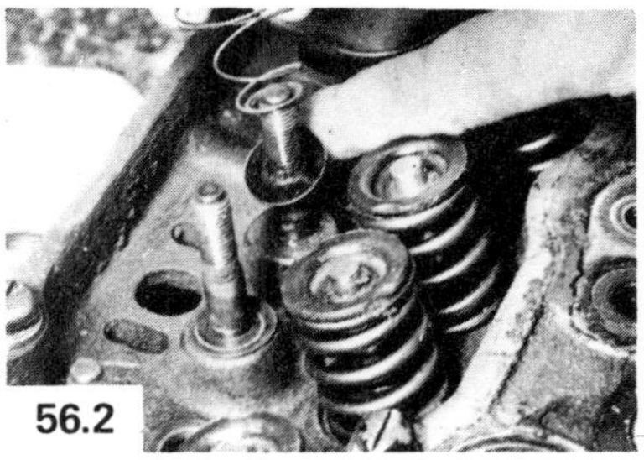

Replacing the rocker assembly

section and that the two bearing surfaces of the interior of the arm and the ball face, are clean and lubricated with engine oil (photo).
4 Oil the stud thread with SAE 90 oil and fit the nut with the self-locking collar uppermost. Screw it down until the locking collar is on the stud (photo).
5 Replace the pushrods through the head in the holes in line with each valve and rocker stud. It is easy to drop the pushrods inadvertently and if they fall at an angle, the lower end could get past the tappet and drop down into the crankcase. This would mean certain removal of the head and possible removal of the sump in order to retrieve it. It would be advisable, therefore, to push the top end through a small 'collar' of stiff cardboard, or hold it in a bulldog clip so that it cannot drop through. When the lower end is felt to be firmly seated in the tappet recess the clip or collar may be removed (photo).
6 Next screw down the stud nut so that the top of the pushrod engages in the recess in the rocker arm and approximately 1/8 in. of the stud protrudes above the top of each nut. It is as well to check the efficiency of the self-locking ring at this stage by checking that at least 3 lb/ft torque is needed to turn the nut. If it should be less the nuts could turn with vibration and should be renewed.

57 Valve clearances - adjustment

1 The valve clearances should be set by placing a .010 in feeler gauge of the correct size (see Specifications) between the valve stem and rocker arm and turning the nut until the clearance is correct. This operation is carried out when each valve is closed, and the appropriate cam is at its lowest point. The initial settings can be made with the engine cold but recheck them when the engine is warm (photo).
2 To ensure that each valve is in the correct position for checking the clearance, proceed as follows:
3 Set the crankshaft pulley marker to the TDC pointer cast in the timing case (the upper one - see Chapter 4), and the distributor rotor at No 1 spark plug lead position.
4 If the engine is being reassembled after an overhaul, it would be best to leave this job until it is back in the car. It will be easier to turn the engine over as needed.
5 With the No 1 piston on compression at TDC both valves for that cylinder (nos 1 and 2) may be set to their correct clearances as described in paragraph 1.
6 The firing order being 1,3,4,2, the crankshaft may then be rotated ½ revolution clockwise and No 3 piston will be at TDC so that valves Nos 5 and 6 may be adjusted. A further ½ revolution and No 4 piston is at TDC for valves Nos 7 and 8 to be checked and finally the last half revolution presents No 2 piston at TDC for valves Nos 3 and 4 to be checked. The table following shows the piston/valve relationship, and which are exhaust and inlet valves:

	Valve No	Piston No
Exhaust	1	1
Inlet	2	1
Inlet	3	2
Exhaust	4	2
Exhaust	5	3
Inlet	6	3
Inlet	7	4
Exhaust	8	4

7 Another way of setting the valves to check the clearance, is to do them in pairs, regardless of crankshaft setting. The table below shows the valves linked as pairs, and when either one of a pair is fully open (valve spring compressed) the other is fully closed and the clearance may be checked.

Valve fully open	Check and Adjust
Valve No 8	Valve No 1 exhaust
Valve No 6	Valve No 3 inlet
Valve No 4	Valve No 5 exhaust
Valve No 7	Valve No 2 inlet
Valve No 1	Valve No 8 exhaust
Valve No 3	Valve No 6 inlet
Valve No 5	Valve No 4 exhaust
Valve No 2	Valve No 7 inlet

If the order of the left hand column is followed, the minimum amount of engine turning will be necessary.

58 Inlet manifold - replacement

1 Fit a new aluminium gasket so that it lines up with the ports in the top of the cylinder head (photo).
2 Refit the manifold and replace the three bolts holding it down (photo).
3 The centre one will need holding with pliers in order to place it easily into the centre locating hole (photo).

59 Exhaust manifold - replacement

1 Place a new gasket in position, having made sure that the mating faces of both the manifold and the cylinder head are perfectly clean and flat.
2 Replace the manifold and when reconnecting the exhaust pipe fit a new gasket. Make sure that the inlet manifold brace is also refitted.

60 Crankcase ventilation valve

Some engines are fitted with a positive ventilation valve in the inlet manifold for the action of crankcase gases for combustion. This valve should be examined for any sign of deterioration in the spring or valve seat. Renew the unit if necessary. It is most important that the valve operates correctly and does not leak.

61 Final engine reassembly

1 All the components removed (see Section 8) should be replaced on the engine where possible before the engine is replaced in the car, as it is generally more easily done at bench level. One possible exception is the carburettor which is somewhat vulnerable and could be damaged when replacing the engine. If a pre-engaged starter motor is fitted it should be noted that it cannot be fixed with the engine mounting bracket in position when the engine is in the car. Even though the valve clearances have not been finally set, replace the rocker cover for protection. You need not fit the new gasket yet, as the cover will have to be removed once more (photo).

62 Engine replacement

Engine replacement is generally speaking a straightforward reversal of the removal sequence. The following hints and tips will, however be found useful:
1 If the engine and gearbox have both been removed together, it is best to replace them together. This will mean fitting the engine to the gearbox on the bench and will make the mating of the gearbox input shaft to the clutch that much easier (see Chapter 6). It will also obviate the need to disconnect the propeller shaft, which is necessary if the gearbox is removed and replaced separately from the engine. As the engine is lowered into the car, the front of the propeller shaft

Replacing the pushrods

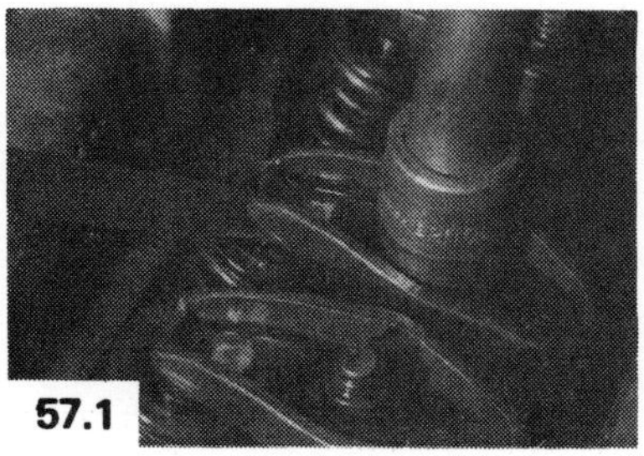

Adjusting the valve clearances

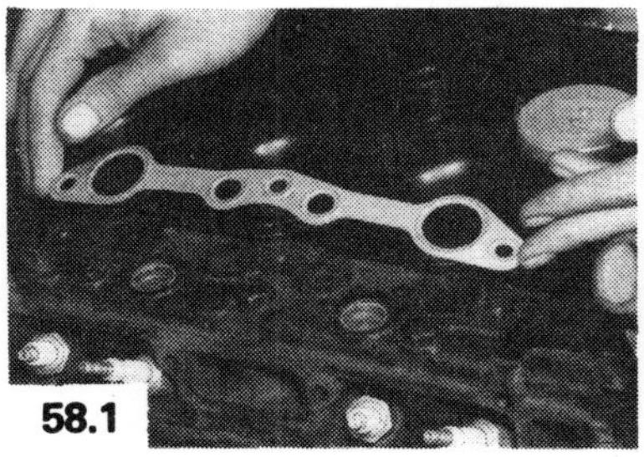

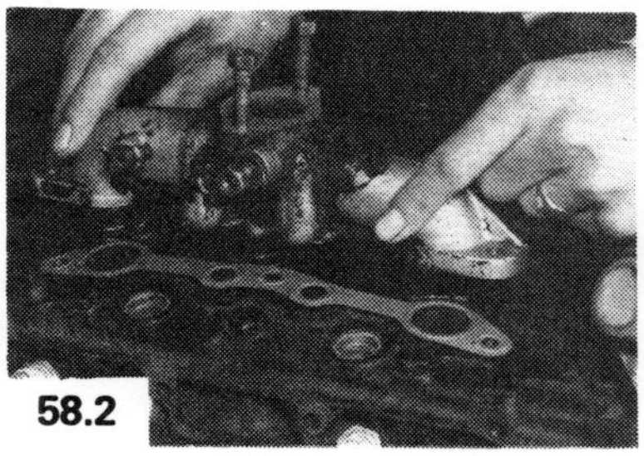

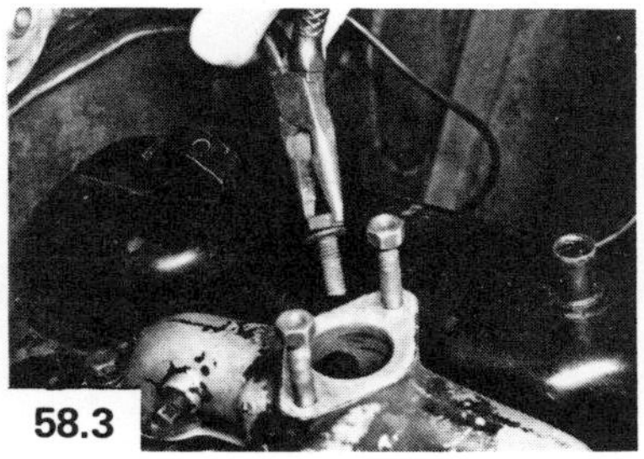

Replacing the gasket, inlet manifold and centre bolt

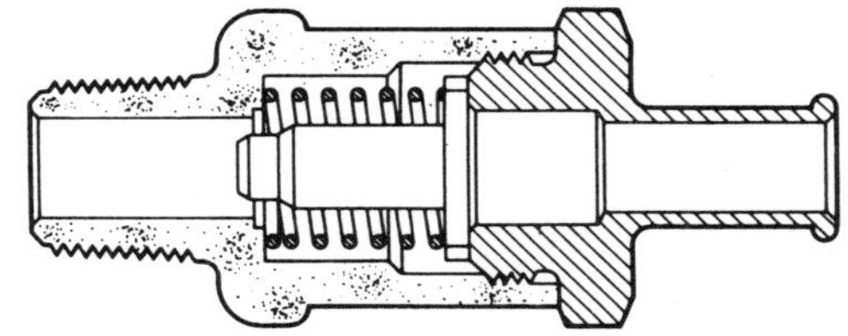
Fig 1:19 Crankcase ventilation valve - cross section

Replacing the rocker cover

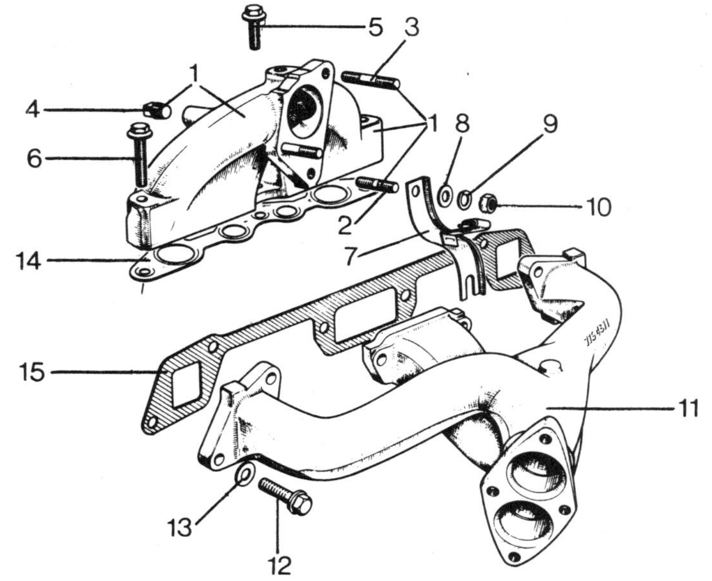

Fig 1:20 EXHAUST AND INLET MANIFOLDS – 1159 cc '90' ENGINE

1 Inlet manifold and studs assembly
2 Brace mounting stud
3 Carburettor mounting stud
4 Blanking plug
5 Inlet manifold mounting bolt
6 Inlet manifold mounting bolt
7 Manifold brace
8 Plain washer
9 Lockwasher
10 Nut
11 Exhaust manifold
12 Exhaust manifold mounting
13 Plain washer
14 Inlet manifold gasket
15 Exhaust manifold gasket

must be fed into the rear of the gearbox by someone underneath the car. It is also important when reconnecting the engine to the gearbox, that the bracing plate is correctly fitted between the bellhousing and sump flange as described in Chapter 6, Section 2, paragraph 16. This is important for the rigidity of the assembly.

2 It is helpful to replace the exhaust manifold before the engine is put back. It acts as a good sling attachment. Note also the remarks concerning starter motor replacement regarding engine mounting (see Chapter 10 - Starter motor - removal and replacement).

3 When the engine is being put back into the car make sure that it is watched every inch of the way to ensure that no pipes or wires get caught up or damaged. If, for any reason the engine will not go where you want it to, look and see why. Do not force anything. As soon as the engine is re-located on its forward mountings it must still be supported at its rear end due to the fact that the gearbox is not yet installed. This can be done by placing a block of wood behind the cylinder head.

4 If the engine is separated from the gearbox, smear a little grease on the tip of the gearbox input shaft.

5 Always fit new oil and air cleaner elements after an overhaul.

6 The bonnet will need two pairs of hands to support it when refitting it. Fix the bracket bolts and nuts just tight enough to hold it and then close the bonnet to ensure it is correctly lined up and central, before tightening them.

7 The following final check list should ensure that the engine starts safely and with the minimum of delay:

a) Fuel lines to pump and carburettor - connected and tightened.
b) Water hoses connected and clipped.
c) Radiator and engine drain taps closed.
d) Water system replenished.
e) Sump drain plug fitted and tight.
f) Oil in engine.
g) Oil in gearbox and level plug tight.
h) LT wires connected to distributor and coil.
i) Spark plugs tight.
j) Tappet clearances set correctly.
k) HT leads connected securely to distributor, spark plugs and coil.
l) Rotor arm replaced in distributor and pushed fully home.
m) Choke and throttle linkages connected.
n) Braided earthing cable, engine to frame reconnected.
o) Starter motor lead connected
p) Fan belt fitted and correctly tensioned.
q) Alternator leads connected
r) Battery charged and leads connected to clean terminals.

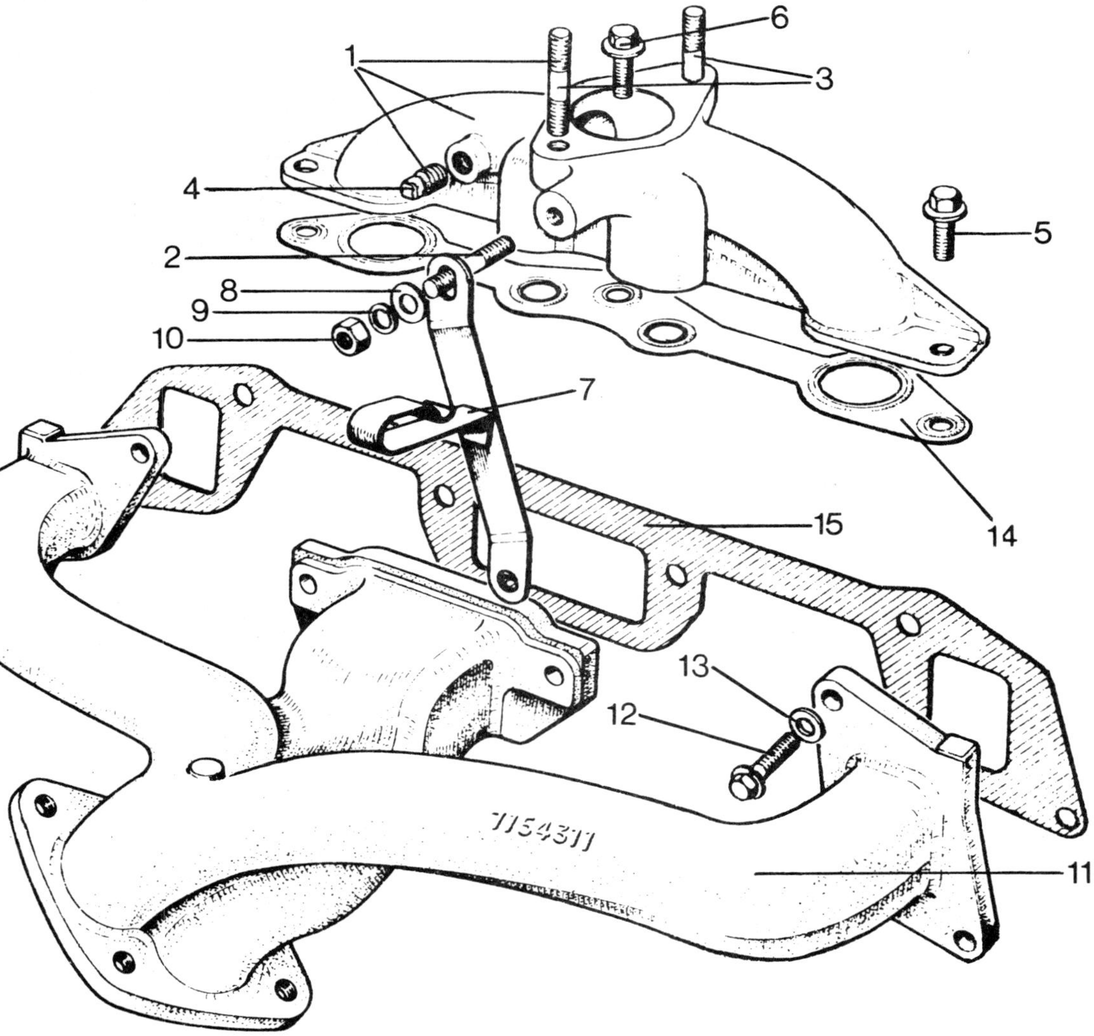

Fig 1:21 EXHAUST AND INLET MANIFOLDS – 1159 cc STANDARD ENGINE

1 Inlet manifold assembly
2 Brace stud
3 Carburettor stud
4 Plug
5 Manifold to head bolt
6 Centre manifold bolt
7 Brace
8 Washer
9 Lockwasher
10 Nut
11 Exhaust manifold
12 Screw
13 Washer
14 Inlet manifold gasket
15 Exhaust manifold gasket

Fault Finding Chart – Engine

Symptom	Reason/s	Remedy
Engine will not turn over when starter switch is operated	Flat battery Bad battery connections Bad connections at solenoid switch and/or starter motor	Check that battery is fully charged and that all connections are clean and tight.
	Starter motor jammed (inertia type)	Turn the square headed end of the starter motor shaft with a spanner to free it. Where a pre-engaged starter is fitted rock the car back and forth with a gear engaged. If this does not free pinion remove starter.
	Defective solenoid	Bridge the main terminals of the solenoid switch with a piece of heavy duty cable in order to operate the starter.
	Starter motor defective	Remove and overhaul starter motor.
Engine turns over normally but fails to fire and run	No spark at plugs	Check ignition system according to procedures given in Chapter 4.
	No fuel reaching engine	Check fuel system according to procedures given in Chapter 3.
	Too much fuel reaching the engine (flooding)	Check the fuel system as above.
Engine starts but runs unevenly and misfires	Ignition and/or fuel system faults	Check the ignition and fuel systems as though the engine had failed to start.
	Incorrect valve clearances	Check and reset clearances.
	Burnt out valves Blown cylinder head gasket	Remove cylinder head and examine and overhaul as necessary.
	Worn out piston rings Worn cylinder bores	Remove cylinder head and examine pistons and cylinder bores. Overhaul as necessary.
Lack of power	Ignition and/or fuel system faults	Check the ignition and fuel systems for correct ignition timing and carburettor settings.
	Incorrect valve clearances	Check and reset the clearances.
	Burnt out valves Blown cylinder head gasket	Remove cylinder head and examine and overhaul as necessary.
	Worn out piston rings Worn cylinder bores	Remove cylinder head and examine pistons and cylinder bores. Overhaul as necessary.
Excessive oil consumption	Oil leaks from crankshaft rear oil seal, timing cover gasket and oil seal, rocker cover gasket, oil filter gasket, sump gasket,sump plug washer	Identify source of leak and renew seal as appropriate.
	Worn piston rings or cylinder bores resulting in oil being burnt by engine. Smoky exhaust is an indication	Fit new rings or rebore cylinders and fit new pistons, depending on degree of wear.
	Worn valve guides and/or defective valve stem seals.	Remove cylinder heads and recondition valve stem bores and valves and seals as necessary.
Excessive mechanical noise from engine	Wrong valve to rocker clearances	Adjust valve clearances
	Worn crankshaft bearings Worn cylinders (piston slap).	Inspect and overhaul where necessary
	Slack or worn timing chain and sprockets	Adjust chain and/or inspect all timing mechanism.
Unusual vibration	Fan blade broken off	Break off another fan blade to balance fan until renewal is possible
	Broken engine/gearbox mounting	Renew mounting
	Misfiring on one or more cylinders	Check ignition system

NOTE: When investigating starting and uneven running faults do not be tempted into a snap diagnosis. Start from the beginning of the check procedure and follow it through. It will take less time in the long run. Poor performance from an engine in terms of power and economy is not normally diagnosed quickly. In any event the ignition and fuel systems must be checked first before assuming any further investigation needs to be made.

Chapter 2 Cooling system

Contents

Specifications

Type of system	Pressurised, pump assisted circulation with thermostat temperature control	
Capacity		
Without heater	9.2 pints	
With heater	10.2 pints	
Radiator	Flow capacity is 5 gallons (Imp) in 22 seconds (maximum) with a constant 2 foot head through a 1¼ inch bore pipe. Leak test pressure is 20 lbs/sq in Filler cap valve opens at between 13½ to 17½ lbs/sq in	
Fan belt	V pulley drive. Tension permits depression of ¼ in between generator and fan pulleys under thumb pressure.	
Thermostat	Western Thomson (valve opens upwards)	AC (valve opens downwards)
Opening temperature	85°–89°C	80°–84°C
Fully open temperature	102°C	98°C
Fully open lift from flange	.51 in (min)	.48 in (min)
Water pump		
Pulley flange fit on spindle	.0009 in to .0026 interference	
Rotor fit on spindle	.0004 in to .0021 interference	

1 General description

The engine cooling water is circulated by a thermo syphon, water pump assisted system, and the coolant is pressurised. This is both to prevent the loss of water down the overflow pipe with the radiator cap in position, and to prevent premature boiling in adverse conditions.

The radiator cap is, in effect, a safety valve designed to lift at a pressure of 13½-17½ lbs/sq in which means that the coolant can reach a temperature above 212°F (100°C) before it lifts the cap. It then boils off, steam escaping down the overflow pipe. When the temperature/pressure decreases, the cap reseats until the temperature/pressure builds up again. In addition there is a vacuum valve in the cap which permits air to enter the system when it cools down (Fig 2:1).

It is, therefore, important to check that the radiator cap fitted is of the correct specification (the relief pressure is stamped on the top) and in good condition, and that the spring behind the sealing washer has not weakened. Most garages have a device in which radiator caps can be tested.

The system functions in the following fashion: Cold water in the bottom of the radiator circulates up the lower radiator hose to the water pump, where it is pushed round the water passages in the cylinder block, helping to keep the cylinder bores and pistons cool.

The water then travels up into the cylinder head and circulates round the combustion spaces and valve seats absorbing more heat, and then, when the engine is at its proper operating temperature, travels out of the cylinder head, past the open thermostat into the upper radiator hose and so into the radiator head tank.

The water travels down the radiator where it is rapidly cooled by the in-rush of cold air through the radiator core, which is created by both the fan and the motion of the car. The water, now cold, reaches the bottom of the radiator, whereupon the cycle is repeated.

When the engine is cold the thermostat (a valve which opens and closes according to the temperature of the water) maintains the circulation of the same water in the engine, excluding that in the radiator.

The cooling system comprises the radiator, top and bottom water hoses, heater hoses (if heater/demister fitted), the impeller water pump, (mounted on the front of the engine it carries the fan blades and is driven by the fan belt), the thermostat and the two drain taps.

Only when the correct minimum operating temperature has been reached, as shown in the specification, does the thermostat begin to open, allowing water to return to the radiator.

2 Cooling system - draining

1 With the car on level ground drain the system as follows:
2 If the engine is cold, remove the filler cap from the radiator by turning the cap anticlockwise. If the engine is hot, having just been run, then turn the filler cap very slightly until the pressure in the system has had time to disperse. Use a rag over the cap to protect your hand from escaping steam. If, with the engine very hot, the cap is released suddenly, the drop in pressure can result in the water boiling. With the pressure released, the cap can be removed.
3 If antifreeze is in the radiator drain it into a clean bucket or bowl for re-use.
4 Detach the bottom hose from its union on the radiator, and remove the block drain plug. The cylinder block drain plug will be found on the right hand side of the block.
5 When the water has finished running, probe the cylinder block drain plug orifice to dislodge any sediment which may be blocking the orifice and preventing all the water draining out. **Note:** Draining the radiator will not drain the cylinder block fully - this is why a separate plug is incorportated in the block.

3 Cooling system - flushing

1 With time the cooling system will gradually lose its efficiency as the radiator becomes choked with rust scales, deposits from water and other sediment. To clean the system out, remove the radiator cap and detach the bottom hose from its union on the radiator. Leave a hose running in the radiator cap orifice for ten to fifteen minutes.
2 In very bad cases the radiator should be reverse flushed. This can be done with the radiator in position. Detach the bottom hose from the radiator, and remove the radiator cap; then, place a garden hose in the bottom hose orifice of the radiator. Water under pressure, is then forced up through the radiator and out of the header tank filler orifice.
3 The hose is then removed and placed in the filler orifice and the radiator washed out in the usual fashion.

4 Cooling system - filling

1 Replace the bottom hose and the cylinder block drain plug.
2 Fill the system slowly to ensure that air locks are minimised. If a heater is fitted it will be necessary to disconnect the outlet hose at the connector and fill up through the pipe to force air out of the heater matrix.
3 Do not fill the system higher than within ½ in.of the filler orifice. Overfilling will merely result in wastage, which is especially to be avoided when antifreeze is in use.
4 Only use antifreeze mixture with an ethylene glycol base. Vauxhall also use an anti-corrosion inhibitor - adding 1/8 pint each month for the first six months is recommended. Thereafter, it should be added each time the coolant is changed. If the inhibitor is used it is strongly recommended that Vauxhall antifreeze is also used to prevent possible chemical incompatibility.
5 Replace the filler cap and turn it firmly clockwise to lock it in position.

5 Radiator - removal, inspection, cleaning and replacement

1 Drain the cooling system as described in Section 2.
2 Undo the clip which holds the top water hose to the header tank and pull it off (photo).
3 Unclip the hose from the bottom of the radiator.
4 Remove the bolts (three each side) which hold the side plates and radiator to the body frame (photo).
5 Lift the radiator complete with plates from the car (photo).
6 With the radiator removed from the car any leaks can be soldered up or repaired with a substance such as 'Cataloy'. An unfortunate fault on some models is the fragile nature of the upper and lower radiator tank sections where the hose pipe unions are soldered in. With time they have been known to fracture due to the rocking motion of the flexibly mounted engine - or, of course, with rough handling when removing the flexible rubber hoses. If this occurs, remove the radiator and resolder the joint, running in a reinforcing fillet of solder. This reduces the flexibility of the thin tank material at the joint.

Clean out the inside of the radiator by flushing as described in Section 3. When the radiator is out of the car, it is well worthwhile to invert it for reverse flushing. Clean the exterior of the radiator by hosing down the matrix (honeycomb cooling material) with a strong water jet to clear away embedded dirt and insects which will impede the air flow.
7 If it is thought that the radiator may be partially blocked, it is possible to test it. Five gallons of water poured through a 1¼ in diameter pipe from a height of 2 ft above the filler cap should pass through the radiator in 22 seconds. If there are obvious indications of blockage a good proprietary chemical product such as 'Radflush' should be used to clear it.
8 Inspect the radiator hoses for cracks, internal or external perishing, and damage caused by overtightening of the securing clips. Replace the hoses as necessary. Examine the radiator hose securing clips and renew them if they are rusted or distorted. The drain taps should be renewed if leaking, but ensure the leak is not caused by a faulty washer behind the tap. If the tap is suspected, try a new washer first to see if this clears the trouble.
9 Replacement is a straightforward reversal of the removal procedure.

6 Thermostat - removal, testing and replacement

1 To remove the thermostat, partially drain the cooling system (four pints is enough), loosen the upper radiator hose at the thermostat elbow end and pull if off the elbow.
2 Unscrew the two set bolts and spring washers from the thermostat housing and lift the housing and paper gasket away (photo).
3 Remove the thermostat and suspend it by a piece of string in a saucepan of cold water together with a thermometer. Neither the thermostat nor the thermometer should touch the bottom of the saucepan, to ensure a false reading is not given.
4 Heat the water, stirring it gently with the thermometer to ensure temperature uniformity, and note when the thermostat begins to open. The temperature at which this should happen is given in the 'Specifications' (Fig 2:2).
5 Discard the thermostat if it opens too early. Continue heating the water until the thermostat is fully open. Then let it cool down naturally. If the thermostat will not open fully in boiling water, or does not close down as the water cools, then it must be exchanged for a new one. If a thermostat is unserviceable it is better to run without one rather than with one which is faulty.
6 If the thermostat is stuck open when cold, this will be apparent when removing it from the housing.
7 Replacing the thermostat is a reversal of the removal procedure. Remember to use a new paper gasket between the thermostat housing elbow and the thermostat. Renew the thermostat elbow if it is badly eaten away. Ensure that the thermostat jiggle pin is located at the highest point of the thermostat housing.

7 Water pump - removal and replacement

1 Partially drain the cooling system as described in Section 2.

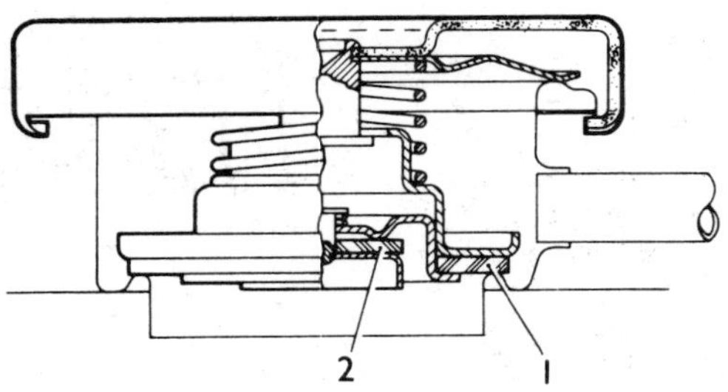

Fig 2:1 CROSS SECTION OF RADIATOR CAP
1 Pressure relief valve   2 Anti-vacuum valve

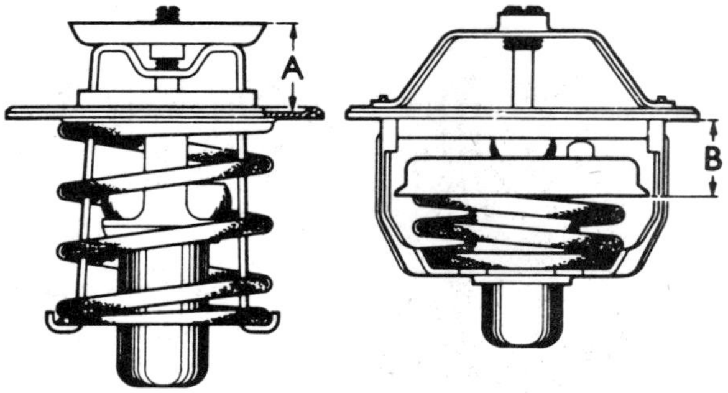

Fig 2:2 TWO TYPES OF THERMOSTAT
Dimension 'A' = .51 inch (fully open)
Dimension 'B' = .48 inch (fully open)

Fig 2:3 Water distribution tube withdrawn from cylinder head after removal of pump

5.2

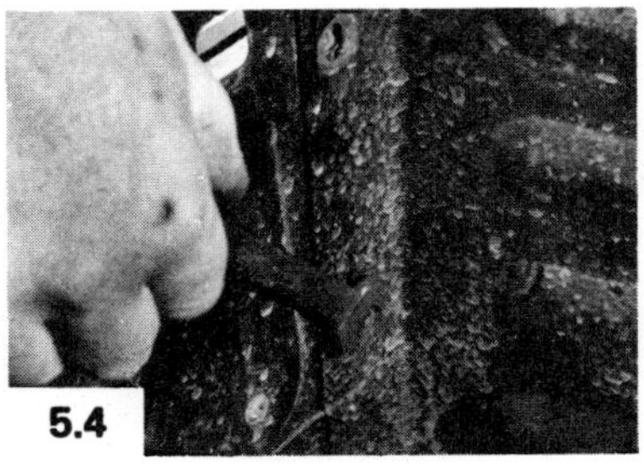

5.4

5.5

Removing the radiator assembly

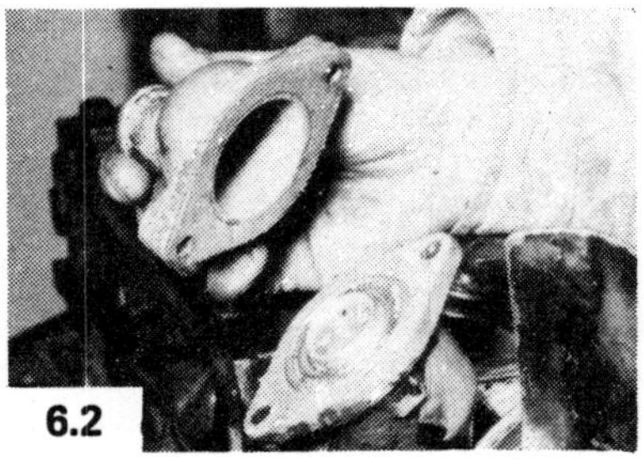

6.2

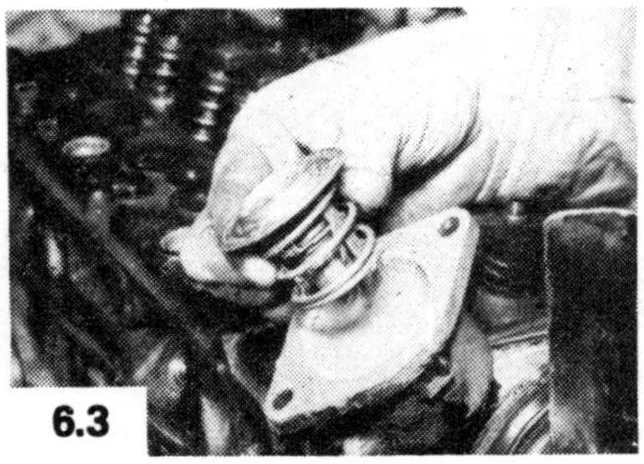

6.3

Removing the thermostat

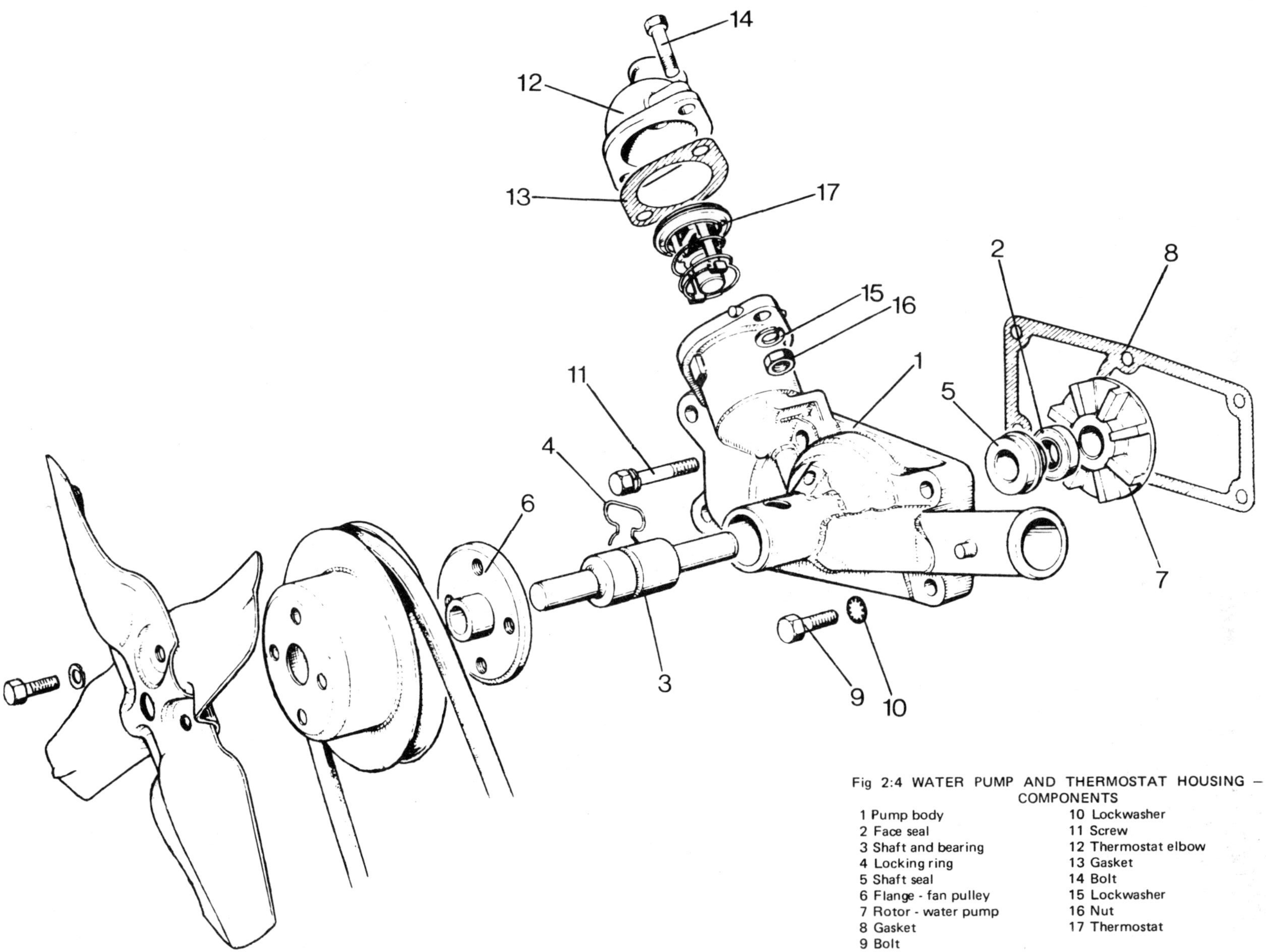

Fig 2:4 WATER PUMP AND THERMOSTAT HOUSING – COMPONENTS

1 Pump body
2 Face seal
3 Shaft and bearing
4 Locking ring
5 Shaft seal
6 Flange - fan pulley
7 Rotor - water pump
8 Gasket
9 Bolt
10 Lockwasher
11 Screw
12 Thermostat elbow
13 Gasket
14 Bolt
15 Lockwasher
16 Nut
17 Thermostat

2 Undo the clips which hold the hoses to the water pump and pull the hoses off.
3 Remove the fan belt.
4 Undo the six bolts which hold the pump body to the cylinder block. Lift the water pump away and remove the gasket. Whilst the pump is removed it is a good idea to check the condition of the water distributor tube in the cylinder head behind the pump. The tube can be withdrawn for cleaning or renewal (Fig 2:3).
5 Replacement is a straightforward reversal of the removal sequence NOTE: The fan belt tension must be correct when all is reassembled. If the belt is too tight undue strain will be placed on the water pump and alternator bearings, and if the belt is too loose it will slip and wear rapidly as well as giving rise to low electrical output from the alternator.

8 Water pump - dismantling and reassembly

1 Having removed the assembly from the engine complete with fan, remove the fan and drive pulley by unscrewing the four bolts and washers.
2 In order to get at the seal the rotor must be drawn off the shaft. There is no satisfactory way of doing this other than by using a puller - preferably one with two split claws so that the strain can be put onto the rotor astride the vanes which are the strongest part. If the rotor is a particularly tight fit any other way of attempting to remove it will probably break it (Fig 2:5).
3 With the rotor off, the seal can be withdrawn. Examine the seal seat on the body of the pump for damage or pitting.
4 If the shaft/bearing assembly needs to be renewed (due to excessive play in the bearings) it can now be removed. First lift out the locking ring and then heat the body of the pump in water to 82°C (180°F). The shaft can then be drifted out complete with flange, at the flange end of the body. When out, press or drift off the flange.
5 Reassembly sequence is in the reverse order, but care must be taken to fit everything back in certain positions.
6 If the shaft assembly is being renewed, first of all press the flange on to the shorter end of the new shaft.
7 Ensure that the flange boss is towards the end of the shaft and that the outer face of the flange is 3.46 in (88 mm) from the end of the body (Fig 2:6).
8 Reheat the pump body and install the shaft and bearing so that the groove in the bearing coincides with the groove in the body bore. Refit the locking ring (Fig 2:7).
9 Smear the face of the new seal and also around the body bore with the recommended grease, and install the seal. Note that the seal has a sleeve in the centre and this must be engaged properly. There are corresponding pips and grooves.
10 Press on the rotor (vanes inwards) so that the clearance between the flat face of the rotor and the pump body is .004 in (1 mm). Fit a new gasket when replacing the pump on the block.

9 Antifreeze coolant solution

1 Where temperatures are likely to drop below freezing point, the coolant system must be adequately protected by the addition of antifreeze. Even if you keep the engine warm at night it is possible for water to freeze in the radiator with the engine running in very cold conditions - particularly if the engine cooling is being adequately dealt with by the heater radiator. The thermostat stays closed and the radiator water does not circulate.
2 It is best to drain the coolant completely and flush out the system first.
3 The table below gives the details of the antifreeze percentage to be used (on the basis of the cooling system capacity of 10¼ pints).

%	Quantity	Complete protection
25	2½ pints	−11°C
35	3½ pints	−19°C
45	4½ pints	−29°C

4 Mix the required quantity of antifreeze with 4 pints of water and fill the system. Top up with water and then run the engine up to normal temperature with the heater turned on.

10 Fan belt - removal, replacement and adjustment

1 If the fan belt is worn, or has stretched unduly, it should be replaced. The most usual reason for replacement is breakage in service and every wise motorist will carry a spare.
2 Even though the belt may have broken and fallen off, go through the removal routine which is first of all to loosen the two alternator pivot bolts and the nut on the adjusting link (brace), then move the alternator inwards. Take the old belt off the three pulleys (Fig 2:8).
3 Put a new belt over the pulleys.
4 The alternator must now be used as a tensioner in effect, by pulling it away from the engine and locking it in the required position. This can call for some sustained effort unless the pivot bolts are slackened only a little so that the alternator is quite stiff to move. A lever between the alternator and block can help, but do not apply any pressure to the rear end shield or the cover may break. Always tighten the front long securing bolt first.
5 The tension of the belt midway between the alternator and water pump pulleys should be ¼ inch under thumb pressure. If in doubt it is better to be a little slack than tight. Only slipping will occur if it is too slack. If too tight, damage can be caused by excessive strain on the pulley bearings.
6 When the adjustment is right, tighten all the mounting bolts.
7 With a new belt, check the tension 250 miles after fitting.
8 Periodic checking of the belt tension is necessary and there is no hard and fast rule as to the most suitable interval, because fan belts do not necessarily stretch or wear to a pre-determined schedule. Assuming most owners check their own oil and water regularly, it is suggested as a good habit to check the fan belt tension every time the bonnet goes up. It takes only a second.

11 Water temperature gauge - fault diagnosis and rectification

1 If no reading is shown on the electrically operated water temperature gauge when the engine is hot and the ignition switched on, either the gauge, the sender unit, or the wiring in between is at fault. Alternatively, No 2 fuse may have blown. If the fuel gauge is also not working, check that the instrument voltage stabiliser is in order as described in Chapter 10.
2 Check the fuse and, if satisfactory, pull off the wire from the sender unit in the cylinder head and connect it to earth.
3 Switch on the ignition and check if the gauge is working. The needle should rise to the 'H' (hot) or 130°C mark which indicates that the sender unit must be renewed. To do this simply undo it and fit a replacement item.
4 If the sender unit is apparently working correctly, check the gauge and, if necessary, replace it. Details of how to remove the gauge are given in Chapter 10.

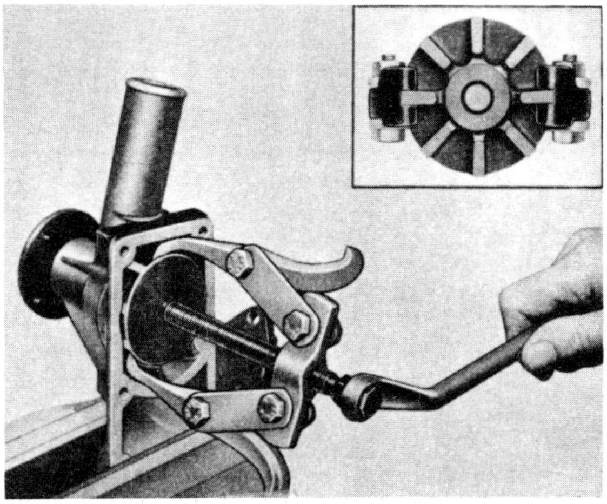

Fig 2:5 Using a puller to draw the water pump impeller from the shaft. Inset shows the puller claws on the ridged part of the impeller

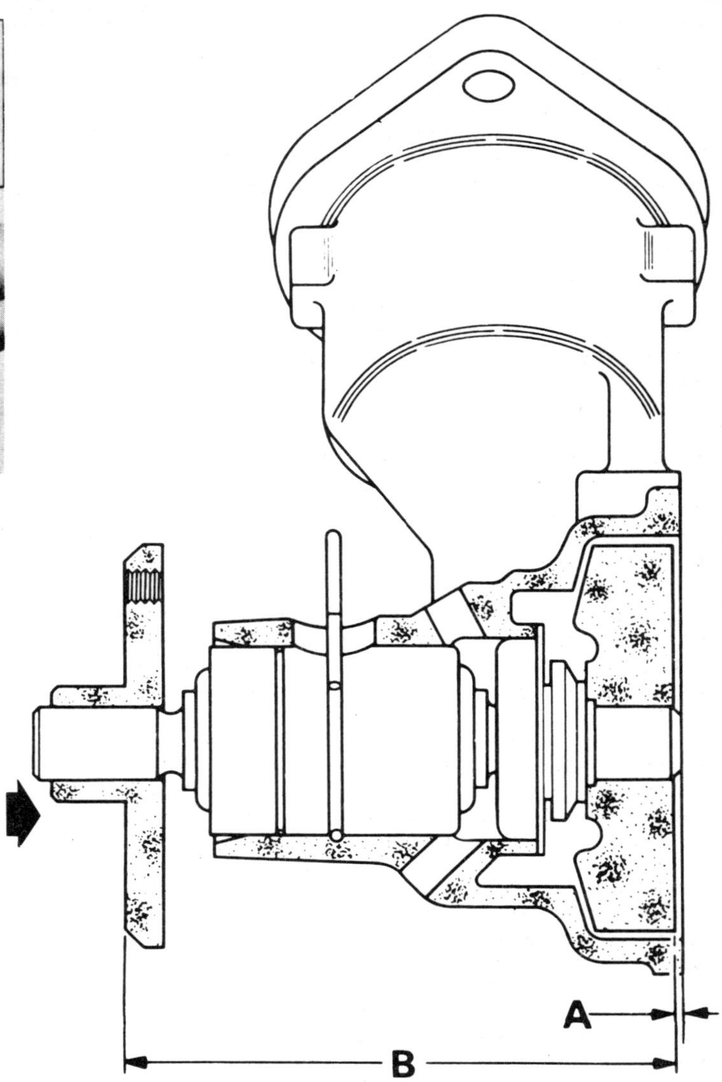

Fig 2:6 CROSS SECTION OF WATER PUMP TO SHOW IMPORTANT ASSEMBLY DIMENSIONS
A = .044—.046 inch B = 3.46 inch

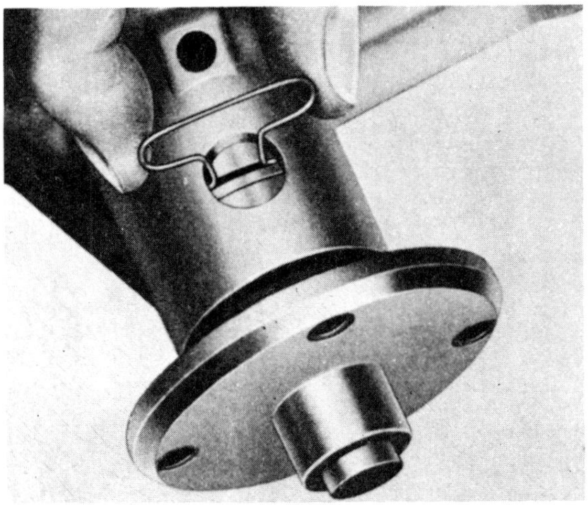

Fig 2:7 Position of water pump shaft assembly locking ring

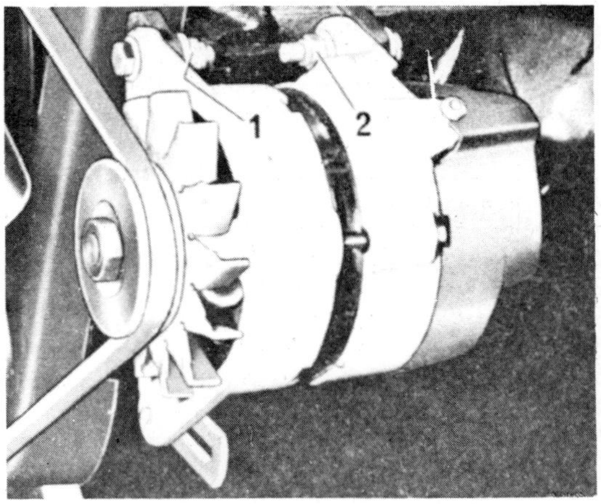

Fig 2:8 ALTERNATOR MOUNTING BOLTS FOR SLACKENING WHEN ADJUSTING FAN BELT
1 Front lug bolt (tighten first) 3 Adjusting brace clamp bolt
2 Rear lug bolt

Fault Finding Chart – Cooling System

Symptom	Reason/s	Remedy
Loss of coolant	Leak in system	Examine all hoses, hose connections, drain taps and the radiator and heater for signs of leakage when the engine is cold, then when hot and under pressure. Tighten clips, renew hoses and repair radiator as necessary
	Defective radiator pressure cap	Examine cap for defective seal or spring and renew if necessary.
	Overheating causing rapid evaporation due to excessive pressure in system forcing vapour past radiator cap	Check reasons for overheating.
	Blown cylinder head gasket causing excess pressure in cooling system forcing coolant past radiator cap overflow	Remove cylinder head for examination.
	Cracked block or head due to freezing	Strip engine and examine. Repair as required
)verheating	Insufficient coolant in system	Top up.
	Water pump not turning properly due to slack fan belt	Tighten fan belt.
	Kinked or collapsed water hoses causing restriction to circulation of coolant	Renew hose as required.
	Faulty thermostat (not opening properly)	Fit new thermostat.
	Engine out of tune	Check ignition setting and carburettor adjustments.
	Blocked radiator either internally or externally	Flush out cooling system and clean out cooling fins.
	Cylinder head gaskets blown forcing coolant out of system	Remove head and renew gasket.
	New engine not run-in	Restrict engine speed until run-in.
ngine running too cool	Missing or faulty thermostat	Fit new thermostat

Chapter 3 Fuel system and Carburation

Contents

Specifications

Fuel Pump

Make and type	AC 'YD'
Delivery pressure	2½–3½ lb in^2
Diaphragm spring load when compressed to ½ in	4½ lbs approx

Carburettors

Make	Zenith 301Z			
	1159 cc		1256 cc	
Air cleaner element	Gauze	Paper	Std	LC
Identification number	3345	3346	3398	3447
Choke	22 mm	22 mm	24 mm	24 mm
Main jet	100	107.5	105	105
Correction jet	180	175	165	175
Economy jet	60	–	80	65
Pilot jet	50	55	55	45
Pump injector	50	50	50	55
Needle valve	1.6 mm	1.6 mm	1.6 mm	1.6 mm
Needle valve washer thickness	1.0 mm	1.0 mm	1.0 mm	1.0 mm
Fuel level	23 mm below face of float chamber with float removed			
Econostat outlet	–	1 mm with .75 mm air bleed		
Idling speed	700 to 750 rpm			

Note. 1159 cc engines with paper element filters have positive crankcase ventilation. Positive crankcase ventilation is standard on 1256 cc engines

High altitude: Main jet	- 5000 - 7000 ft	97.5	105	102.5	102.5
	- 7000 - 10,000 ft	95	102.5	100	100
	- 10000 - 15000 ft	95	102.5	100	100
Correction jet	- 5000 - 7000 ft	180	175	165	175
	- 7000 - 10000 ft	180	175	165	175
	- 10000 - 15000 ft	185	180	170	175

Carburettors - Extra performance 1159 cc engines

Make	Zenith/Stromberg 150CDS - 3350 150CDST (automatic transmission) - 3352
Metering needle (standard)	5BN
Air valve spring colour	Blue
Fast idle cam	SA F5 (automatic transmission)
Cold start needle	J6
Needle valve	1.75 mm
Valve washer thickness	1.6 mm
Float setting	Invert carburettor. Highest point of floats 15.5 - 16.5 mm above face of body with valve seated

Jet adjuster setting - nominal	Two turns down from flush with bridge
High altitude needles - 5000 - 7000 ft	5BQ
- 7000 - 10000 ft	5BR
- 10000 - 15000 ft	5BS

Extra performance 1159 cc engine (European exhaust emission control)

Make	Zenith/Stromberg 150CD-SETV
Number (manual)	3393
(automatic)	3356
Metering needle	B5AY
Jet orifice	2.29 mm
Jet initial setting	2½ turns down from flush from bridge
Air valve spring colour	Blue
Fast idle cam	A6
Cold start needle	F7
Needle valve (fuel inlet)	1.75 mm
Needle valve washer thickness	1.6 mm
Float setting as for 150CDS type carburettor	
Idling speed (manual in neutral)	775 to 825 rpm
(automatic in 'D')	775 to 825 rpm
C O level	2.5 to 3.5%

Carburettor - Automatic transmission 1256 cc

Make	Zenith 34 IVET
Identification number	3400
Main jet	80
Compensatory jet	105
Idling jet	45
Pump jet	40
Needle valve	1.75 mm
Needle valve washer thickness	2 mm
High altitude jet settings:	
Main jet - 5000 - 7000 ft	77.5
- 7000 - 10000 ft	75
- 10000 - 15000 ft	70
Compensating jet - 5000 - 7000 ft	102.5
- 7000 - 15000 ft	100

1 General description

The fuel system consists of a seven gallon fuel tank mounted in the boot of the car; mechanical fuel pump bolted to the left hand side of the cylinder block near the bellhousing, actuated by an eccentric on the camshaft and lever arm; and one of the following carburettors: Zenith 301Z, Zenith/Stromberg 150CD or Zenith 34IV.

Fuel is carried from the tank to the carburettor by lengths of metal piping joined by a flexible hose and clipped to the left hand longitudinal subframe member. An air filter of either oil wetted gauze or paper element type is fitted to the carburettor.

2 Air filter element - removal and servicing

1 The air cleaner can be detached from the carburettor as a unit or the top cover removed solely to get access to the filter element. The former method is preferable, as the inside can be cleaned thoroughly without risk of dirt falling into the carburettor (photo).

2 To clean the filter element, proceed as follows:

a) (Paper tvpe). Lightly tap the end surfaces on a hard object and continue until dust and debris stop falling from the element. Do not try to brush, wash or use compressed air on a paper type element.

b) (Gauze type). Rinse in paraffin, blow out with an air pump or shake dry, dip element in clean engine oil and allow excess to drain off (Fig 3:1).

3 Thoroughly clean the interior of the air filter, including the intake tube and base plate.

4 Ensure any gaskets are correctly positioned (paper element). Replace the element on the cover base and snap the cover back in position (Fig 3:2).

3 Zenith 301Z carburettors - description

The carburettor fitted on the standard engine is a fixed choke design and incorporates an economy unit to correct fuel mixture at certain intermediate engine speeds. The principle of operation is as follows: At full throttle opening with the choke flap open, the depression (low pressure) in the choke tube draws a fuel/air mixture from the main discharge beak.

This fuel/air mixture has been emulsified in the emulsion tube below the discharge beak. The fuel has reached the emulsion tube, via the reserve well, from the main jet in the float chamber.

When the engine is cold and the choke flap is closed, the throttle flap is automatically slightly opened a predetermined amount.

The choke tube depression draws principally on the discharge beak and, therefore, a very rich mixture reaches the engine, as air from the main air inlet has been closed off.

At idling speed, with the throttle shut, there is no depression at the main discharge beak. It is now concentrated at the idling discharge orifice on the engine side of the throttle flap. Fuel from the main reserve well is drawn via the pilot jet to this orifice, taking the requisite amount of air for the mixture through the pilot air bleed and bypass orifice. The volume of the mixture supplied is controlled by the idling mixture control screw.

As soon as the throttle is opened further, the bypass orifice is then also subject to depression, so instead of feeding air in one direction to the idling discharge orifice, it now delivers fuel/air mixture in

Removal of air cleaner from Zenith carburettor

Fig 3:1 Dismantled air cleaner with wire gauze element fitted to top cover

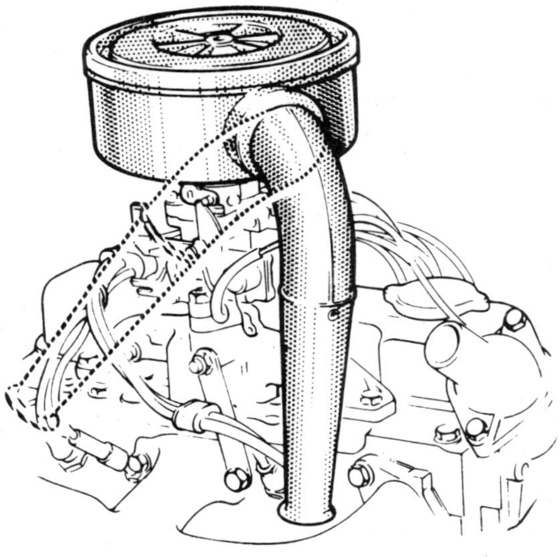

Fig 3:2 Air cleaner unit showing alternative position of air intake pipe in summer

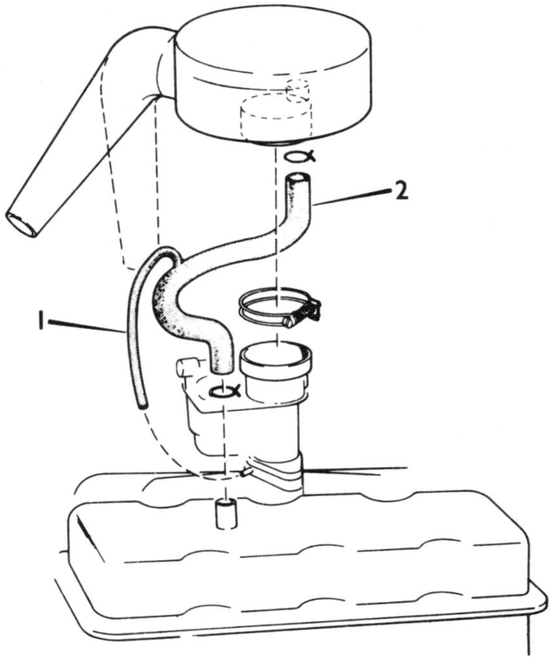

Fig 3:3 POSITIVE CRANKCASE VENTILATION HOSE CONNECTION
1 Branch hose 2 Main hose

the other direction until the throttle is open sufficiently for the main discharge beak to take over.

The economy unit augments the fuel flow from the main jet automatically through the economy jet when the choke tube depression is low. At cruising speeds, when the depression is high a diaphragm operated valve shuts off the economy jet.

There is also an accelerator pump which delivers a metered jet of neat fuel into the choke tube whenever the accelerator pedal is operated quickly. This gives the richer mixture necessary for rapid acceleration. The fuel is drawn from the float chamber into the pump chamber via a non-return valve at the bottom of the float chamber. When the pump is operated quickly (ie sudden accelerator pedal operation) the pump release valve is forced shut under pressure and the fuel passes through the injector. If the pump is operated slowly, the pressure is insufficient to close the release valve, so fuel passes through it back to the float chamber rather than out of the injector. Models fitted with positive crankcase ventilation are basically similar but have no economy unit fitted. Instead there is an econostat device. Venting of the float chamber at idling speed is external and controlled by a plunger type vent controlled by a leaf spring.

Manual gearbox cars equipped with exhaust emission control have a Zenith 301ZE carburettor. This is similar to the 301Z except for a sealed calibration screw - no adjustment being possible.

4 Zenith 301Z carburettor - removal, dismantling and reassembly

1 Slacken the clip holding the air filter unit to the carburettor air intake.

2 Slacken the clip at the bottom of the hose connection from the filter to the rocker cover and lift off the air filter unit complete.

3 Disconnect the throttle cable from the lever by detaching the clip washer from the clevis pin. Also unhook the spring from the other end of the throttle spindle (photos).

4 Slacken off the choke cable retaining bolt, unclip the cable from the bracket and move the cable to one side (photos).

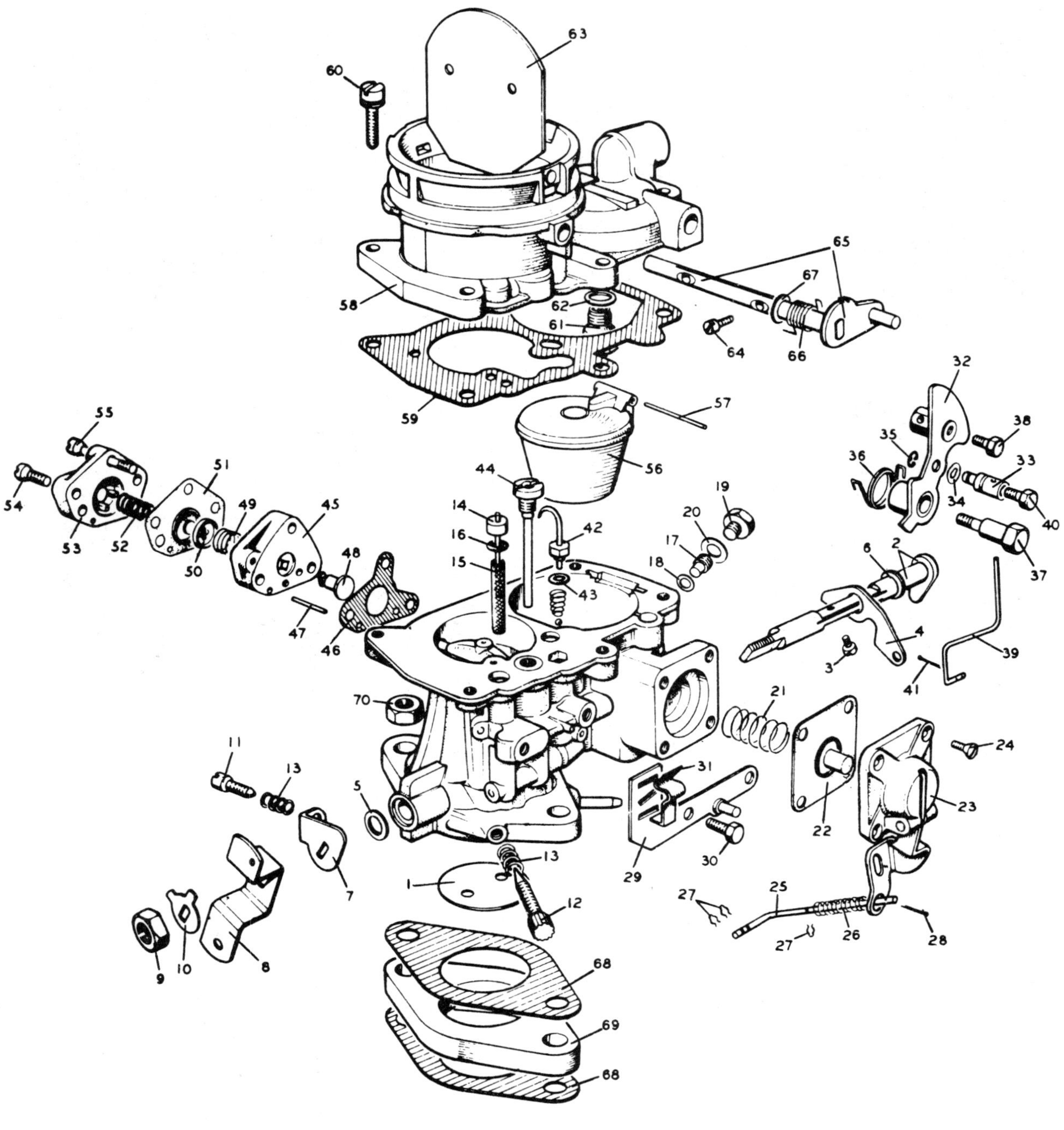

Fig 3:4 ZENITH 301Z CARBURETTOR – EXPLODED VIEW

1 Throttle flap
2 Spindle
3 Screw
4 Lever (floating)
5 Sealing washer
6 Distance washer
7 Throttle stop plate
8 Throttle lever
9 Spindle end nut
10 Lockwasher
11 Slow running screw
12 Volume control screw
13 Spring
14 Pilot jet
15 Filter gauze
16 Sealing ring
17 Main jet
18 Fibre washer
19 Sealing plug
20 Washer
21 Spring
22 Pump diaphragm
23 Pump cover
24 Cover screw
25 Pump control rod
26 Control rod spring
27 Circlip
28 Split pin
29 Choke bracket
30 Screw
31 Clip
32 Choke cam
33 Choke cam swivel
34 Washer
35 Circlip
36 Cam return spring
37 Choke cam pivot
38 Screw
39 Interconnection rod
40 Screw
41 Split pin
42 Injector
43 Sealing ring
44 Emulsion tube
45 Economy unit
46 Gasket
47 Economy unit jet
48 Economy unit valve
49 Spring
50 Washer
51 Economy unit diaphragm
52 Diaphragm spring
53 Cover
54 Cover screw
55 Unit screw
56 Float
57 Float spindle
58 Float chamber cover
59 Gasket
60 Cover screw
61 Needle valve
62 Washer
63 Choke flap
64 Screw
65 Spindle
66 Return spring
67 Washer
68 Gasket
69 Heat insulator
70 Nut

4.3a

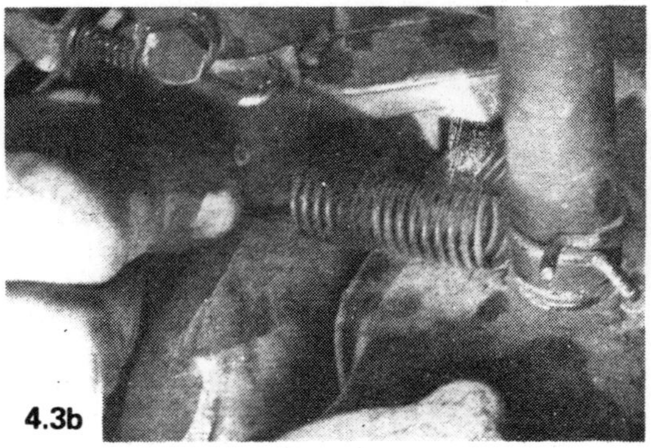

4.3b

Disconnecting the throttle cable and spring from a Zenith carburettor

4.4a

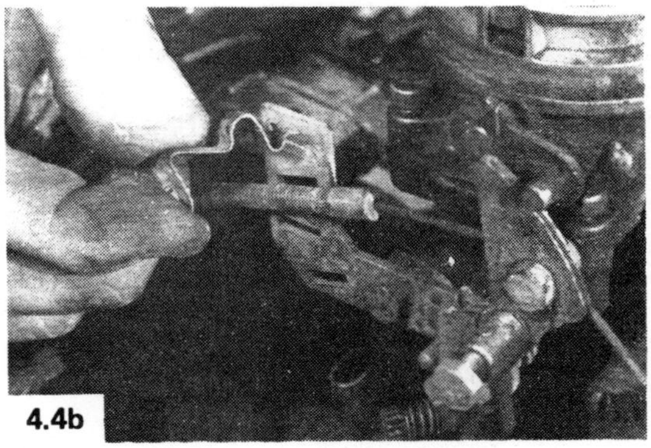

4.4b

Disconnecting choke cable from a Zenith carburettor

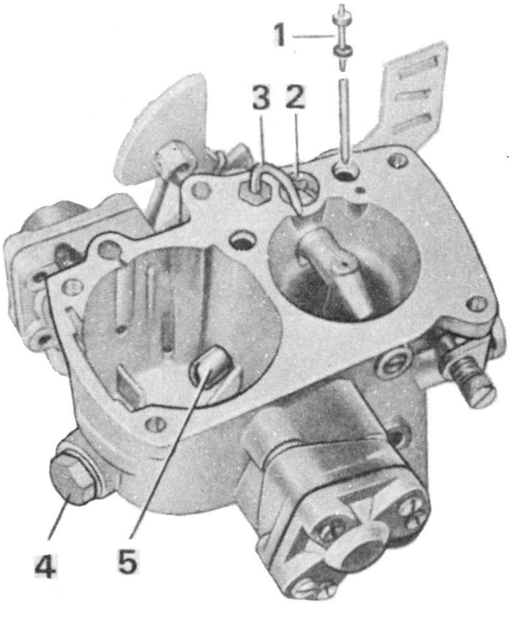

Fig 3:5 ZENITH 301Z CARBURETTOR – DETAILS OF INTERNAL COMPONENTS

1 Pilot jet and filter tube
2 Emulsion tube
3 Injector tube
4 Main jet access plug
5 Main jet

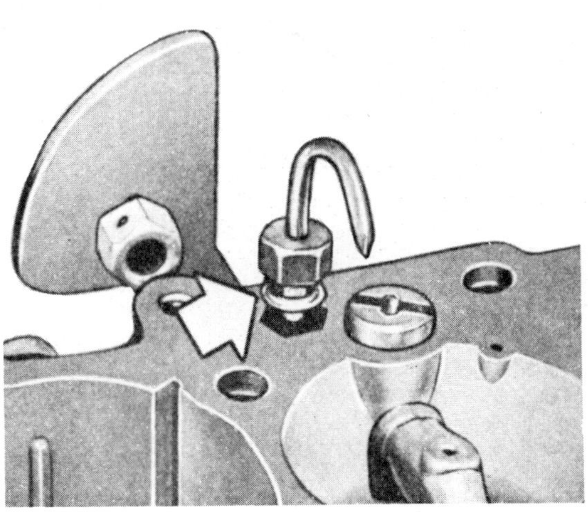

Fig 3:6 ZENITH 301Z CARBURETTOR

Accelerator pump injector pipe. Note sealing ring arrowed

5 Disconnect the fuel pipe from the carburettor by undoing the union from the float chamber cover.
6 Remove the two nuts holding the carburettor to the inlet manifold. There is very little clearance above the manifold studs so the nuts should be undone together and the carburettor lifted, so that eventually the nuts will clear the tops of the studs. It is impossible to remove only one nut with the other tightened down. Remove the carburettor, heat block and gaskets from the manifold.
7 Dismantling. Remove the five screws holding the float chamber cover to the body. Unscrew the needle valve and packing washer, (retain the washer) remove arm spindle and lift out the needle valve actuating arm and float. Handle the float carefully as it is fragile.
8 Pull the pump injector and sealing ring from the body and remove the spring and ball underneath from the drilling. Do not lose the ball. Unscrew the correction jet and emulsion tube, and pilot jet. Remove the sealing plug and washer and unscrew the main jet and washer through the orifice (Fig 3:6).
9 Disconnect the control rod and spring from the pump lever. Undo the four screws and remove the pump cover, diaphragm and return spring (Fig 3:7).
10 To remove the economy unit, undo the three long screws and then the two short screws holding the subassembly together. Remove the cover, diaphragm spring, diaphragm, valve washer and spring body and vacuum tube, gasket and valve. The economy jet can then be unscrewed from the body (Fig 3:8).
11 Inspect the jets, needle valve, actuating arm and float top for excessive wear or damage, also the diaphragms for any signs of fatigue, or tears. Also shake the float to ensure it does not contain any petrol. NOTE: It is suggested that all jets, diaphragms, float needle valve actuating arm and needle valve are renewed. This may seem a needless expense, but it is obvious that a certain degree of wear will have taken place in all these components and this will lower performance and economy. Therefore, the small outlay for these new parts will soon be recouped in lower running costs. Even if the other parts are not renewed, the needle valve assembly certainly should be, as a worn needle valve will allow excessive fuel into the carburettor, making the mixture over-rich. This will not do the engine or your pocket any good!
12 The carburettor body, and all components, should be washed in clean petrol, and then blown dry. If an air jet is not available in any form, the parts can be wiped dry with a soft non-fluffy rag. Jets can be blown through by mouth to clear obstructions, or poked through with a nylon bristle. DO NOT scour the carburettor body or components with a wire brush of any variety, or poke through the jets with wire. The reason for this is that the carburettor is a precision instrument made of relatively soft metal (ie body - aluminium, jets - brass) and wire will scratch this material, possibly altering the performance of the carburettor.
13 Reassembly is a straightforward reversal of the dismantling procedure, but note the following points:
a) New gaskets and sealing washers should be used. Note the normal thickness of the fibre washer under the needle valve is 1.6 mm.
b) When refitting the pump injector ensure that a sealing ring is fitted. Also when replacing the ball and spring in the drilling, ensure that the spring does not actually touch the ball and that its top (wide end) is just below the bottom of the hexagon recess.
c) Ensure that the economy jet is screwed into the carburettor body before refitting the economy unit.
d) The pump control rod and spring should be refitted to the appropriate hole in the pump lever, so that the large end of the spring contacts the lever.
f) The strangler flap return spring should be refitted in the centre notch of the spindle lever (Fig 3:9).
g) Never use jointing compound on carburettor gaskets.
14 Before replacing the carburettor, adjust the choke to throttle setting as follows: Close the throttle flap by unscrewing the throttle stop screw. Place a number 61 drill, or a 1.0 mm piece of wire between the throttle flap and bore (Fig 3:10). Hold the strangler cam against the stop pin and this will allow the strangler flap to be closed by the spindle spring. Loosen the swivel bolt and move the rod through it until the floating lever touches the pump rod lever. Retighten the swivel bolt and remove the drill or wire. The adjustment is now completed and the carburettor can be assembled to the manifold.

5 Zenith 301Z carburettor - replacement

1 Thoroughly clean the carburettor and manifold flanges and both sides of the insulating block.
2 Push one gasket over the manifold studs on to the flange, followed by the insulating block and the second gasket. Replace the carburettor. Replace the two mounting washers and nuts together. It will be necessary to hold the carburettor up the studs until the nuts can be caught onto the stud threads. Do not overtighten the nuts (photo).
3 Reconnect the choke control, petrol pipe, vacuum pipe and the throttle controls, not forgetting the return spring on the carburettor, (Fig 3:11).
4 Now check that the throttle flap opens fully when the pedal is depressed. If not, the cable must be adjusted.

6 Zenith 301Z carburettor - adjustment

1 It is assumed that the throttle flap to choke setting has already been adjusted correctly as described earlier, and that the carburettor is installed.
2 First check that the throttle flap is correctly operated by the accelerator cable. It should be vertical when the pedal is fully depressed and at the same time it must be ascertained that the pedal is at the limit of its downwards travel when the throttle is wide open. If it is not, then excessive strain will be imposed on the carburettor. Slacken the clamp which holds the wire to the operating lever, then press the accelerator pedal to the floor and move the arm so that the flap is in the open position. Re-clamp the cable. Check that the throttle shuts when the accelerator pedal is raised. Adjustment can also be made on the outer cable but care must be taken to guard against the possibilities mentioned.
3 The accelerator pump stroke is adjustable for summer and winter motoring. This adjustment allows a richer mixture during acceleration in the winter to compensate for the mixture's partial loss of vapourising ability and, in summer, lessens the delivery through the injector, so preventing over-richness (Fig 3:12).
4 To alter the setting, remove the split pin, lift out the pump rod from the inner hole and place it in the outer one. Replace the split pin.
5 Slow running adjustment is set with the volume control screw. With the engine warm, set the throttle stop screw to a fast idle speed. Then turn the volume control screw in whichever direction is necessary to obtain the fastest, smoothest tickover. Reduce the tickover speed with the throttle stop screw and, if necessary, make any further minor adjustments to maintain the smoothness of tickover using the volume control screw. It should be remembered that not only the carburettor is responsible for smooth idling: distributor settings are also important and should be checked if the carburettor adjustment is insufficient. A full check against engine fault diagnosis, in Chapter 1, may be necessary if a satisfactory tickover speed and smoothness cannot be obtained (Fig 3:13).
6 The level of the fuel in the float chamber is important in correct operation of the carburettor.
a) Check first that the needle valve is in good condition.
b) Run the engine to ensure that the chamber is full and make sure the car is standing level.
c) Remove the top cover of the carburettor and carefully remove the float.
d) Measure the level of the fuel below the edge of the chamber which should be 23 mm.

If the level is wrong, then it should be altered by increasing the

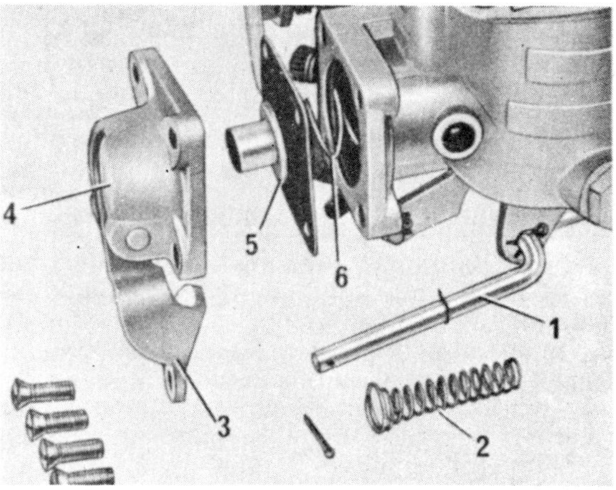

Fig 3:7 ZENITH 301Z CARBURETTOR – ACCELERATOR PUMP DETAILS

1 Actuating rod
2 Rod spring
3 Pump lever
4 Cover
5 Diaphragm
6 Return spring

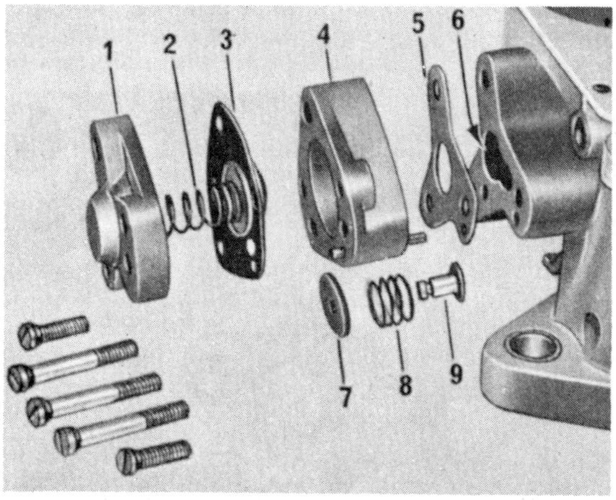

Fig 3:8 ZENITH 301Z - DETAIL OF ECONOMY UNIT

1 Cover
2 Spring
3 Diaphragm
4 Valve body
5 Gasket
6 Jet recess
7 Valve washer
8 Valve spring
9 Valve

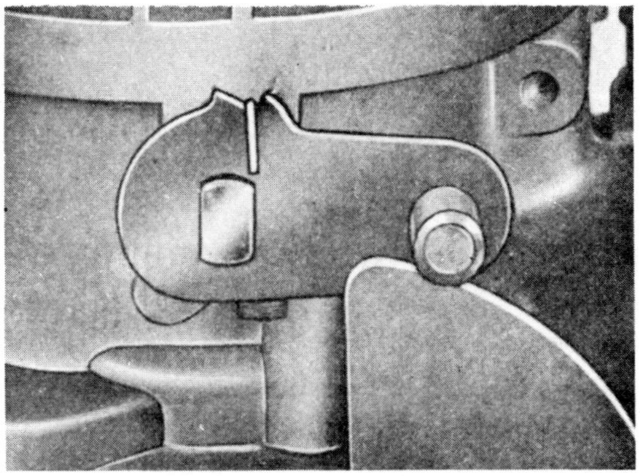

Fig 3:9 ZENITH 301Z - Correct position of choke flap lever return spring

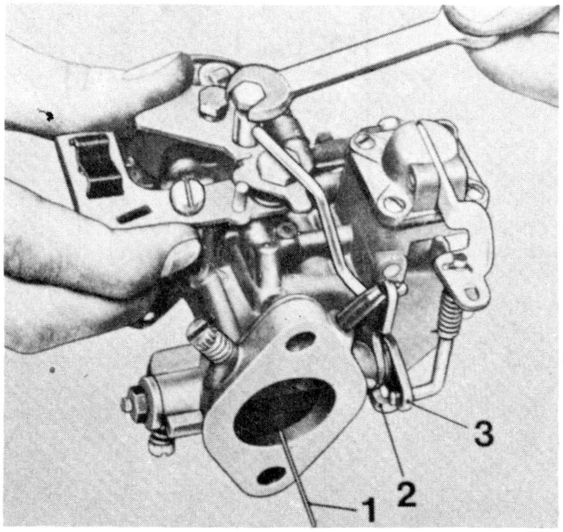

Fig. 3:10 ZENITH 301Z - SETTING THE CHOKE OPERATING LEVER RELATIVE TO THE THROTTLE FLAP (FAST IDLE)

1 1.0 mm dia setting rod
2 Floating lever
3 Pump rod lever

Fitting Zenith carburettor manifold gaskets and insulator block

washer thickness under the needle valve to lower the lever, and vice versa to raise it. The cover should be replaced and the engine run each time an alteration is made so that the result may be checked. Do NOT attempt to alter the fuel level by messing about with the float.

7 On engines with positive crankcase ventilation systems, the carburettor float chamber vent plunger valve setting must be correct. To check this requires removal of the carburettor. With a suitable rod or blade, set the gap between the edge of the throttle flap at .8 mm (.032 in), and then adjust the spring screw so that the vent plunger is just closed. With the throttle quite shut the plunger should be open (Fig 3:14 and 3:15).

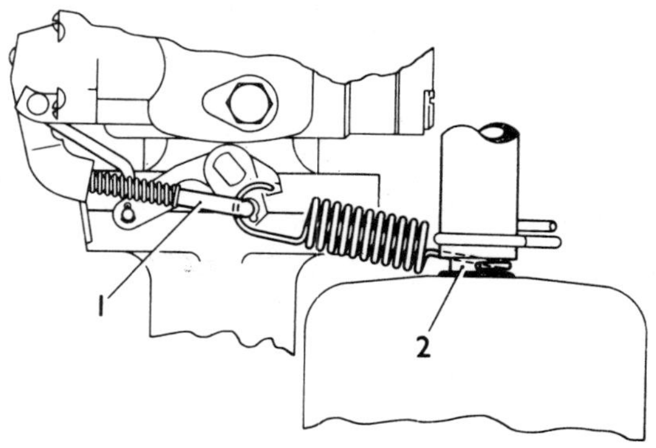

Fig 3:11 ZENITH 301Z CARBURETTOR
Throttle cable return spring fitted between rod (1) and crankcase/ventilation pipe connections on inlet manifold

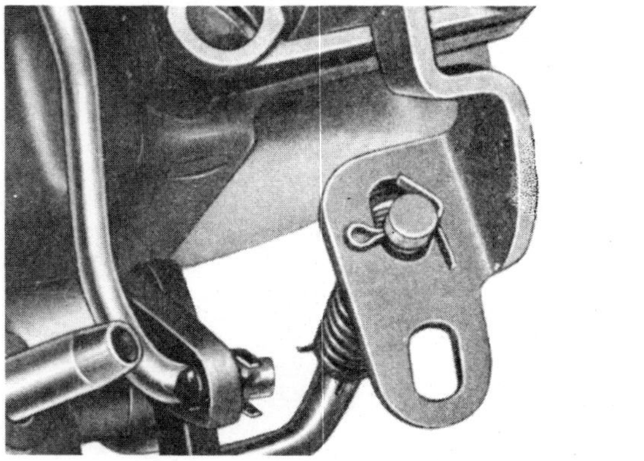

Fig. 3:12 ZENITH 301Z - Accelerator pump rod setting holes for winter and summer (lower)

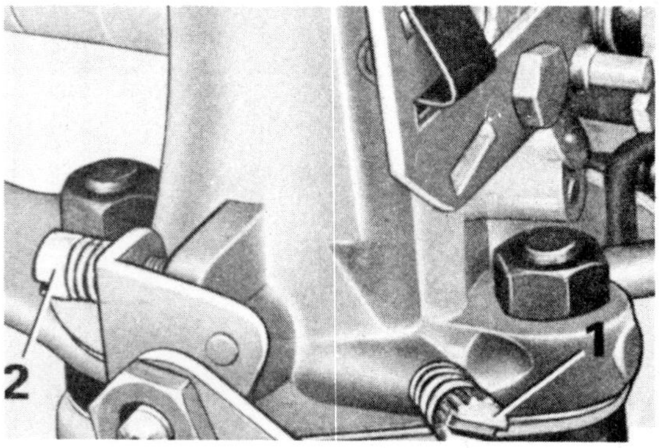

Fig. 3:13 ZENITH 301Z
1 Volume control screw 2 Throttle stop screw

7 Stromberg 150 CD carburettor - description

1159 cc SL models are fitted with a single Stromberg 150CD (constant depression) carburettor. It has a single horizontal variable choke and is quite different in principle of operation to the Zenith.

1159 cc extra performance models built with exhaust emission control are equipped with a Stromberg 150CD-SETV carburettor. Apart from having a wire-sealed jet adjuster nut, the unit is very similar to that used on later 1256 cc models which is fully described in Chapter 13.

The air intake is choked by the cylindrical air valve which can move vertically.

To the base of the air valve a tapered needle is fitted. This runs in and out of a jet orifice through which fuel can be drawn from the float chamber, which is underneath the main body. Suction from the engine inlet manifold passes through a hole in the base of the air valve to the suction chamber. This suction acts on the diaphragm to which the air valve and metering needle are attached, and raises them. This increases the air flow through the choke tube and the fuel flow through the jet as the tapered needle withdraws. As the air valve rises, so the concentration of suction through the valve hole is reduced, and the valve reaches a point of equilibrium, balanced against throttle opening and air valve height.

Sudden acceleration demands would apparently cause the air valve to rise sharply, thus tending to weaken the mixture. In fact, the rise of the valve is damped by an oil controlled piston. Thus, when the throttle is opened suddenly, the initial suction is concentrated at the fuel jet and the quantity of air let through to reduce the mixture richness to normal, occurs slightly later as the piston rises. The taper of the metering needle is obviously the main controlling feature of the carburettor's performance and this controls the fuel/air mixture at all heights of the valve. At the same time the height of the air valve is nicely balanced, according to throttle opening, in conjunction with the metering needle. The jet itself is adjustable by raising or lowering, thus altering the position of the jet orifice in relation to the taper of the needle.

For cold starts, there is a device mounted on the side, consisting of a rotating disc with a number of holes drilled in it. When the disc is moved by the control lever, a passage way is opened up and, depending on the number of holes uncovered, so the flow is restricted. The fuel from the float chamber can then be drawn direct into the choke tube - supplementing that from the main jet.

8 Stromberg 150 CD carburettor - adjustment

1 As there is no separate idling jet, the mixture for all conditions is supplied by the main jet and variable choke. Thus the strength of the mixture throughout the range depends on the height of the jet in the carburettor body and, when the idling mixture is correct, the mixture will be correct throughout the range. A slotted nut at the base of the **carburettor increases or decreases the strength of the mixture. Viewed from below, turning the nut clockwise raises the jet and weakens the mixture. Turning the jet anticlockwise enriches the mixture. The idling speed is controlled by the throttle stop screw.**

2 To adjust a Stromberg 150 CD carburettor from scratch it is assumed that the fast idle cam has been set correctly as described in the reassembly procedure.

3 Run the engine until it is at its normal working temperature and then remove the air cleaner. Insert a .002 in feeler gauge between the air valve and carburettor body and screw in the jet adjusting screw until it touches the air valve and the feeler is freed. The feeler should

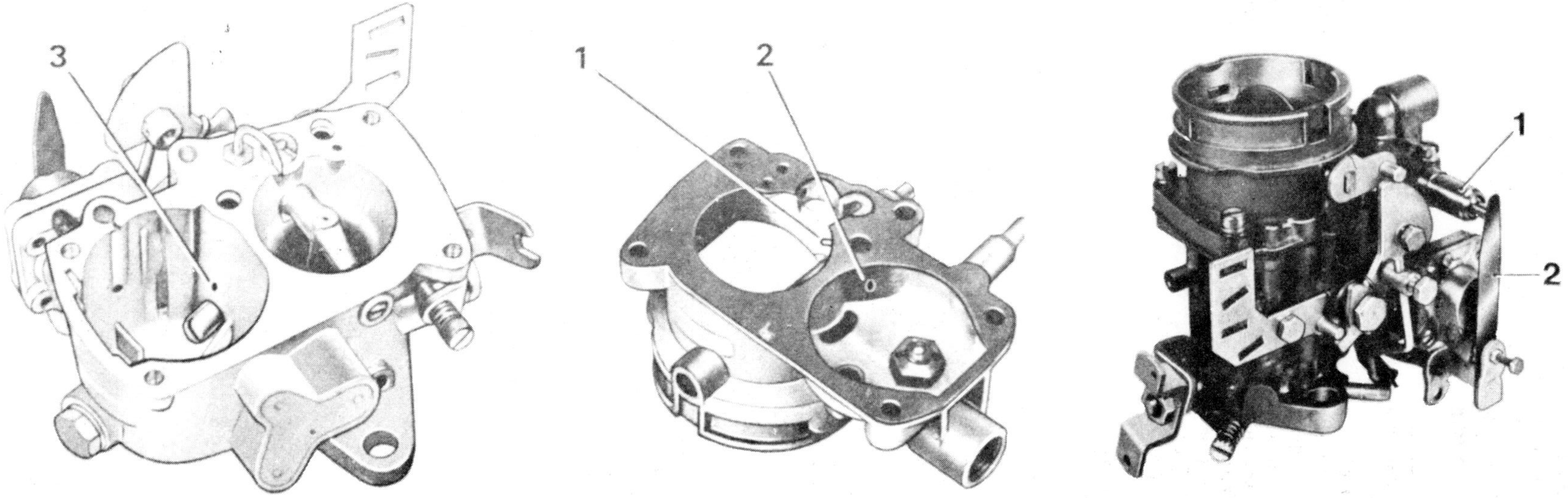

Fig 3:14 ZENITH 301Z CARBURETTOR
Differences on models fitted with positive crankcase ventilation

1 Fuel discharge tube 2 Air bleed 3 Drilling 4 Float chamber vent plunger valve 5 Plunger valve spring

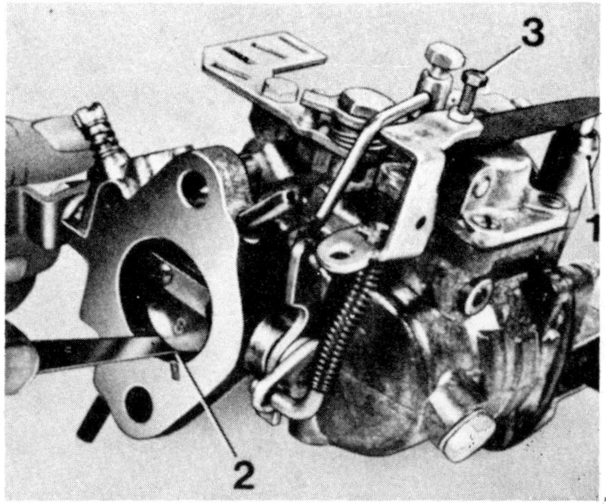

Fig 3:15 ZENITH 301Z CARBURETTOR (POSITIVE CRANKCASE VENTILATION)
Valve adjustment

1 Vent plunger valve 3 Adjusting screw
2 Throttle flap gap .8mm (.032 in)

Fig 3:16 (RIGHT) Stromberg 150 CDS carburettor - general view

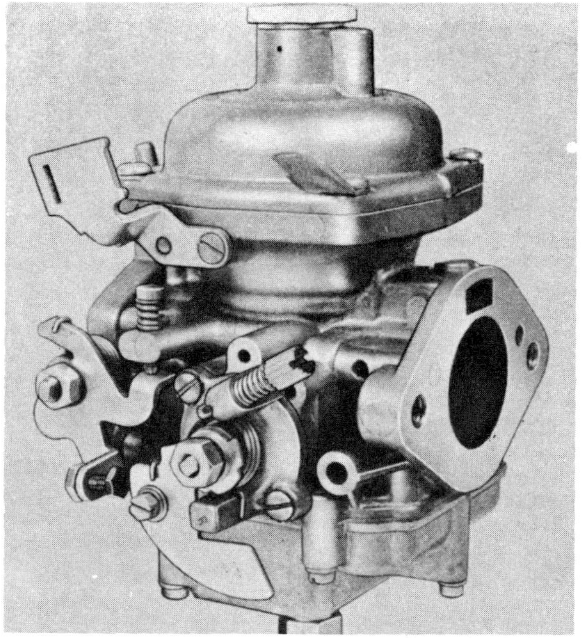

now be withdrawn and the adjuster unscrewed two complete revolutions. This will give an approximate setting. Start the engine and adjust the throttle stop screw so that the engine runs fairly slowly and smoothly (about 750 rpm) without vibrating excessively on its mountings. To get the engine to run smoothly at this speed, it may be necessary to turn the jet adjuster nut a small amount in either direction, but not more than ½ turn.

3 To test if the correct setting has been found, lift the air valve piston 1/32 in with the lifting pin. This is a very small amount and care should be taken to lift the piston only fractionally. If the engine speed increases and stays so, then the mixture is too rich. If it hesitates or stalls, it is too weak. Re-adjust the jet adjusting nut and recheck. All is correct when the engine speed rises momentarily and then drops when the air valve is lifted the specified 1/32 in. Make sure that the air valve damper is correctly filled with engine oil. With the plunger and air filter removed, the level should be ¼ inch below the top of the air valve guide. Lift the air valve with a finger through the air intake, during the topping up operation.

4 Replace the air cleaner.

5 **Note: On later carburettors, the air valve has a recessed underside and the feeler blade method for initial setting is unsuitable. Instead, rotate the jet adjuster nut until the jet is seen to be below the bridge. Now screw the adjuster clockwise and at the same time raise and release the air valve repeatedly (air valve damper removed). Insert a finger into the carburettor air intake to do this. The sound of the valve contacting the bridge will change when the jet is raised sufficiently. When this point is reached, turn the jet one quarter turn down for final setting.**

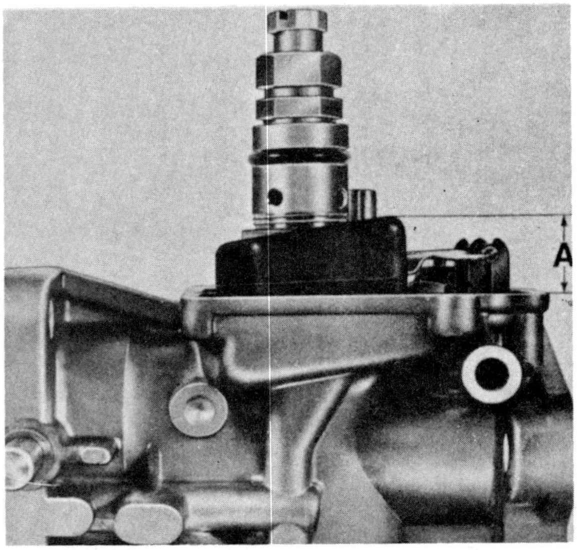

Fig 3:17 STROMBERG 150 CDS CARBURETTOR
Setting height of float. A = 15.5–16.5 mm

Fig 3:18 Stromberg 150CDS cold start device removed to show disc and fuel orifices

9 Stromberg 150 CD carburettor - float chamber fuel level setting

1 Take off the air cleaner and then remove the carburettor from the engine.

2 Slacken the jet bush retainer and undo the screws which hold the float chamber to the base of the carburettor. Remove the float chamber.

3 Turn the carburettor body upside down and accurately measure the highest point of the floats which should be 15.5-16.5 mm above the edge of the chamber. During this operation ensure that the needle is against its seating. To reset the level, carefully bend the tag which bears against the end of the needle (Fig 3:18).

4 When replacing the float chamber, ensure that the head of the fulcrum pin on which the float pivots will abut the lug in the float chamber casting.

10 Stromberg 150 CD carburettor - dismantling and reassembly

1 Take off the air cleaner, disconnect the choke and accelerator controls at the carburettor, also the vacuum advance and retard pipe, and undo the nuts and spring washers holding the carburettor in place. Remove the carburettor.

2 With the carburettor on the bench, undo and remove the damper cap and plunger. Then undo the four screws which hold the suction chamber cover in place and lift off the cover.

3 The air valve, complete with needle and diaphragm, is then lifted out. Handle the assembly with the greatest of care as it is very easy to knock the needle out of true.

4 The bottom of the float chamber is removed by undoing five screws and the washers which hold it in place. Take out the pin and remove the float assembly.

5 If wished, the needle may be removed from the air valve by undoing the grub screw.

6 To remove the diaphragm from the piston, simply undo the four screws and washers which hold the diaphragm retaining ring in place.

7 The jet and associated parts are removed after the jet locking nut has been undone. The cold start device can be removed from the side of the main body after undoing the two securing screws (Fig 3:19).

8 On reassembly there are several points which should be noted particularly. The first is that if fitting a new needle to the piston, ensure it has the same markings as the old stamped on it, and fit it so that the needle shoulder is perfectly flush with the base of the piston. This can be done by placing a metal ruler across the base of the valve and pulling the needle out until it abuts the rule.

9 Thoroughly clean the piston and its cylinder in paraffin.

10 Refit the jet and associated parts, lift the piston and tighten the jet assembly.

11 Turn the mixture adjusting nut clockwise until the tip of the jet

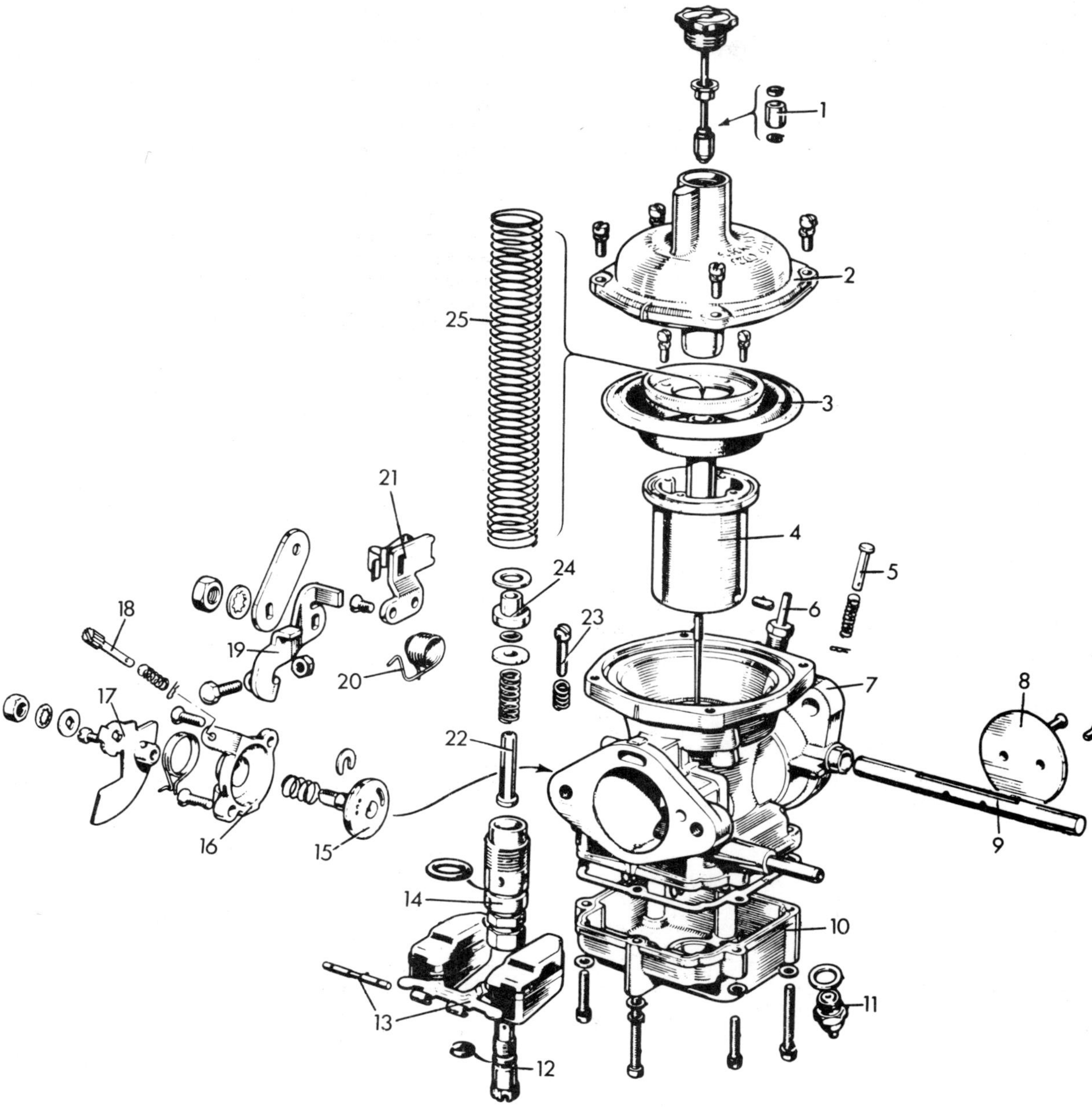

Fig 3:19 STROMBERG 150 CDS CARBURETTOR – EXPLODED VIEW

1 Piston damper plunger
2 Top cover
3 Diaphragm and retainer plate
4 Air valve and metering needle
5 Air valve lifting pin
6 Manifold vacuum take off union
7 Main carburettor body
8 Throttle flap
9 Throttle spindle
10 Float chamber
11 Needle valve
12 Jet adjusting screw
13 Float and pivot pin
14 Jet retainer
15 Cold start regulation disc
16 Cold start unit cover
17 Fast idle cam
18 Cold start selector pin
19 Throttle spindle lever with cold start stop screw
20 Throttle spindle return spring
21 Cable clip bracket
22 Jet
23 Throttle stop screw
24 Jet bush
25 Air valve spring

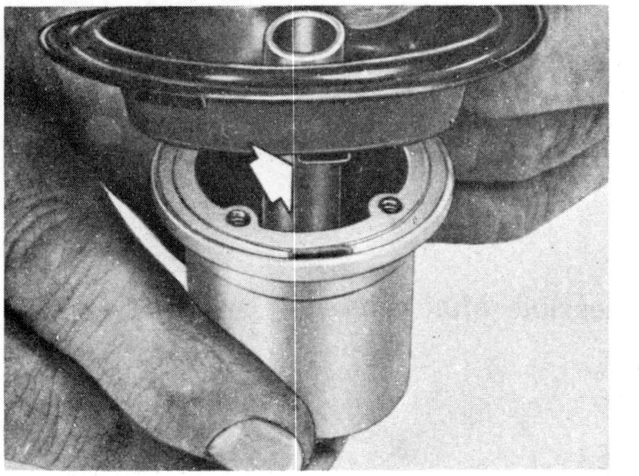

Fig 3:20 STROMBERG 150 CDS CARBURETTOR
Showing engagement of diaphragm tongues into respective slots in air valve and body

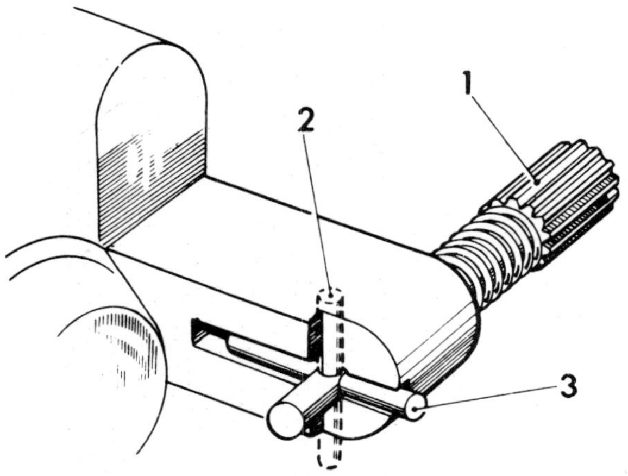

Fig 3:21 COLD START SETTING PIN
1 Spring loaded stop spindle 2 Normal setting position 3 Cold (below −18°C) setting position

just stands proud into the choke tube. Now loosen the jet bush retainer about one turn so as to free the bush.

12 Allow the piston to fall. As it descends, the needle will enter the orifice and automatically centralise it. With the needle still in the orifice, tighten the jet assembly slowly, frequently raising and dropping the piston ¼ in to ensure the orifice bush has not moved. Finally check that the piston drops freely without hesitation and hits the bridge with a soft metallic sound.

13 Make sure that the holes in the diaphragm line up with the screw holes in the piston and retaining ring, and that the diaphragm is correctly positioned with the upper and lower tongues engaged in the slots. Reassembly is otherwise a straightforward reversal of the dismantling sequence (Fig 3:20). When installing the float, make sure that the flat face of the float is at the bottom.

14 The cold start device can be set to operate in two positions. The two-position stop can be set to control the maximum amount to which the unit may rotate. With the cross-pin in the slot the device may rotate to its full extent, but this is not usually needed unless the temperature drops below 18°C. The fast idle cam operates against the throttle lever via an adjusting screw, which may be set to ensure a suitably fast idling speed under cold start situations (Fig 3:21).

15 To adjust the fast idle, the cold start cam must be set correctly in relation to the throttle flap. To do this put the cold start setting pin in the normal (vertical) position and press the cam against the end of the pin. Then move the adjusting screw on the throttle control lever against the cam until there is a .8 mm (.032 in) gap between the throttle flap and body. The gap can be measured with a suitably sized drill (Fig 3:22).

11 Stromberg 150 CDST carburettor - description and general

1 The 150 CDST carburettor is used on those 1159 cc high performance engines fitted with automatic transmission systems. It is identical to the other version with the exception of having an automatic cold start device (Fig 3:23).

This device is controlled by the temperature of the engine coolant. A bi-metal spring thermostat controls a tapered needle which regulates fuel flowing into a mixing chamber. In addition, a pushrod and piston, controlled by inlet manifold vacuum, override the action of the thermostat under certain engine speed conditions.

2 It is not recommended that this item is tampered with but in the event of the owner deciding to investigate because of malfunction the following points should be noted on reassembly:

3 When replacing the fast idle cam and thermostat lever, the lever peg should engage in the taper needle slot and the spade in the flat in the vacuum kick rod (Fig 3:24).

4 The interconnecting spring should be fitted so that the inner end goes over the long arm of the thermostat lever and the other end hooks into the hole in the fast idle cam afterwards (Fig 3:25).

5 The clearance between the fast idle screw and cam (on the base circle) should be .020 in when the thermostat lever is rotated fully anticlockwise. The fast idle screw can be adjusted as required and resecured. (Fig 3:26).

6 When fitting the water jacket and cover make sure that the square loop on the thermostat spring engages the peg on the thermostat lever (Fig 3:27).

7 Correct positioning of the thermostat after assembly is indicated by line up marks on the housing and rims. No variations should be made (Fig 3:28).

8 The only adjustment after assembly is to the idling screw which is set when the engine has reached normal running temperature.

Adjustment should never be made to the fast idle screw except as described in paragraph 5 (Fig 3:30).

9 If the fast idle cam should stick in the open position for some reason, it can be moved back manually. Remove the plug in the housing and insert a piece of 3/16 in rod and push it back (Fig 3:29).

12 Zenith 34 IVET carburettor - description and adjustments

1 The Zenith 34 IVET carburettor is used on the 1256 cc engine where automatic transmission is fitted. It is of the double choke tube downdraught type with an acceleration pump, diaphragm type economy device and a water heated automatic choke. It comprises a float chamber, top cover and emulsion block (screwed to the top cover) wherein all the jets are housed. The accelerator pump is housed in the emulsion block also.

2 If the carburettor is not dismantled the only adjustment made should be the throttle stop screw. Do not tamper with the fast idle adjustment screw as this is only set when the carburettor has been dismantled.

The complete dismantling of the carburettor is not detailed here although the more ambitious can follow the exploded drawings - one or two important points and settings in connection with the automatic choke are also illustrated.

13 Fuel tank - removal and replacement

1 Remove the battery from the car as a safety measure, also ensure there are no open flames in the vicinity. DO NOT smoke during

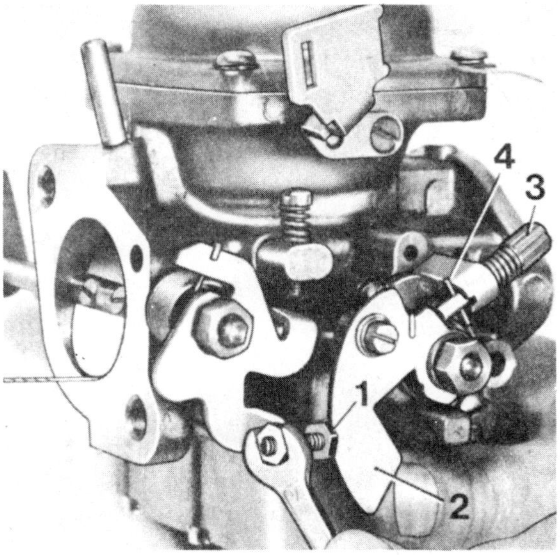

Fig 3:22 STROMBERG 150 CDS CARBURETTOR
Adjustment of fast idle cam
1 Adjustment screw
2 Cold start cam
3 Cold start setting spindle
4 Setting spindle pin

Fig 3:24 STROMBERG 150 CDST CARBURETTOR AND COLD START DEVICE
1 Thermostat lever peg
2 Thermostat lever spade
3 Vacuum kick rod

Fig 3:25 STROMBERG 150 CDST CARBURETTOR
Cold start device. Location of interconnecting spring over long lug of thermostat lever (arrowed). Other end hooks into fast idle cam

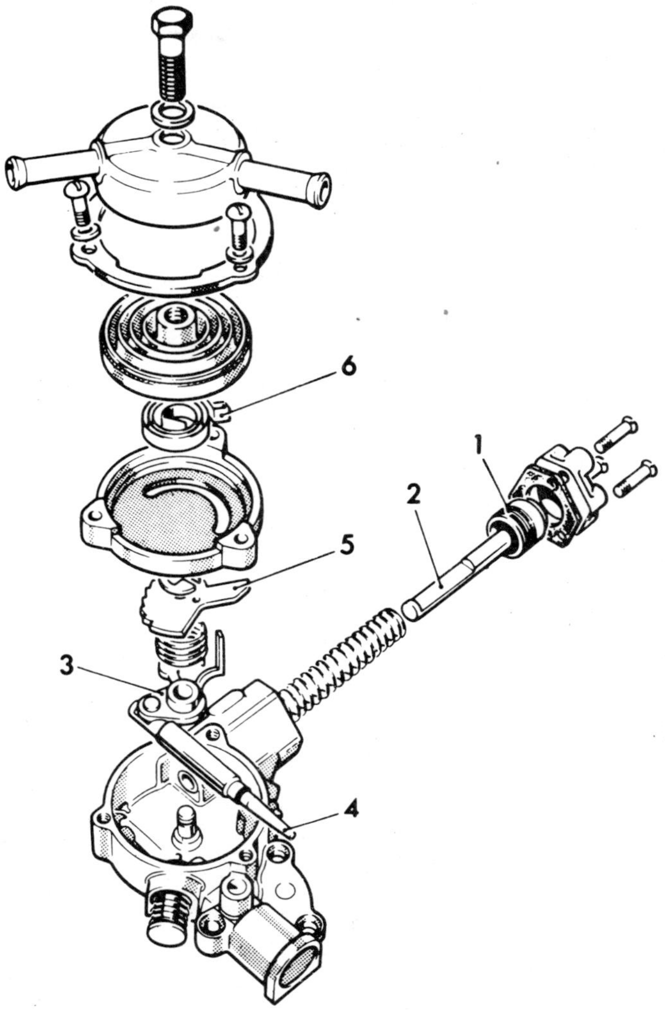

Fig 3:23 STROMBERG 150 CDST CARBURETTOR – COLD START DEVICE – EXPLODED VIEW
1 Vacuum piston
2 Kick rod
3 Thermostat lever
4 Tapered fuel needle
5 Fast idle cam
6 Thermostat coil

Fig 3:26 STROMBERG 150 CDST CARBURETTOR
Cold start device. Fast idle cam stop setting
A = .020 inches

Fig 3:27 STROMBERG 150 CDST CARBURETTOR – COLD START DEVICE
Engagement of thermostat and lever pin (arrowed)

Fig 3:28 STROMBERG 150 CDST CARBURETTOR – COLD START DEVICE
Line up of thermostat and housing marks

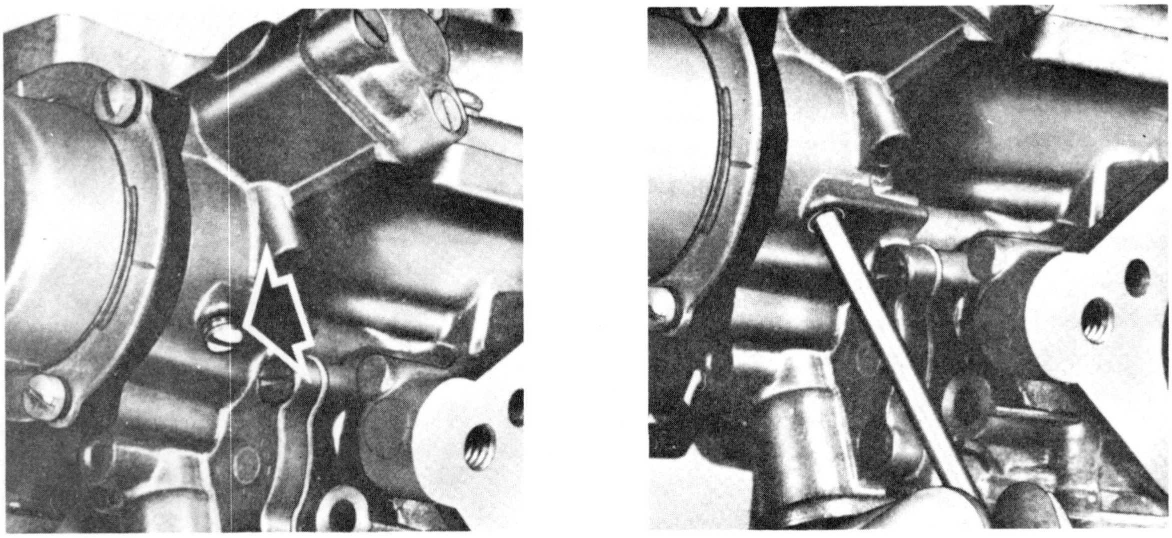

Fig 3:29 Stromberg 150 CDST carburettor - cold start device - manual resetting of thermostat

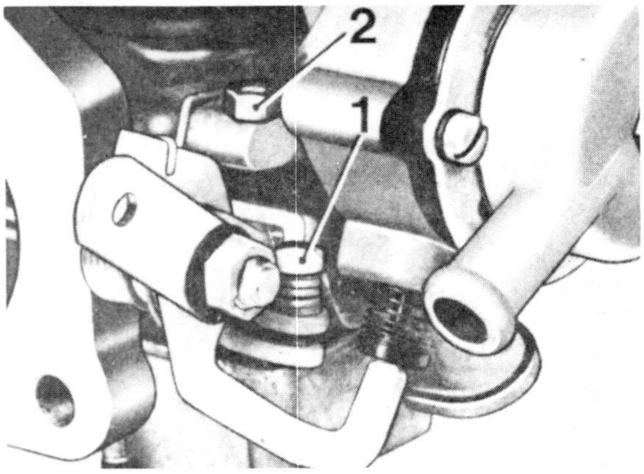

Fig 3:30 STROMBERG 150 CDST CARBURETTOR
1 Throttle stop screw
2 Fast idle stop screw (to be adjusted only when cold start device is dismantled)

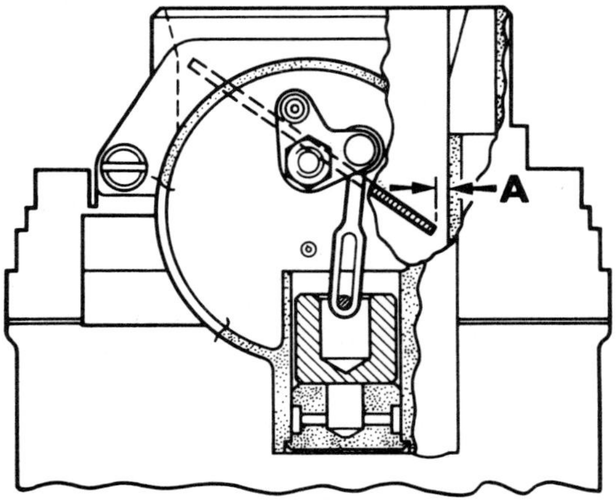

Fig 3:31 ZENITH 34 IVET CARBURETTOR
Setting of choke flap when piston is at bottom of stroke. Dimension A = 2.5 mm

Fig 3:32 ZENITH 34 IVET CARBURETTOR
Float setting. With cover inverted, gasket in position and needle valve closed dimension 'A' = 30.5–31.5 mm. Adjust by bending float arm tag on needle valve

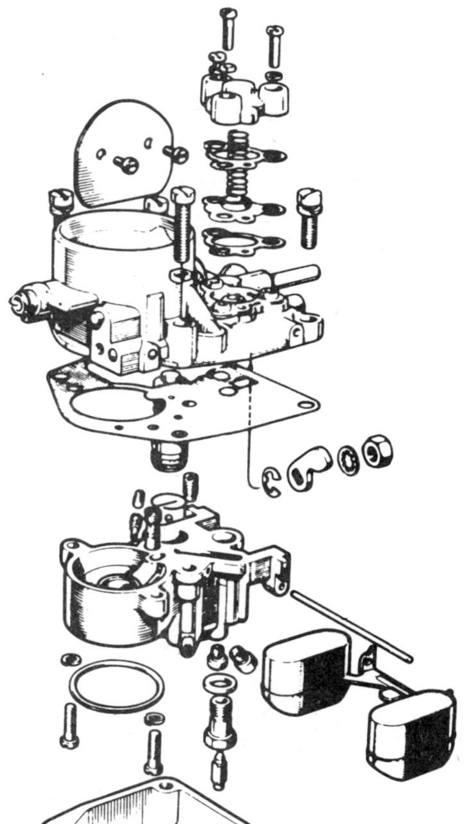

Fig 3:33 ZENITH 34 IVET CARBURETTOR
Exploded view (less auto choke assembly)

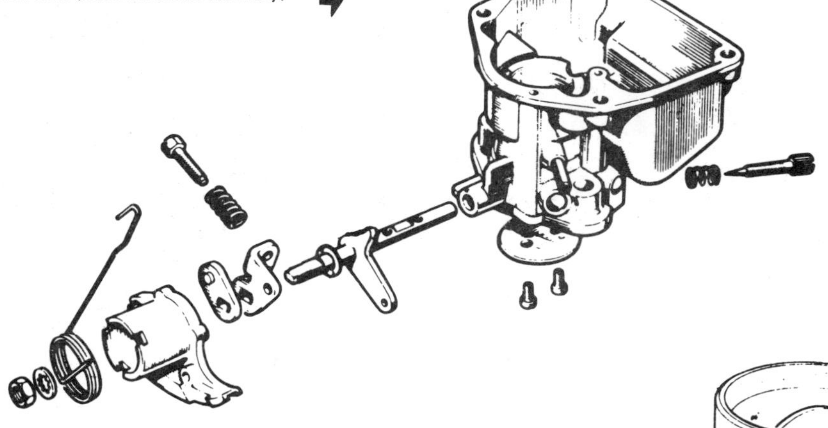

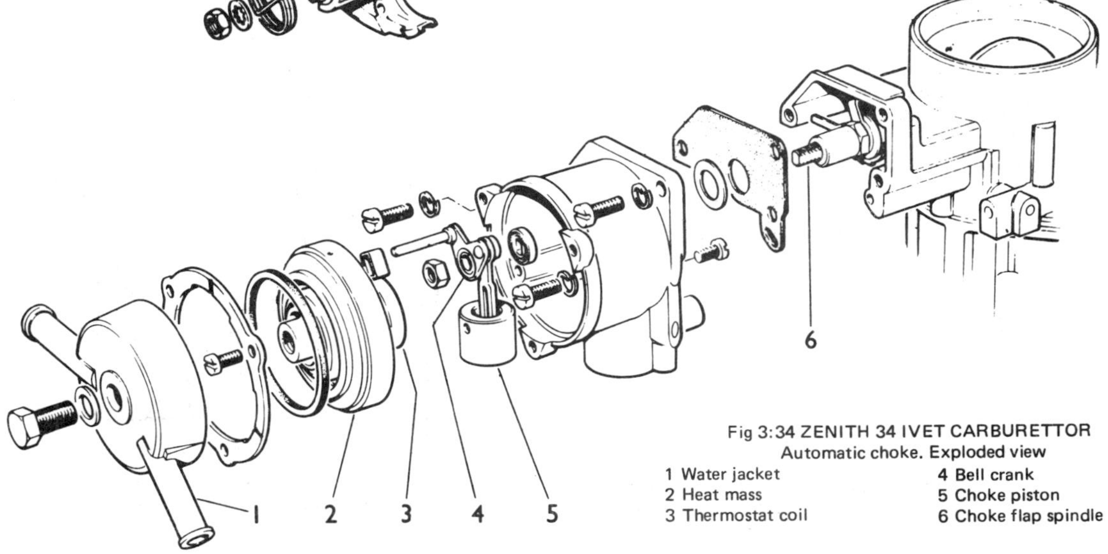

Fig 3:34 ZENITH 34 IVET CARBURETTOR
Automatic choke. Exploded view

1 Water jacket
2 Heat mass
3 Thermostat coil
4 Bell crank
5 Choke piston
6 Choke flap spindle

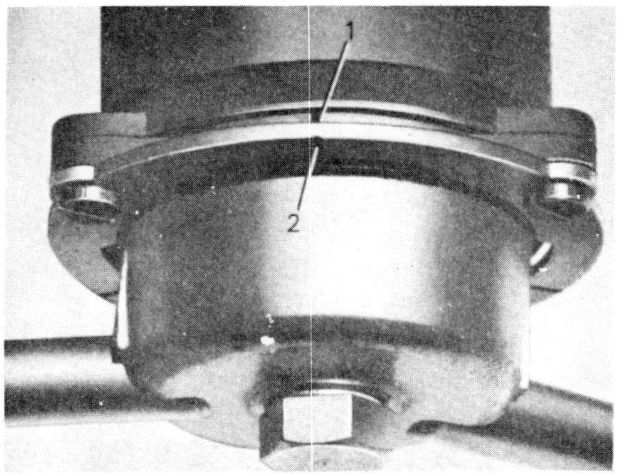

Fig 3:35 ZENITH 34 IVET
Thermostat is set by lining up mark on heat mass (1) with mark on clamping ring (2)

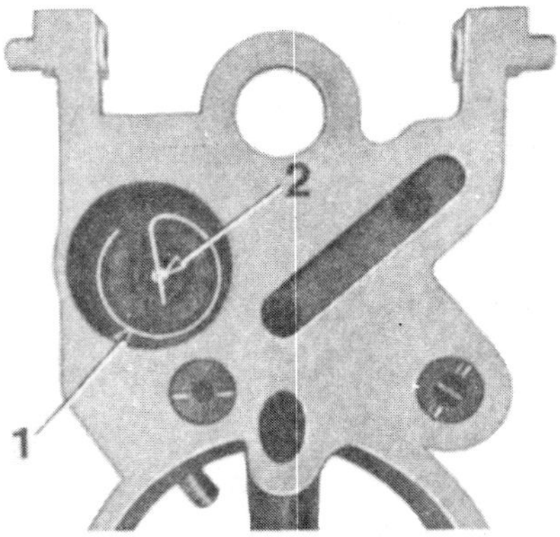

Fig 3:36 ZENITH 34 IVET CARBURETTOR
Accelerator pump non-return valve ball (2) is held by spring clip (1) locked at bottom of recess

Fig 3:37 ZENITH 34 IVET CARBURETTOR
Automatic choke piston is prevented from coming out by screw in chamber (arrowed)

removal and replacement of the tank.

2 The tank outlet pipe is on the front of the tank and should be disconnected by removing the clip and pulling off the flexible pipe connection. If the tank is to be drained this should be done now by attaching a longer piece of pipe to the outlet and syphoning the contents out.

3 From inside the boot lift the floor mat (and floor panel on estate cars) and the tank mounting screws are accessible. Before lifting the tank out, disconnect the lead from the gauge sender unit. Take care not to strain the filler neck when manoeuvring it through the grommet in the body

4 On cars fitted with 12 gallon tanks the gauge sender unit is mounted on the front together with the fuel feed pipe. These tanks are secured by bolts from underneath the car (Fig 3:40).

5 Replacement is a reversal of the removal procedure but the flange should be sealed with a suitable compound to prevent water coming up into the luggage compartment.

6 Repairs to the fuel tank to stop leaks are best carried out using resin adhesives and hardeners as supplied in most accessory shops. In cases of repairs being done to large holes, fibre glass mats or perforated zinc sheet may be required to give area support. If any soldering, welding or brazing is contemplated, the tank must be steamed out to remove any traces of petroleum vapour. It is dangerous to use naked flames on a fuel tank without this, even though it may have been lying empty for a considerable period.

14 Fuel pump - removal

1 Disconnect the two fuel lines from the pump and make sure that the line from the fuel tank is blocked by clamping or plugging it.

2 Undo the two retaining nuts holding the pump to the engine. The forward nut is not very accessible and will require a box spanner in order to remove it.

3 Lift the pump away from the mounting studs and lift away the insulating block.

15 Fuel pump - dismantling

1 Remove the top cover securing screw and lift off the cover and filter screen. Do not lose the nylon spacer collar round the screw between the cover and the screen: this ensures that the screen is held down in position.

2 Mark the relationship between centre and base sections of the pump body and then remove the five securing screws and lift off the centre section (Fig 3:41).

3 To release the diaphragm, depress the centre and turn it 90°. This will release the diaphragm pull rod from the stirrup in the operating link. Lift out the seal and seal retainer (Fig 3:41)

4 Do not remove the rocker arm and pivot pin from the body base unless there are signs of excessive wear - in which case it would probably be more economical to obtain an exchange pump.

5 To remove the valve assemblies from the body centre section, they must be prised out carefully past the stakes which locate them. Remove the sealing ring fitted behind each valve.

16 Fuel pump - inspection, reassembly and replacement

1 Examine the diaphragm for signs of splitting or cracking and renew it, if in any doubt.

2 If the valves are suspected of malfunctioning, replace them.

3 The filter screen should be intact with no signs of enlarged holes or broken strands.

4 Renew the oil seal.

5 Clean up the recesses where the valves have been staked into the body to ensure that when replaced the valves will seat neatly.

6 To refit the valves, fit new gaskets first and then press them

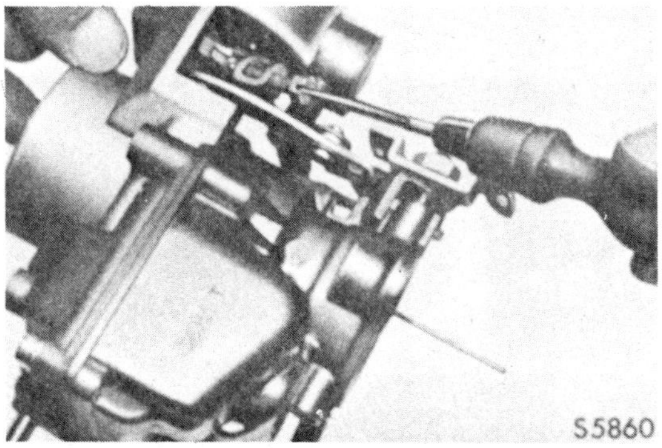

Fig 3:38 ZENITH 34 IVET CARBURETTOR
Fast idle adjustment is made with 1.3 mm gap (No 55 drill) between throttle flap and choke tube and with choke flap closed. Adjust screw against top step of cam.

Fig 3:39 ZENITH 34 IVET CARBURETTOR
Correct position of throttle control lever spring

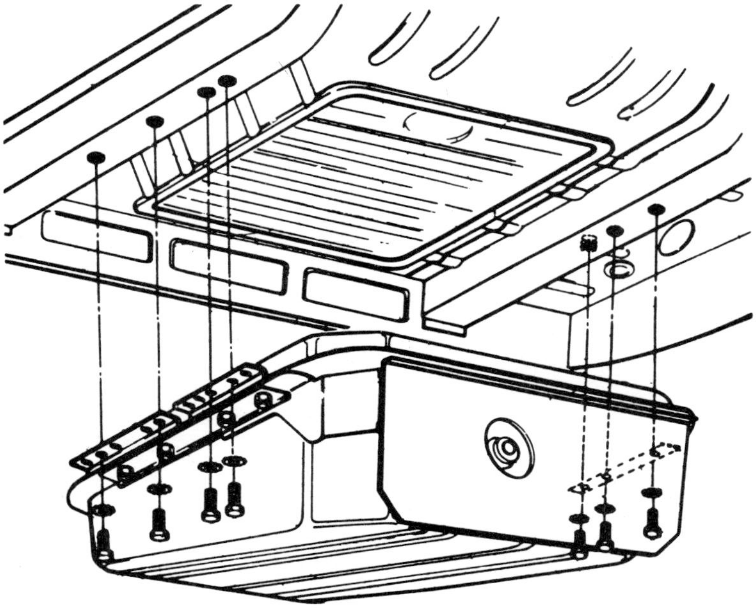

Fig 3:40 Fuel tank - 12 gallon

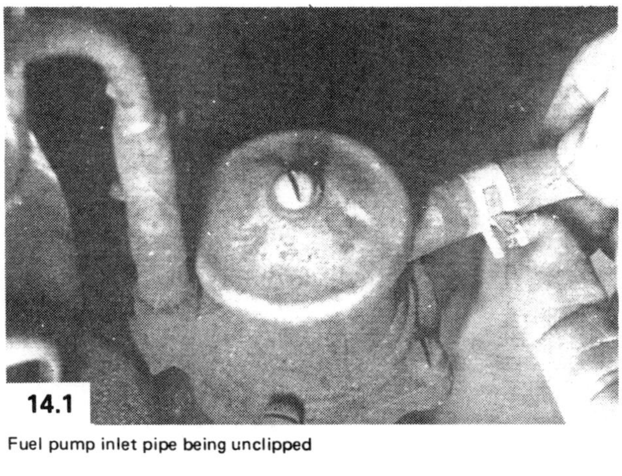

Fuel pump inlet pipe being unclipped

carefully home, preferably with a tube that will locate round their rims. If this is not available, press round each rim with a non-metallic article, a little at a time so that they bed down squarely. Then stake the body in six positions round each valve to hold them in position.

7 To refit the diaphragm, first put a new oil seal, followed by the retainer, into the body base. Put the diaphragm pull rod through the seal and the groove in the rocker arm link. Then turn the diaphragm anticlockwise 90° so that it lines up with the screw holes, and the lug on the body aligns with the tab on the diaphragm.

8 Move the rocker arm until the diaphragm is level with the body flanges and hold the arm in this position. Reassemble the two halves of the pump, ensuring that the previously made marks on the flanges are adjacent to each other.

9 Insert the five screws and lockwashers and tighten them down finger tight.

10 Move the rocker arm up and down several times to centralise the diaphragm, and then with the arm held down, tighten the screws securely in a diagonal sequence.

11 Replace the gauze filter in position. Fit the cover sealing ring; fit the cover; and insert the bolt with the fibre washer under its head. Do not overtighten the bolt but ensure that it is tight enough to prevent any leaks.

12 Fuel pump replacement is a straightforward reversal of the removal procedure. However, note the following points:

a) The fuel pump should be assembled to the engine block with new gaskets.

b) Ensure that the pump operating arm is resting on the camshaft, and not under it.

c) Do not overtighten the pump retaining nuts.

d) Test that the pump is working, by disconnecting the pipe feed at the carburettor, holding a container under it and getting someone to turn the engine. The fuel should spurt out in intermittent jets.

17 Fuel gauge sender unit - fault finding

1 If the fuel gauge does not work correctly, the fault is either in the sender unit in the fuel tank, the gauge in the instrument panel, or the wiring.

2 To check the sender unit, first disconnect the green/black wire from the unit at the connector on the tank. With the ignition on, the gauge should read 'full'. With the same lead connected to earth the gauge should read 'empty'. If BOTH of these situations are correct then the fault (if any) lies in the sender unit.

3 If the gauge does not read 'full' with the wire disconnected from the sender unit, the wire should then be disconnected from the gauge unit (having removed the instrument panel as described in Chapter 10). If the gauge now reads 'full' then the fault lies in the wire from the gauge to the sender unit.

4 If not, the gauge is faulty and should be replaced (for details see Chapter 10).

5 With the wire disconnected from the sender unit and earthed, if the gauge reads anything other than 'empty', check the rest of the circuit as described in Chapter 10.

6 To remove the sender unit, first remove the boot floor cover and disconnect the lead from the sender. The six screws should be removed and the unit can then be lifted out together with the float.

7 Replacement is a straightforward reversal of the removal procedure. Fit a new gasket, and position it so that the terminal blade points to the front of the car. After the wire is connected, bend the terminal blade down flush with the tank.

18 Exhaust system

1 The exhaust system incorporates two silencers and is supported at the front where it bolts to the manifold; and underneath by two flexible hangers. One supports the rear silencer and the other the tail pipe.

2 If any one section of the system needs replacement, it is often easiest and (in the long run) cheapest, to replace the whole lot.

3 Make sure that the attitude and clearance of the silencers is correct and a gap of at least one inch exists between them and the floor panel, and always fit a new manifold to pipe gasket.

Some early 1256 cc engines were fitted with a single exhaust downpipe.

With twin outlet pipes from the manifold, the centre two nuts should be tightened up first.

19 Accelerator cable and pedal

1 Should the accelerator cable break, it may be removed by first removing the whole pedal assembly from the floor of the car. The clevis pin can then be taken out of the cable end. The other end of the cable is detached from the carburettor. The cable outer is removed by undoing the clamping nuts at each end and detaching it from its mountings.

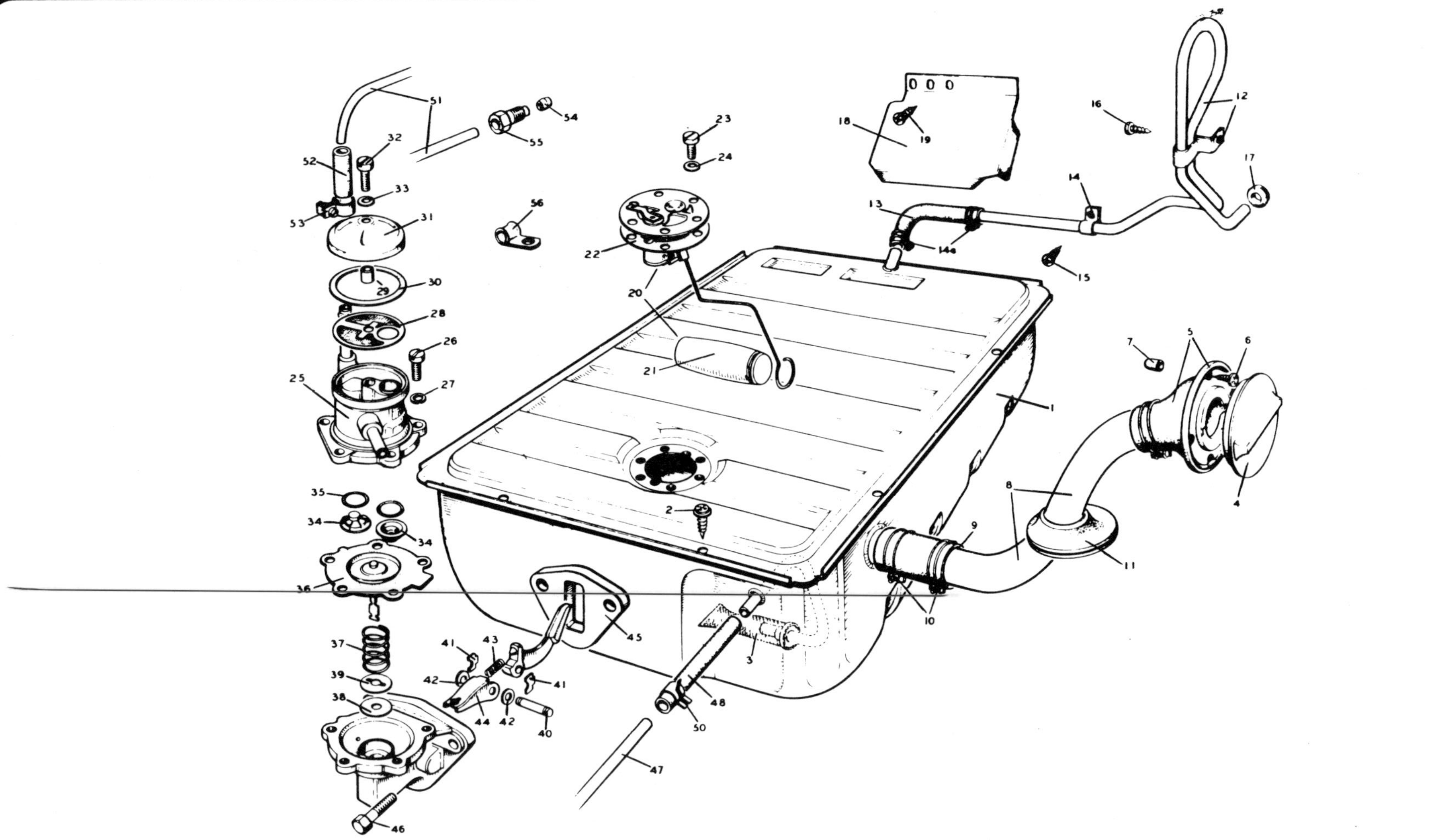

Fig 3:41 FUEL PUMP, TANK AND GAUGE SENDER UNIT COMPONENTS – EXPLODED VIEW

1 Tank
2 Mounting screw
3 Tank filter
4 Filler cap
5 Filler neck
6 Screw
7 Thread protector
8 Filler pipe
9 Hose
10 Hose clip
11 Grommet
12 Vent pipe
13 Hose
14 Fixing clip
15 Screw
16 Screw
17 Grommet
18 Finisher plate
19 Fixing screw
20 Fuel gauge sender unit
21 Float
22 Gasket
23 Screw
24 Washer
25 Pump upper body
26 Screw
27 Lockwasher
28 Filter screen
29 Spacer collar
30 Cover gasket
31 Cover
32 Cover screw
33 Screw gasket
34 Valve
35 Gasket
36 Diaphragm
37 Diaphragm spring
38 Oil seal
39 Oil seal retainer
40 Rocker arm pin
41 Pin retainer
42 Spacer washer
43 Spring
44 Link
45 Heat insulator
46 Bolt
47 Delivery pipe to pump
48 Connecting hose
50 Hose clip
51 Pipe to carburettor
52 Connecting hose
53 Hose clip
54 Pipe sleeve
55 Union nut
56 Pipe clip

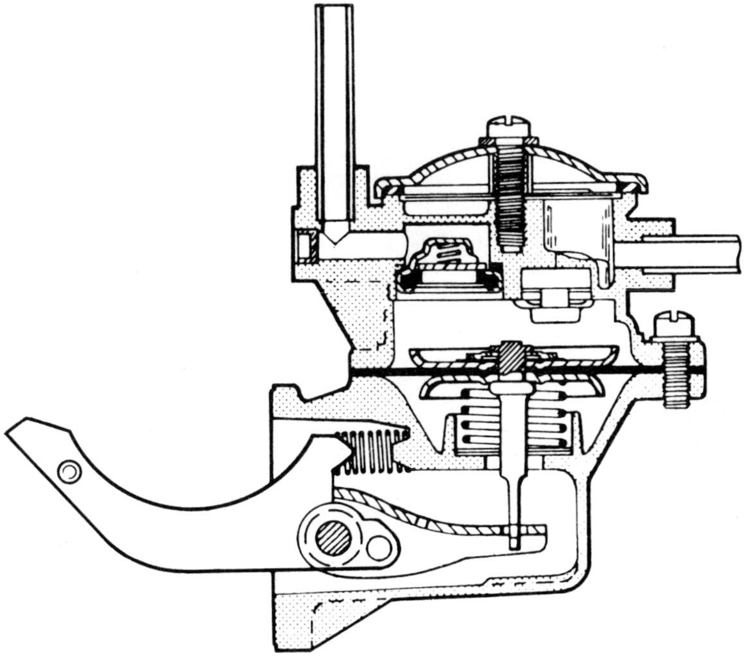

Fig 3:42 Fuel pump - cross section view

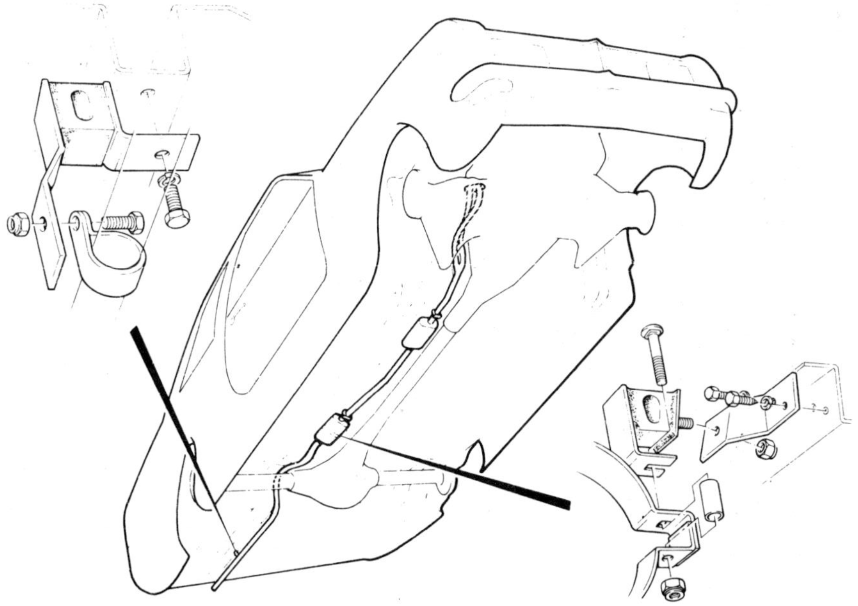

Fig 3:43 Exhaust system - detail of mounting brackets

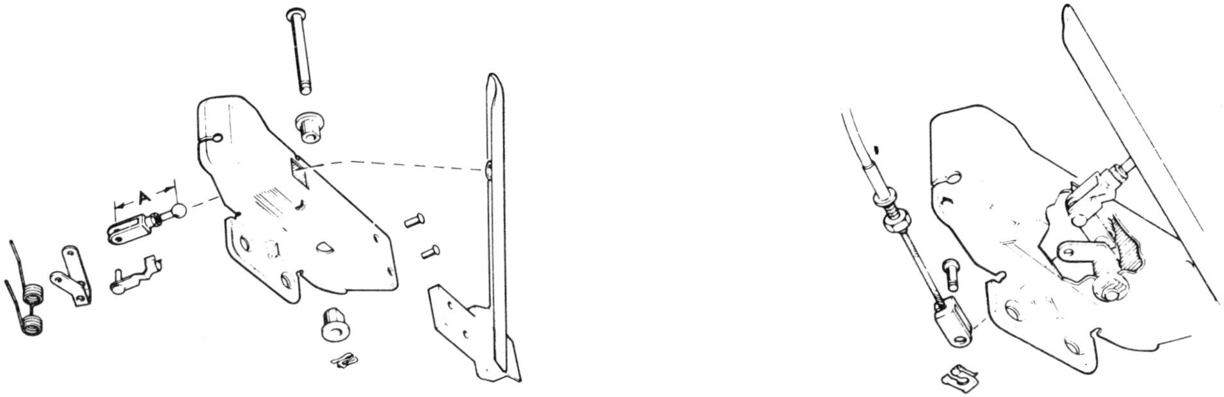

Fig 3:44 Accelerator pedal assembly - exploded and assembled views - 'A' = 2 inches

Fault Finding Chart – Fuel System and Carburation

Unsatisfactory engine performance and excessive fuel consumption are not necessarily the fault of the fuel system or carburettor. In fact they more commonly occur as a result of ignition and timing faults. Before acting on the following it is necessary to check the ignition system first. Even though a fault may lie in the fuel system it will be difficult to trace unless the ignition is correct. The faults below, therefore, assume that this has been attended to first (where appropriate).

Symptom	Reason/s	Remedy
Smell of petrol when engine is stopped	Leaking fuel lines or unions	Repair or renew as necessary.
	Leaking fuel tank	Fill fuel tank to capacity and examine carefully at seams, unions and filler pipe connections. Repair as necessary.
Smell of petrol when engine is idling	Leaking fuel line unions between pump and carburettor.	Check line and unions and tighten or repair.
	Overflow of fuel from float chamber due to wrong level setting, ineffective needle valve or punctured float.	Check fuel level setting and condition of float and needle valve, and renew if necessary.
Excessive fuel consumption for reasons not covered by leaks or float chamber faults	Worn jets (Zenith)	Renew jets
	Over-rich jet setting (Stromberg)	Adjust jet
	Sticking strangler flap (Zenith)	Check correct movement of strangler flap.
Difficult starting, uneven running, lack of power, cutting out	One or more jets blocked or restricted	Dismantle and clean out float chamber and jets.
	Float chamber fuel level too low or needle valve sticking	Dismantle and check fuel level and needle valve.
	Fuel pump not delivering sufficient fuel	Check pump delivery and clean or repair as required.
	Air valve piston not operating correctly (Stromberg)	Dismantle and examine. Clean and repair as required.

Chapter 4 Ignition system

Contents

Specifications

Spark plugs

Standard	AC 42 X LS
Electrode gap	.030
HT leads - suppressor type	4000 - 8000 ohms resistance per foot

Coil

Make	AC-Delco
Current consumption	.78 amps at 1000 rpm (distributor)
Primary coil resistance	1.3 - 1.5 ohms at 20°C
Resistor coil resistance	2 ohms at 20°C

Distributor

Make	Delco-Remy D202
Rotation	Anticlockwise
Firing sequence	1, 3, 4, 2
Contacts points gap - new	.022 in
- used	.020 in
Contact spring arm tension	17 - 21 ozs
Cam dwell angle (up to engine no. 1639738)	35° - 37°
Cam dwell angle (engine no 1639738 onwards)	49° - 51°
Points breaker plate rotation load	10 - 16 ozs
Mainshaft - diameter	.4895 in to .4900 in
- clearance in bushes	.003 in to .0013 in
- end float - initial	.002 in to .005 in
- maximum	.010 in
Upper thrust washer thickness	.029 in to .033 in
Lower thrust washer thickness	.062 in to .066 in

Ignition timing

Static advance - 1159 cc and 1256 cc		
low compression	9° BTDC	
high compression (pre-1974 models)	4½° BTDC	
high compression (1974 models onwards)	9° BTDC	
Static advance - 1159 cc - high performance	9° BTDC	
Vacuum advance - 1159 cc and 1256 cc		
Standard engine	Vacuum (Hg)	Distributor degrees
	5	0 - 1½
	7	1 - 4½
	9	3½ - 7½
	11+	5½ - 7½
1159 cc High performance engine	2	0
	4	0 - 5
	6	5¼ - 7¾
	8	7½ - 9½
	10+	9 - 11

Centrifugal advance
Cut in speed 300 - 425 rpm (distributor)

Distributor rpm	Distributor degrees
400	0 - 1¾
600	2½ - 5
800	5¾ - 8¼
1000	8¾ - 11½
1200	9¾ - 12½
1400	10¾ - 13½
1600	11¾ - 14½
1800	13 - 15½
2000+	14 - 16½

NB Tickover speed in 1256 cc engines is above 400 rpm.

Special distributor for 1159 cc emission control engine
475 - 600 rpm (distributor)

Distributor rpm	Distributor degrees
600	0 - 3
800	4½ - 7
1000	9 - 11
1200	10 - 12
1400	11 - 13
1600	12 - 14
1800	13 - 15
2000	14 - 16
2200	14 - 16
2400	14 - 16
2600+	14 - 16

1 General description

In order that the engine can run correctly it is necessary for an electrical spark to ignite the fuel/air mixture in the combustion chamber at exactly the right moment in relation to engine speed and load. The ignition system is based on feeding low tension voltage from the battery to the coil where it is converted to high tension voltage. The high tension voltage is powerful enough to jump the spark plug gap in the cylinders many times a second under high compression pressures, providing that the system is in good condition and that all adjustments are correct.

The ignition system is divided into two circuits, the low tension circuit and the high tension circuit.

The low tension (sometimes known as the primary) circuit consists of the battery, lead to the control box, lead to the ignition switch, lead from the ignition switch to the low tension or primary coil windings (terminal SW), and the lead from the low tension coil windings (coil terminal CB) to the contact breaker points and condenser in the distributor.

The high tension circuit consists of the high tension or secondary coil windings, the heavy ignition lead from the centre of the coil to the centre of the distributor cap, the rotor arm, and the spark plug leads and spark plugs.

The system functions in the following manner: High tension voltage is generated in the coil by the interruption of the low tension circuit. The interruption is effected by the opening of the contact breaker points in this low tension circuit. High tension voltage is fed via the carbon brush in the centre of the distributor cap to the rotor arm of the distributor.

The rotor arm revolves anticlockwise at half engine speed inside the distributor cap, and each time it comes in line with one of the four metal segments in the cap, which are connected to the spark plug leads, the opening and closing of the contact breaker points causes the high tension voltage to build up, jump the gap from the rotor arm to the appropriate metal segment, and so via the spark plug lead to the spark plug, where it finally jumps the spark plug gap before going to earth.

The ignition is advanced and retarded automatically, to ensure the spark occurs at just the right instant for the particular load at the prevailing engine speed.

The ignition advance is controlled both mechanically and by a vacuum operated system. The mechanical governor mechanism comprises two weights, which move out from the distributor shaft as the engine speed rises, due to centrifugal force. As they move outwards they rotate the cam relative to the distributor shaft, and so advance the spark. The weights are held in position by two springs and it is the tension of the springs which is largely responsible for correct spark advancement.

The vacuum control consists of a diaphragm, one side of which is connected via a small bore tube to the carburettor, and the other side to the contact breaker plate. Depression in the inlet manifold and carburettor, which varies with engine speed and throttle opening, causes the diaphragm to move, so moving the contact breaker plate, and advancing or retarding the spark. A fine degree of control is achieved by a spring in the vacuum assembly.

2 Contact breaker points - adjustment

1 To adjust the contact breaker points to the correct gap, first pull off the two clips securing the distributor cap to the distributor body, and lift away the cap. Clean the cap inside and out with a dry cloth. It is unlikely that the four segments will be badly burned or scored, but if they are, the cap will have to be renewed.
2 Check the carbon brush located in the top of the cap to make sure that it is not broken or missing.
3 Gently prise the contact breaker points open to examine the condition of their faces. If they are rough, pitted or dirty, it will be necessary to remove them for resurfacing, or for replacement points to be fitted.
4 Presuming the points are satisfactory, or that they have been cleaned and replaced, measure the gap between the points by turning the engine over until the contact breaker arm is on the peak of one of the four cam lobes.
5 A .020 in feeler gauge should now just fit between the points (.022 on new points).
6 If the gap varies from this amount, slacken the contact securing screw (photo).
7 Adjust the contact gap by inserting a screwdriver in the nick in the side of the fixed plate and lever it in the required direction (photo).
8 Replace the rotor arm and distributor cap and clip the cap into place.

3 Contact breaker points - removal and replacement

1 Remove the distributor cap (photo).
2 Remove the rotor arm by pulling it straight up. Do not pull it by the contact spring. If it is tight, lever it carefully from underneath with a screwdriver (photo).
3 Remove the contact points holding screw.
4 Pull the condenser lead clip and coil lead clip from the nylon insulator holding the end of the moving contact spring.
5 Lift the complete contact set assembly off the pivot pin of the mounting plate.
6 If the condition of the points is not too bad, they can be reconditioned by rubbing the contacts clean with fine emery cloth or a fine carborundum stone. It is important that the faces are rubbed flat and parallel to each other so that there will be complete face to face contact when the points are closed. One of the points will be pitted and the other will have deposits on it.
7 It is necessary to completely remove the built-up deposits, but not necessary to rub the pitted point right down to the stage where

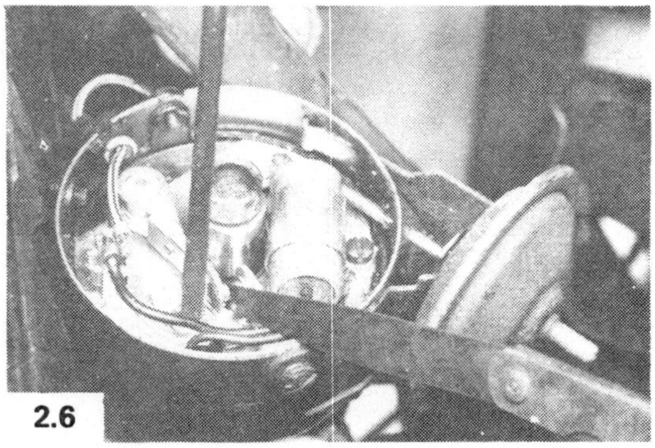
2.6
Slackening distributor fixed contact securing screw

2.7
Adjusting distributor contact gap

3.1
Removing distributor cap

3.2
Removing distributor rotor arm

all the pitting has disappeared, though obviously if this is done it will prolong the time before the operation of refacing the points has to be repeated.

8 Thoroughly clean the points before refitting them. Locate the fixed contact plate over the base of the pivot pin and then fix the moving contact into position so that the end of the spring fits over the centre boss of the nylon insulator. Replace the fixing screw. Press in the condenser and coil lead tags to the nylon insulator behind the spring (Fig 4:2 and photos).

9 Adjust the gap as described in Section 2.

4 Condenser - removal, testing and replacement

1 The purpose of the condenser (sometimes known as capacitor) is to ensure that when the contact breaker points open there is no sparking across them which would waste voltage and cause rapid deterioration of the points.

2 The condenser is fitted in parallel with the contact breaker points. If it develops a short circuit, it will cause ignition failure as the points will be prevented from interrupting the low tension circuit.

3 If the engine becomes very difficult to start or begins to misfire whilst running and the breaker points show signs of excessive burning, then the condition of the condenser must be suspect. A further test can be made by separating the points by hand with the ignition switched on. If this is accompanied by a bright spark at the contact points it is indicative that the condenser has failed.

4 Without special test equipment the only sure way to diagnose condenser trouble is to replace a suspected unit with a new one and note if there is any improvement.

5 To remove the condenser from the distributor, remove the distributor cap and the rotor arm.

6 Pull out the condenser lead clip from the nylon insulator where it fits behind the spring.

7 Undo the mounting bracket screw and remove the condenser.

8 Replacement is simply a reversal of the removal process. Take particular care that the condenser wire cannot short circuit against any portion of the breaker plate.

5 Distributor - removal and replacement

1 To remove the distributor complete with cap from the engine, begin by pulling the plug lead terminals off the four spark plugs. Free the HT lead from the centre of the coil to the centre of the distributor by undoing the lead retaining cap from the coil.

2 Pull off the pipe holding the vacuum tube to the distributor vacuum advance and retard take off pipe.

3 Disconnect the low tension wire from the coil.

4 Undo and remove the bolt which holds the distributor clamp plate to the crankcase and lift out the distributor (photo).

5 NOTE: If it is not wished to disturb the timing then under no circumstances should the clamp pinch bolt, which secures the distributor in its relative position in the clamp, be loosened. Providing the distributor is removed without the clamp being loosened from the distributor body, the timing will not be lost.

6 Replacement is a reversal of the above process. When putting the distributor into position, line up the offset tongue of the spindle with the offset slot in the drive shaft. The correct timing will be automatically obtained provided the clamp plate has not been touched. If it is moved, retime the ignition as described in Section 9.

6 Distributor - dismantling

1 The only time when the distributor should be dismantled is when it is wished to recondition it, and certain parts should always be discarded as a matter of course. Ensure that these parts, described in the following text, are available before taking the distributor down.

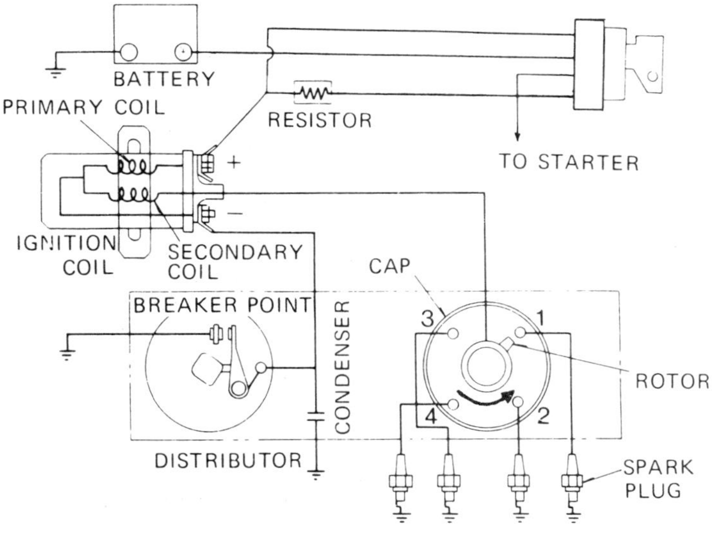

Fig 4:1 Schematic diagram of ignition circuit

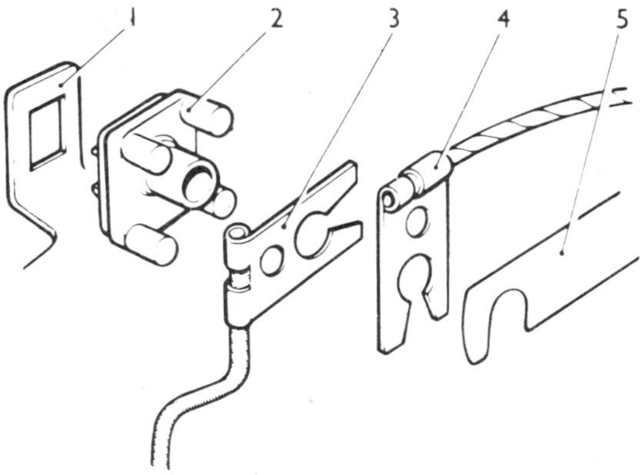

Fig 4:2 DETAIL OF CONNECTION OF CONTACT POINTS AND ASSOCIATED WIRES

1 Breaker base plate lug
2 Nylon insulator
3 Condenser wire
4 LT wire from coil
5 Moving contact spring

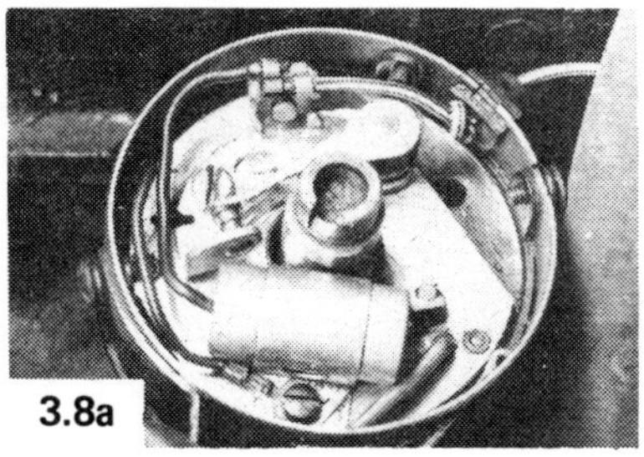
3.8a

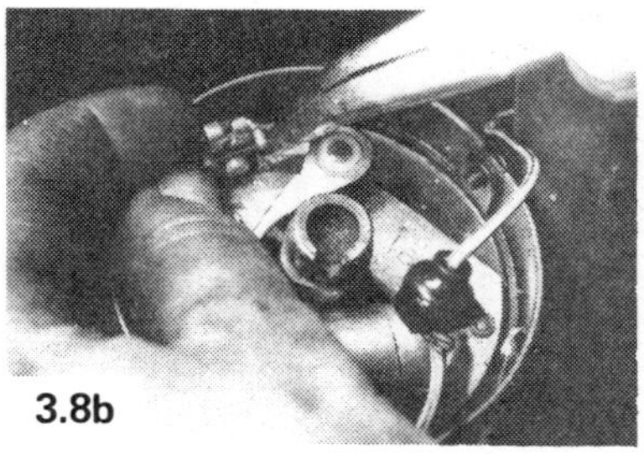
3.8b

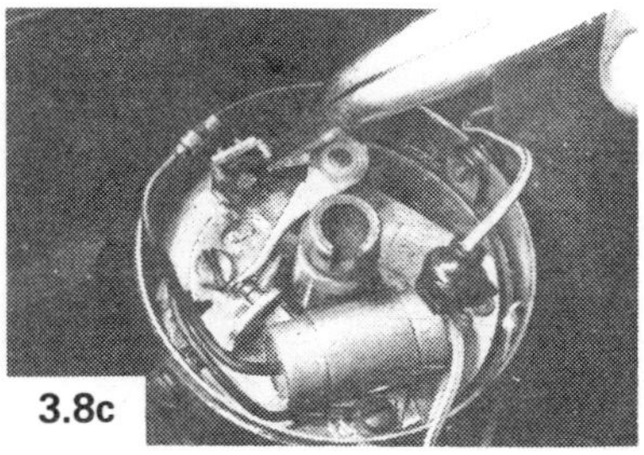
3.8c

Replacing contact points in distributor

5.4

Distributor removed from engine

2 With the distributor removed from the car and on the bench, remove the distributor cap and lift off the rotor arm. If very tight, lever it off gently with a screwdriver.
3 Remove the points from the distributor as described in Section 3.
4 Carefully remove the vacuum unit assembly after undoing the two screws from the side of the distributor body which also serves to partially hold the contact breaker plate in place. Undo the third plate retaining screw and lift out the plate assembly and the condenser.
5 With a fine nosed punch remove the retaining pin from the bottom end of the distributor mainshaft.
6 Take the tagged washer off the mainshaft.
7 Pull the mainshaft out of the housing and throw away the thrust washers and the advance weight springs.
8 Undo the clamp bolt and remove the clamp and oil seal ring from the shank of the housing.

7 Distributor - inspection and repair

1 Check the points as described in Section 2. Check the distributor cap for signs of tracking indicated by a thin black line between the segments. Replace the cap if any signs of tracking are found.
2 If the metal portion of the rotor arm is badly burned or loose, renew the arm. If slightly burnt clean the arm with a fine file. Check that the rotor contact spring setting is between .30 to .34 in (Fig 4:4).
3 Check that the carbon brush is intact in the centre of the distributor cover.
4 Examine the fit of the breaker plate on the bearing plate and also check the breaker arm pivot for looseness or wear, and renew as necessary.
5 Examine the balance weights and pivot pins for wear, and renew the weights or cam assembly if a degree of wear is found.
6 Examine the length of the balance weight springs and compare them with new springs. If they have stretched they must be renewed. It is almost inevitable that they will have stretched and it is best to fit new springs as a matter of course.
7 Check that the mainshaft is not a slack fit in the housing bushes. If it is, the points gap setting will fluctuate according to the degree of slackness. Replace the bushes and also the shaft if the old one is still slack.
8 The diaphragm should be checked visually for proper operation when the engine is running.

8 Distributor - reassembly

1 Reassembly is a straightforward reversal of the dismantling process, but there are several points which should be noted.
2 Lubricate with engine oil the balance weights and other parts of the mechanical advance mechanism, the cam, the mainshaft, and the felts, during assembly.
3 Always use a new upper and lower thrust washer and check the mainshaft end float between the bottom thrust washer and the housing with a feeler gauge. The dimension should be between .002 and .005 in using new thrust washers otherwise .010 in.
4 If a new mainshaft is being fitted check that it is of the correct type as they can vary between the same type of distributor. The figure stamped on the underside of the centrifugal advance mechanism plate indicates the type being used. It will also be necessary to drill the end of the shaft to accept a new pin using a No 30 drill. To do this replace the shaft into the distributor body with the upper thrust washer in place, and put the tabbed retaining washer on the end of the shaft. If you are not too sure of your ability to drill the shaft accurately in order to obtain the specified end float, then it is always better to err on the side of a little too much end float which can be taken up with additional spacer washers. Be sure to drill the hole exactly at right angles to the shaft axis and through the shaft centre line. It is not important about its radial position relative to the tongue on the end of the shaft.
5 Finally set the contact breaker gap to the specified clearance.

9 Ignition - timing

1 If the clamp plate pinch bolt has been loosened and moved on the distributor, the following procedure should be followed:
2 Check the 'Specifications' at the beginning of this Chapter, and see what the timing setting should be according to the model of engine fitted. This setting refers to the position of No 1 piston coming up on its compression stroke and is expressed as the number of degrees the crankshaft has to rotate for that piston to reach top dead centre.
3 Replace the distributor with the clamp bolt slackened but with the clamp securing bolt tightened into the crankcase. The offset torque should be set so that it engages with the offset slot in the drive shaft.
4 Remove No 1 (front) spark plug and turn the engine clockwise (put it in gear and push the car or use a spanner on the crankshaft pulley wheel) with one finger over the plug hole. As soon as pressure is felt stop and then look at the timing position on the pulley wheel. It should be getting near to the two indicator lugs on the crankcase. The first one it reaches is 9° BTDC and between the two is 4½° BTDC (Fig 4:5).
5 Turn the engine and set the pointer on the pulley to the required position on the crankcase.
6 The engine is now set at the correct position with No 1 piston coming up on the compression stroke to TDC. As a basic check, fit the rotor arm and note if it points (more or less) towards the segment on the distributor cap attached to No 1 spark plug lead. If it should be a long way out it means one of two things - either the oil pump drive shaft is incorrectly meshed with the camshaft, or the spark plug leads are mixed up in the distributor cap. If the oil pump drive shaft is exactly 180° out of position then the problem can be solved by altering the plug leads to suit. If otherwise, the distributor can be turned accordingly but there is a limit to how far it can move.
7 Assuming that all is well, turn the distributor body clockwise until the contact points are just about to open. This can be accurately gauged if a 12 volt 6 watt bulb is wired in parallel with the contact points. Switch on the ignition and when the points open the bulb should light.
8 Tighten the distributor clamp bolt.
9 If a stroboscopic light is used for a final static ignition timing check, remove the lead from No 1 plug and then connect the strobe, one wire to the plug and the other to the plug lead. With the engine idling as slowly as possible, shine the strobe light on to the timing case markers (4½° advance) or on the lower marker (9° advance).
10 If the engine speed is increased, then the effect of the vacuum and centrifugal advance controls can be seen and in fact, measured to some extent, in so far as the gap between the two crankcase timing markers represents 9° of crankshaft revolution.

10 Spark plugs and leads

1 The correct functioning of the spark plugs is vital for the correct running and efficiency of the engine. The plugs fitted as standard are listed in the 'Specifications' at the beginning of this Chapter.
2 At intervals of 6000 miles the plugs should be removed, examined, cleaned and, if worn excessively, replaced. The condition of the spark plug will also tell much about the overall condition of the engine.
3 If the insulator nose of the spark plug is clean and white, with no deposits, this is indicative of a weak mixture, or too hot a plug. (A hot plug transfers heat away from the electrode slowly - a cold plug transfers it away quickly).
4 If the tip of the insulator nose is covered with sooty black deposits, then this is indicative that the mixture is too rich. Should

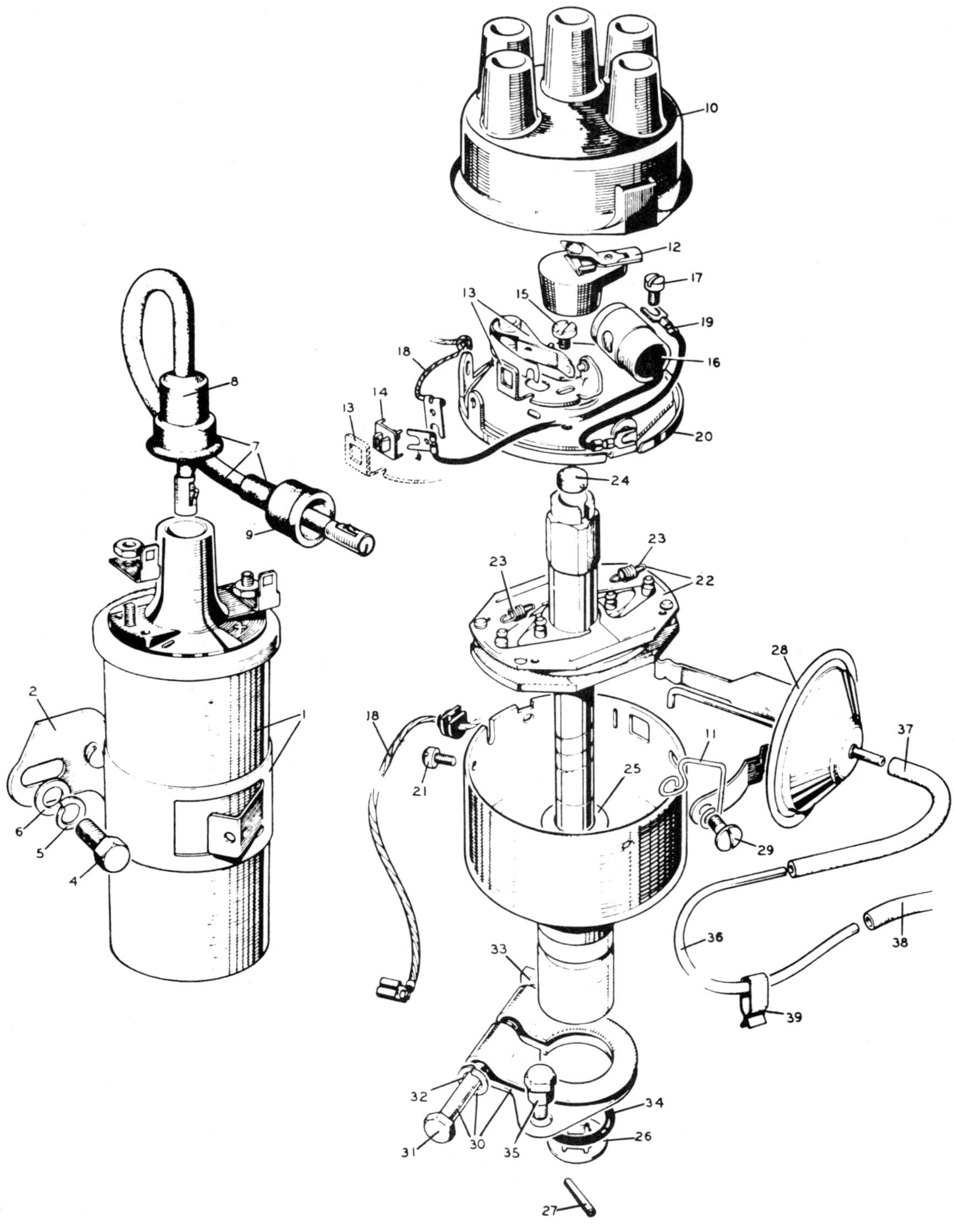

Fig 4:3 DISTRIBUTOR AND COIL – EXPLODED VIEW

1 Coil
2 Mounting bracket
4 Mounting bolt
5 Lockwasher
6 Plain washer
7 HT lead
8 Grommet
9 Grommet
10 Distributor cap
11 Clip
12 Rotor arm
13 Contact points assembly
14 Nylon terminal lug
15 Fixed contact mounting screw
16 Condenser
17 Earth wire screw
18 LT lead from coil
19 Earth wire
20 Contact breaker plate
21 Plate securing screw
22 Main shaft and balance weight assembly
23 Balance weight spring
24 Lubricating feet
25 Upper washer
26 Retaining washer
27 Locating pin
28 Vacuum advance unit
29 Vacuum unit fixing screw
30 Clamping ring assembly
31 Clamping bolt
32 Spring washer
33 Nut
34 Oil seal ring
35 Locating bolt
36 Vacuum pipe
37 Flexible connector pipe
38 Flexible pipe to manifold
39 Clip

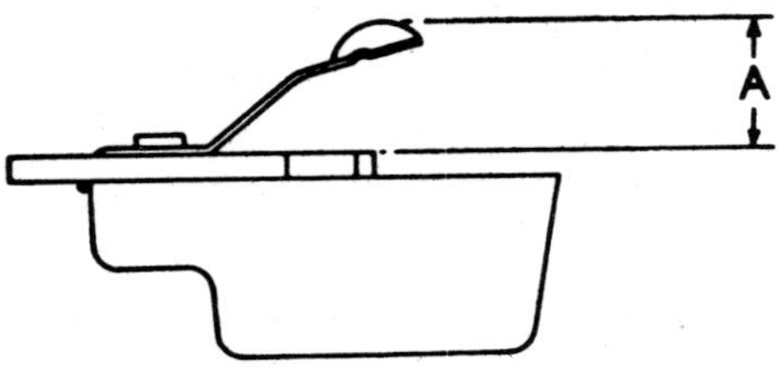

Fig 4:4 ROTOR ARM CONTACT SETTING
A = .30–.34 inch (7–8 mm)

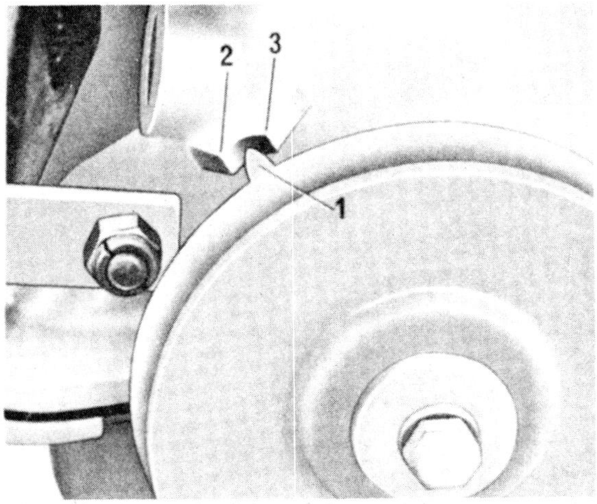

Fig 4:5 IGNITION TIMING – CRANKSHAFT SETTING
1 Crankshaft pulley pointer at 4½° BTDC
2 Indicator lug 9° BTDC
3 Indicator lug TDC

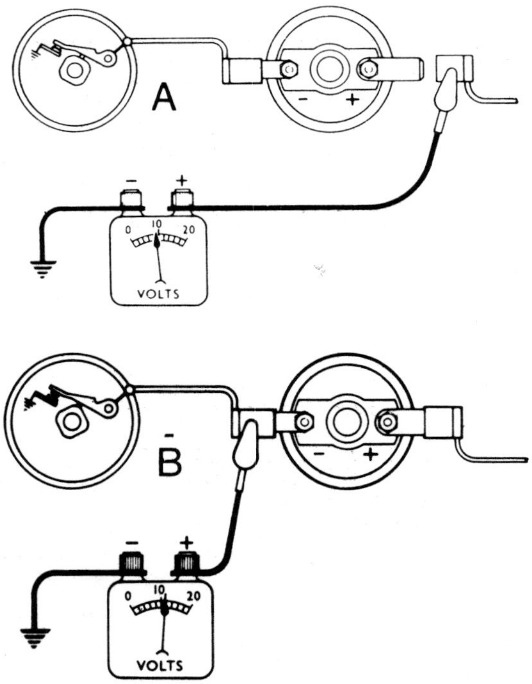

Fig 4:6 LT CIRCUIT TEST CONNECTIONS
A Checking supply to coil
B Checking condition of coil, circuit breaker and condenser

the plug be black and oily, then it is likely that the engine is fairly worn, as well as the mixture being rich.

5 If the insulator nose is covered with light tan to greyish brown deposits, then the mixture is correct and it is likely that the engine is in good condition.

6 If there are any traces of long brown tapering stains on the outside of the white portion of the plug, then the plug will have to be renewed, as this shows that there is a faulty joint between the plug body and the insulator, and compression is being allowed to leak away.

7 Plugs should be cleaned by a sand blasting machine, which will free them from carbon more thoroughly than cleaning by hand. The machine will also test the condition of the plugs under compression. Any plug that fails to spark at the recommended pressure should be renewed.

8 The spark plug gap is of considerable importance, as, if it is too large or too small the size of the spark and its efficiency will be seriously impaired. The spark plug gap should be set to .030 in for the best results.

9 To set it, measure the gap with a feeler gauge, and then bend open, or close, the outer plug electrode until the correct gap is achieved. The centre electrode should never be bent as this may crack the insulation and cause plug failure, if nothing worse.

10 When replacing the plugs, remember to use new plug washers, and replace the leads from the distributor in the correct firing order, which is 1, 3, 4, 2; No 1 cylinder being the one nearest the radiator.

11 The plug leads require no routine attention other than being kept clean and wiped over regularly. At intervals, say twice yearly, pull each lead off the plug in turn and also from the distributor cap. Water can seep down into these joints giving rise to a white corrosive deposit which must be carefully removed from the brass washer at the end of each cable, through which the ignition wires pass.

11 Ignition system - faults and remedies

1 By far the majority of breakdown and running troubles are caused by faults in the ignition system, either in the low tension or high tension circuits.

2 There are two main symptoms indicating ignition faults. Either the engine will not start or fire, or the engine is difficult to start and misfires. If it is a regular misfire, ie the engine is only running on two or three cylinders, the fault is almost sure to be in the secondary, or high tension, circuit. If the misfiring is intermittent, the fault could be in either the high or low tension circuits. If the car stops suddenly, or will not start at all, it is likely that the fault is in the low tension circuit. Loss of power and overheating, apart from faulty carburation settings, are normally due to faults in the distributor, or incorrect ignition timing.

3 If the engine fails to start and the car was running normally when it was last used, first check there is fuel in the petrol tank. If the engine turns over normally on the starter motor and the battery is evidently well charged, then the fault may be in either the high or low tension circuits. First check the HT circuit. NOTE: If the battery is known to be fully charged; the ignition light comes on, and the starter motor fails to turn the engine CHECK THE TIGHTNESS OF THE LEADS ON THE BATTERY TERMINALS and also the secureness of the earth lead to its CONNECTION TO THE BODY. It is quite common for the leads to have worked loose, even if they look and feel secure. If one of the battery terminal posts gets hot when trying to work the starter motor this is a sure indication of a faulty connection to thet terminal.

4 One of the common reasons for bad starting is wet or damp plug leads and distributor. Remove the distributor cap. If condensation is visible internally, dry the cap with a rag and also wipe over the leads. Replace the cap.

5 If the engine still fails to start, check that current is reaching the plugs, by disconnecting each plug lead in turn at the spark plug end, and hold the end of the cable about 3/16 in away from the cylinder block. Spin the engine on the starter motor.

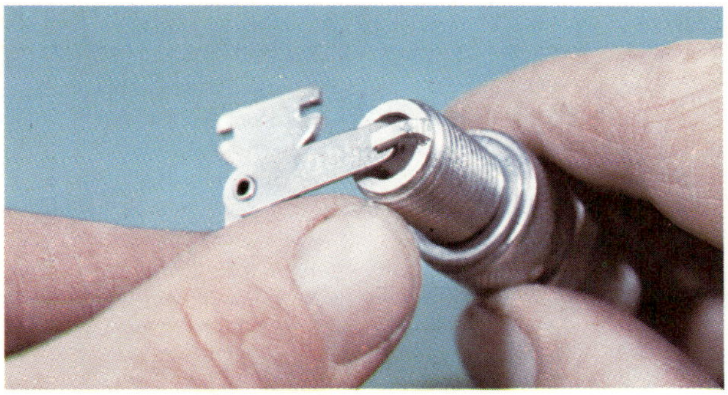

Measuring plug gap. A feeler gauge of the correct size (see ignition system specifications) should have a slight 'drag' when slid between the electrodes. Adjust gap if necessary

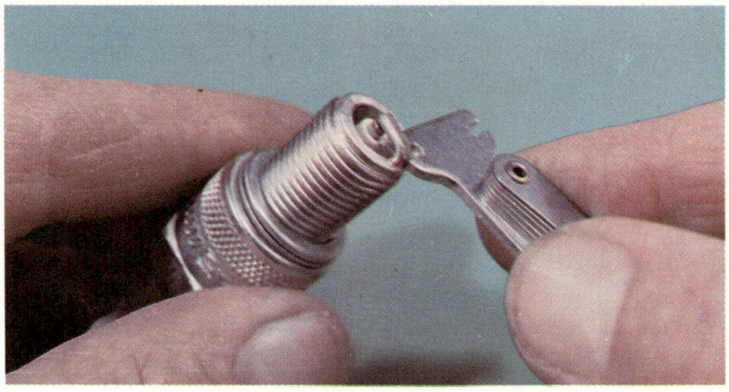

Adjusting plug gap. The plug gap is adjusted by bending the earth electrode inwards, or outwards, as necessary until the correct clearance is obtained. Note the use of the correct tool

Normal. Grey-brown deposits lightly coated core nose. Gap increasing by around 0.001 in (0.025 mm) per 1000 miles (1600 km). Plugs ideally suited to engine and engine in good condition

Carbon fouling. Dry, black, sooty deposits. Will cause weak spark and eventually misfire. Fault: over-rich fuel mixture. Check: carburettor mixture settings, float level and jet sizes; choke operation and cleanliness of air filter. Plugs can be re-used after cleaning

Oil fouling. Wet, oily deposits. Will cause weak spark and eventually misfire. Fault: worn bores/piston rings or valve guides; sometimes occurs (temporarily) during running-in period. Plugs can be re-used after thorough cleaning

Overheating. Electrodes have glazed appearance, core nose very white - few deposits. Fault: plug overheating. Check: plug value, ignition timing, fuel octane rating (too low) and fuel mixture (too weak). Discard plugs and cure fault immediately

Electrode damage. Electrodes burned away; core nose has burned, glazed appearance. Fault: initial pre-ignition. Check: for 'Overheating' but may be more severe. Discard plugs and remedy fault before piston or valve damage occurs

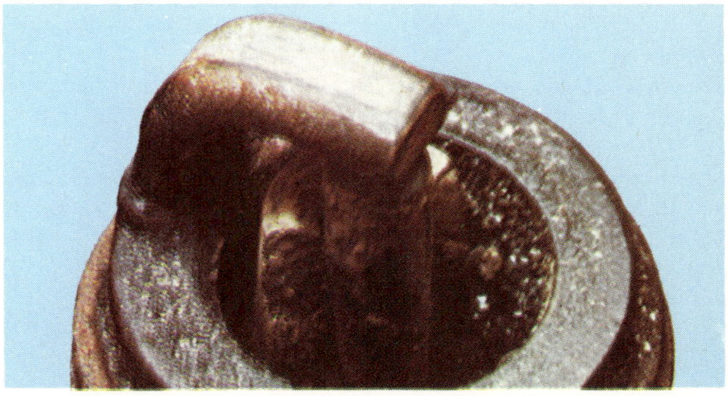

Split core nose (may appear initially as a crack). Damage is self-evident, but cracks will only show after cleaning. Fault: pre-ignition or wrong gap-setting technique. Check: ignition timing, cooling system, fuel octane rating (too low) and fuel mixture (too weak). Discard plugs, rectify fault immediately

6 Sparking between the end of the cable and the block should be fairly strong with a regular blue spark (hold the lead with rubber to avoid electric shocks). If current is reaching the plugs, then remove them and clean and regap them to .030 in. The engine should now start.

7 If there is no spark at the plug leads take off the HT lead from the centre of the distributor cap and hold it to the block as before. Spin the engine on the starter once more. A rapid succession of blue sparks between the end of the lead and the block indicate that the coil is in order and that the distributor cap is cracked, the rotor arm faulty, or the carbon brush in the top of the distributor cap is not making good contact with the spring on the rotor arm. Possibly the points are in bad condition. Clean and reset them as described earlier in this Chapter.

8 If there are no sparks from the end of the lead from the coil, check the connections at the coil end of the lead. If it is in order start checking the low tension circuit.

9 Use a 12v voltmeter or a 12v bulb and two lengths of wire. With the ignition switch on and the points open, test between the low tension wire to the coil (it is marked SW or +) and earth. No reading indicates a break in the supply from the ignition switch. Check the connections at the switch to see if any are loose. Refit them and the engine should run. A reading shows a faulty coil or condenser, or broken lead between the coil and the distributor.

10 Take the condenser wire off the points assembly and with the points open, test between the moving point and earth. If there now is a reading, then the fault is in the condenser. Fit a new one and the fault is cleared.

11 With no reading from the moving point to earth, take a reading between earth and the CB or - terminal of the coil. A reading here shows a broken wire which will need to be replaced between the coil and distributor. No reading confirms that the coil has failed and must be replaced, after which the engine will run once more. Remember to refit the condenser wire to the points assembly. For these tests it is sufficient to separate the points with a piece of dry paper while testing with the points open.

12 The Viva is fitted with a device which boosts the output from the coil when the starter is operated, and battery voltage tends to drop due to load placed upon it. Quite simply, the coil is rated for a continuous 6 volt supply. As the vehicle system is 12 volt a resistor cable is fitted into the LT supply to the coil so that under normal running conditions the coil only receives a 6 volt supply. However, when the starter is operated the system voltage drops. This is usual. In addition to the normal LT feed to the coil, therefore, an additional feed is taken from the starter solenoid switch direct to the coil. This feed only operates when the solenoid starter terminals are closed, ie when the starter is turning. Consequently, for the brief time when the voltage drops from 12 to about 8 volts, 8 volts is fed direct to the 6 volt coil providing a temporary starting boost.

Certain checks are necessary to ensure that:

a) The starter feed is functioning properly - otherwise only about 2 volts would reach the coil on starting.

b) The resistor is in good order - otherwise 12 volts or no volts may reach the 6 volt coil.

The tests detailed are classified by reference to the wiring connections given in Fig 4:6:

A (i) To check the current supply to the coil through the resistor wire, disconnect the resistor wire and connect it to earth via a voltmeter. With contact points closed and ignition switched on the reading should be 6 volts (approx). If not then there is a fault in the wire, the ignition switch, or the feed to the ignition switch.

(ii) With the connections still made, operate the starter motor. The voltage should jump to 9v (approx) whilst the starter is turning.

If the wire is not faulty and the voltage does not rise, then the starter solenoid switch must be faulty and will need renewal.

B (i) To check the primary (LT) coil winding points and condenser, connect the voltmeter to the – (CB) terminal of the coil leaving both coil wires connected also. Switch on the ignition with the contact point OPEN. The voltage should be 12v (approx). If the reading is low or zero, then there is a fault in the coil primary winding or a short circuit at the contact points or condenser.

(ii) With the same connections, close the contact breaker points. With the ignition still on the reading should be 0–.2 volts. If the voltage is greater then contact points are dirty, the wire from the – (CB) coil terminal is broken, or the distributor body to earth resistance is very high. Should indications be that the resistance in the wiring harness is faulty, the whole harness may need replacement as this particular resistance wire is not supplied or serviced separately. However, if the wiring harness is removed and the wires separated, a competent electrician might succeed in fitting a new resistance wire according to the resistance requirement as given in the 'Specifications'.

12 Misfiring - diagnosis and remedies

1 If the engine misfires regularly, run it at a fast idling speed. Pull off each of the plug caps in turn and listen to the note of the engine. Hold the plug cap in a dry cloth or with a rubber glove as additional protection against a shock from the HT supply.

2 No difference in engine running will be noticed when the lead from the defective circuit is removed. Removing the lead from one of the good cylinders will accentuate the misfire.

3 Remove the plug lead from the end of the defective plug and hold it about $^3/_{16}$ in away from the block. Restart the engine. If the sparking is fairly strong and regular the fault must lie in the spark plug.

4 The plug may be loose, the insulation may be cracked, or the points may have burnt away giving too wide a gap for the spark to jump. Worse still, one of the points may have broken off. Either renew the plug, or clean it, reset the gap, and then test it.

5 If there is no spark at the end of the plug lead, or if it is weak and intermittent, check the ignition lead from the distributor to the plug. If the insulation is cracked or perished, renew the lead. Check the connections at the distributor cap.

6 If there is still no spark, examine the distributor cap carefully for tracking. This can be recognised by a very thin black line running between two or more electrodes, or between an electrode and some other part of the distributor. These lines are paths which now conduct electricity across the cap thus letting it run to earth. The only answer is a new distributor cap.

7 Apart from the ignition timing being incorrect, other causes of misfiring have already been dealt with under the section dealing with the failure of the engine to start. To recap - these are that:

a) The coil may be faulty giving an intermittent misfire.

b) There may be a damaged wire or loose connection in the low tension circuit.

c) The condenser may be short circuiting.

d) There may be a mechanical fault in the distributor - a broken driving spindle or contact breaker spring.

8 If the ignition timing is too far retarded, it should be noted that the engine will tend to overheat, and there will be a quite noticeable drop in power. If the engine is overheating and the power is down, and the ignition timing is correct, then the carburettor should be checked, as it is likely that this is where the fault lies.

Chapter 5 Clutch and actuating mechanism

Contents

Specifications

Make	Borg & Beck or Laycock
Type	Diaphragm spring
Diameter	6¼ in
Operating fork free travel	¼ in between fork and cable adjusting nut
Disc hub spring code colour	Pink

Torque Wrench Setting

Clutch cover to flywheel bolts	14 lb/ft - dry threads

1 General description

The clutch consists of an integral pressure plate and diaphragm spring assembly with a single dry plate friction disc between the pressure plate assembly and the flywheel.

The bellhousing on the gearbox encloses the whole unit but only the top half of the bellhousing bolts to the engine. Consequently, there is a semi-circular steel plate bolted to the lower half of the bellhousing to act as a cover.

The clutch is operated mechanically by a Bowden cable direct from the clutch pedal. This actuates a clutch release lever and thrust bearing, the lever pivoting on a ball pin inside the bellhousing and projecting through an aperture in the bellhousing opposite to the pin. Adjustment of free play is effected by a threaded ball joint at the end of the cable where it is attached to the clutch operating lever.

2 Clutch cable - removal and replacement

1 Slacken the locknut at the clutch operating lever end of the cable and remove it and the adjusting ball nut completely from the thread. Remove the sump screw which holds the cable clip locating the outer cable to the side of the engine where the sump joins the crankcase. Then draw the cable through the hole in the bellhousing.

2 Remove the parcel shelf.

3 Withdraw the panel covering the pedal mounting assembly.

4 Pull the clip off the end of the pedal shaft and then draw it to one side far enough to allow the pedal arm to drop down. The end of the cable may then be detached by removing the clevis pin.

5 Unscrew the outer cable locknut and remove the washer.

6 The cable assembly may then be drawn out through the bulkhead. Note the washer fitted on the engine compartment side also.

7 Replacement of the cable is an exact reversal of the removal procedure.

8 If there are signs that the pedal shaft bushes are worn this is a good opportunity to renew them. They can easily be prised out of their locations if the pedal shaft is fully withdrawn. New ones are simply pressed into position.

9 Reset the clutch fork free travel as described in Section 3.

3 Clutch - adjustment

1 The free play in the clutch pedal cannot be determined accurately from the pedal itself. It is necessary to check the gap between the ball joint adjuster and actuating arm.

2 To measure the gap first unhook the return spring from the arm. Then prop, or get someone to hold, the pedal in the fully up position.

3 Move the actuating arm until it can be felt to be up against the clutch and pull the cable to eliminate any end play there may be. Adjust the nut as necessary to obtain the gap of ¼ in between arm and adjuster (Fig 5:2).

4 Too little or no gap will wear out the thrust race prematurely. If very badly adjusted, clutch slip will occur. Too much gap will result in excessive pedal movement before the clutch disengages.

5 Do not forget to refit the return spring. If the spring is broken or disconnected it will immediately be apparent by looseness in the pedal at the top end of its travel. The clutch will still work but the actuating arm will rattle and cause wear on the thrust bearing.

6 Do not forget to tighten the locknut after adjustment is complete (photo).

4 Clutch pedal and shaft - renewal of bushes

1 Should excessive play develop in the movement of the pedals on the shaft, the bushes may be renewed. The shaft can be withdrawn after removing the spring clips at the ends.

2 The pedals can be lowered and the nylon bushes pressed out and new ones fitted.

3 When replacing the shaft use the special grease for lubrication of the nylon bearings such as Duckhams Keenol.

4 Make sure the spring clips are properly located on the ends of the shaft so that they also lock into the mounting plate to prevent the shaft rotating.

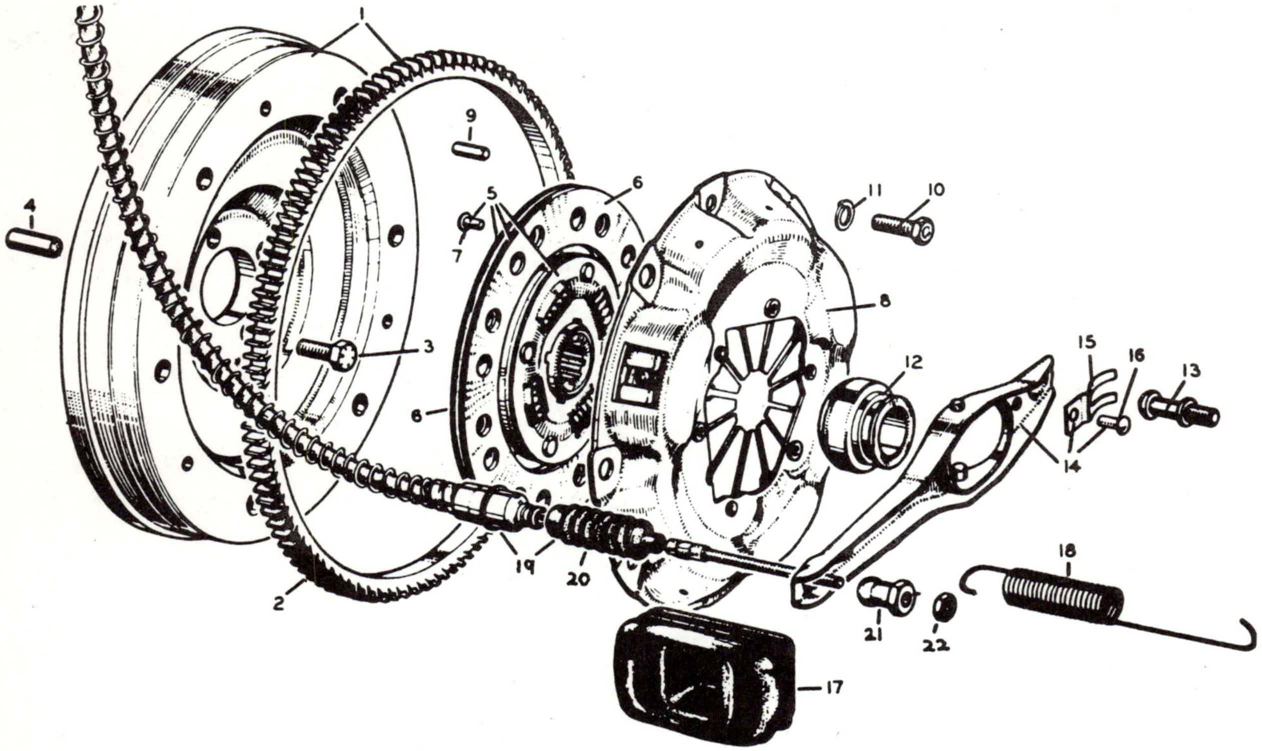

Fig 5:1 CLUTCH ASSEMBLY – EXPLODED VIEW

1 Flywheel and starter ring
2 Ring gear
3 Flywheel mounting bolt
4 Flywheel locating dowel peg
5 Clutch friction disc assembly
6 Friction lining
7 Rivet
8 Clutch pressure plate and cover
9 Clutch cover locating pegs
10 Clutch cover bolt
11 Spring washer
12 Thrust release bearing
13 Operating lever pivot pin
14 Clutch lever assembly
15 Retaining spring
16 Rivet
17 Lever aperture cover
18 Return spring
19 Cable assembly
20 Rubber boot
21 Adjuster nut
22 Locknut

NB Drawing shows lever assembled for left hand drive. On RHD the cable is on the other side

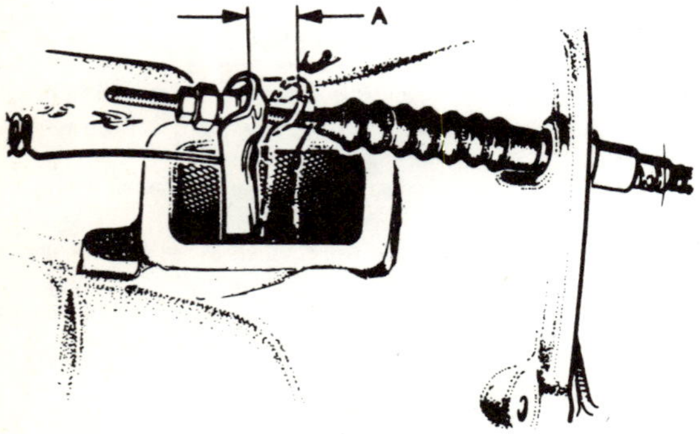

Fig 5:2 CLUTCH ADJUSTMENT – CABLE TO OPERATING LEVER
Dimension 'A' to be ¼ inch (6 mm)

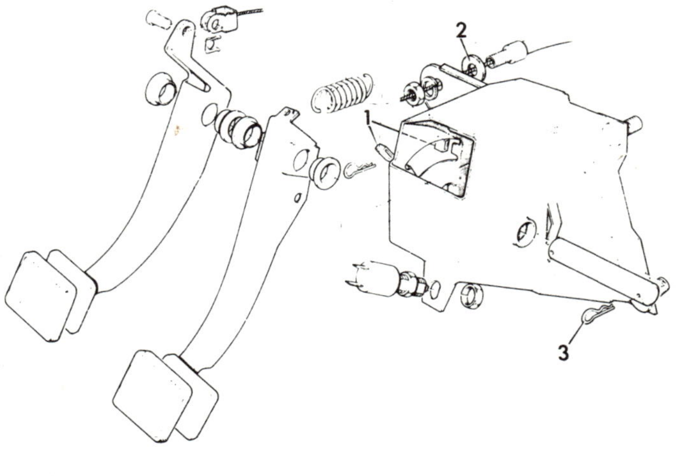

Fig 5:3 CLUTCH PEDAL AND SHAFT ASSEMBLY – EXPLODED VIEW

1 Rubber stop buffers
2 Cable outer - sealing washer
3 Shaft retaining clip

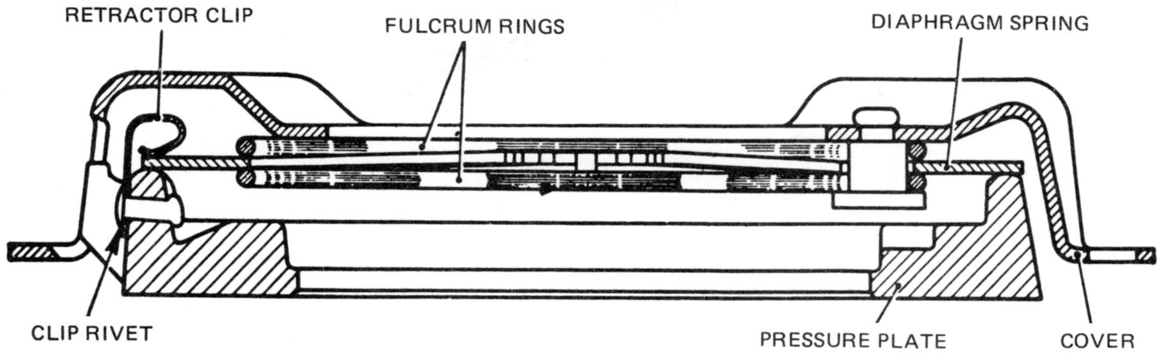

Fig 5:4 Laycock clutch pressure plate assembly - cross section

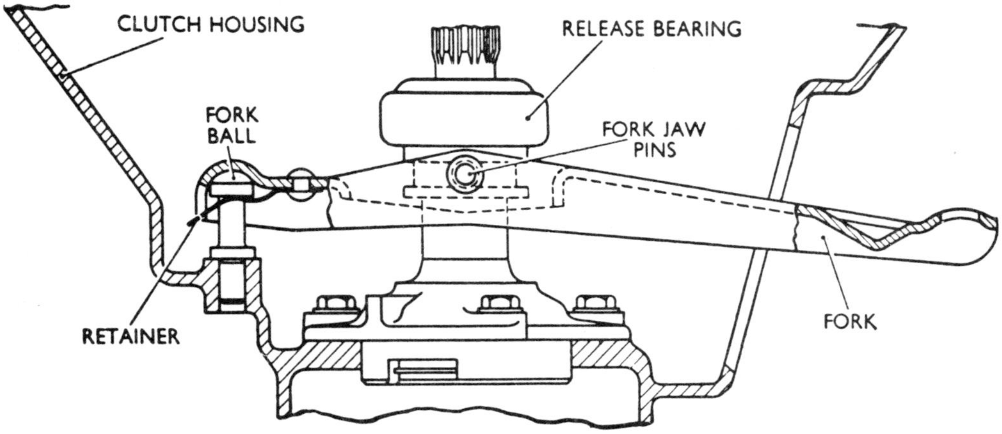

Fig 5:5 Clutch lever and release bearing assembly - cross section

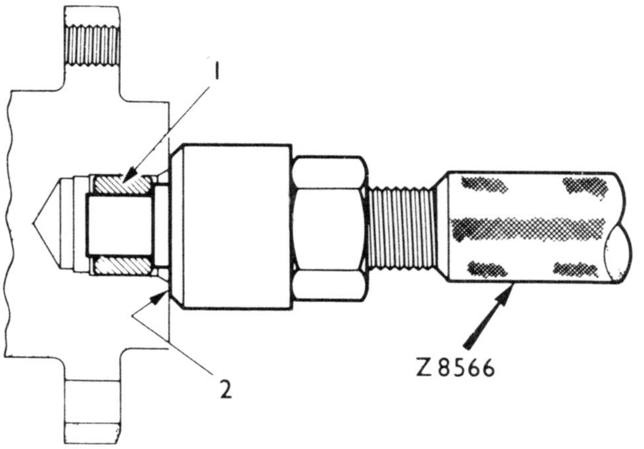

Fig 5:6 CLUTCH PILOT BUSH FITMENT IN FLYWHEEL USING SPECIAL VAUXHALL MANDREL Z8566
1 Bush 2 Flywheel

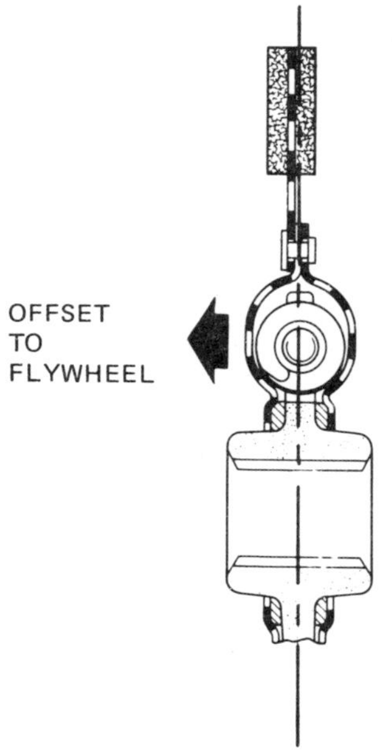

Fig. 5.6a. Position of clutch disc where face is not marked

5 Clutch assembly - removal and inspection

1 Remove the gearbox (see Chapter 6 'Gearbox removal').
2 Mark the position of the clutch cover relative to the flywheel (photo).
3 Slacken off the bolts holding the cover to the flywheel in a diagonal sequence, undoing each bolt a little at a time. This keeps the pressure even all round the diaphragm spring and prevents distortion. When all the pressure is released on the bolts, remove them, lift the cover off the dowel pegs and take it off together with the friction disc which is between it and the flywheel.
4 Examine the diaphragm spring for signs of distortion or fracture.
5 Examine the pressure plate for signs of scoring or abnormal wear.
6 If either the spring or the plate is defective it will be necessary to replace the complete assembly with an exchange unit. The assembly can only be taken to pieces with special equipment and, in any case, individual parts of the assembly are not obtainable as regular spares. Some models are fitted with the Laycock clutch assembly. The friction plates for Borg and Beck and Laycock clutches are now interchangeable, but check first. On these versions the diaphragm spring, driving plate and pressure plate can be separated as they are simply held together by a large retaining ring and no rivets are used. (Fig 5:4). The relative position of the three items should be marked. The retaining ring and anti-rattle springs can then be detached and the pressure points should be greased sparingly with a heavy lubricant, Castrol MS3. Make sure the retaining ring is replaced with the flat sections under the pressure plate lugs and the curved sections against the edge of the diaphragm spring.
7 Examine the friction disc for indications of uneven wear and scoring of the friction surfaces. Contamination by oil will also show as hard and blackened areas which can cause defective operation. If the clearance between the heads of the securing rivets and the face of the friction lining material is less than .025 in it would be worthwhile to fit a new disc. Around the hub of the friction disc are four springs acting as shock absorbers between the hub and the friction area. These should be intact and tightly in position.
8 The face of the flywheel should be examined for signs of scoring or uneven wear and, if necessary, it will have to be renewed and replaced or reconditioned. See Chapter 1 for details of flywheel removal.

6 Clutch assembly - replacement

1 Replacement of the clutch cover and friction plate is the reverse of the removal procedure but not quite so straightforward, as the following paragraphs will indicate.
2 If the clutch assembly has been removed from the engine with the engine out of the car, it is a relatively easy matter to line up the hub of the friction disc with the centre of the cover and flywheel. The cover and friction plate are replaced onto the flywheel with the holes in the cover fitting over the three dowels on the flywheel. The friction plate is supported with a finger while this is being done.
3 Note that the friction plate is mounted with the longer hub of the boss towards the flywheel. Usually the replacement disc is marked 'flywheel side' to prevent a mistake being made (photo). Some early Laycock discs do not have a marked face, in which case the offset side must be towards the flywheel as shown.
4 Replace the cover mounting bolts finger tight sufficiently to just grip the friction plate. Then set the friction plate in position so that the hub is exactly concentric with the centre of the flywheel and the cover assembly. An easy way of doing this is to make a temporary mandrel, using a bar from a socket set, which should fit fairly closely in the flywheel bush. Wrap a few turns of adhesive tape round the bar near the end, which will make a snug fit inside the splined boss of the friction plate. Use this as a centring device. It is most important to get this right when replacing the clutch to an engine which is still in the car. Otherwise difficulty and possibly damage could occur when refitting the gearbox (photo). Chapter 6.
5 Tighten up the cover bolts one turn at a time in a diagonal sequence to maintain an even pressure. Final torque setting should be 14 lb/ft with clean dry bolt threads.

7 Clutch actuating lever and thrust release bearing - removal, inspection and replacement

1 Remove the gearbox as described in Chapter 6, 'Gearbox removal'.
2 Move the lever sideways so that the end over the ball pivot pin is freed by springing back the retaining clip.
3 The lever jaw pins can then be disengaged from the groove in the thrust release bearing and the lever taken off over the end of the input shaft.
4 The clutch release bearing may then be taken off the input shaft.
5 Inspect the pivot pin ball for signs of wear and flats. If necessary it can be removed by driving it out of the bellhousing with a drift. A new one can be driven in with a soft headed hammer.
6 If the release bearing is obviously worn and is noisy it should be replaced. Do not clean the release bearing in any solvent as the ball races have been prepacked with grease and such cleaning would wash it out.
7 Replace the operating lever and release bearing in the reverse order of dismantling. Note that the grooved side of the thrust bearing goes towards the gearbox.
8 Ensure also that the spring retaining clip on the end of the lever fastens securely over the mushroom head of the ball pivot pin.
9 Replace the gearbox as described in Chapter 6, 'Gearbox replacement'.

8 Clutch pilot bush - renewal

1 In the centre of the flywheel is a bushed hole in which runs the input shaft of the gearbox. If the bush is badly worn the clutch operation will be unsatisfactory and out of balance.
2 It is impossible to judge the condition of the bush unless you can obtain a bar of exactly the same diameter as the gearbox shaft to test it. With the engine removed from the car it is possible to measure it but, once again, unless the gearbox is also removed, it is impossible to check the fit of the two parts concerned.
3 The old bush can be removed by finding a bolt which can be forced into the soft metal of the bush. Fit a nut and spacer tube on the bolt, force the bolt into the bush and then draw out the bush by turning the nut. Vauxhall have a special tool for the job (Fig 5:6).
4 Replacement of the bush is best done using a proper mandrel, which will ensure that it is not distorted or damaged when being driven in. It is most important that the shaft is an easy fit in the bush on reassembly.

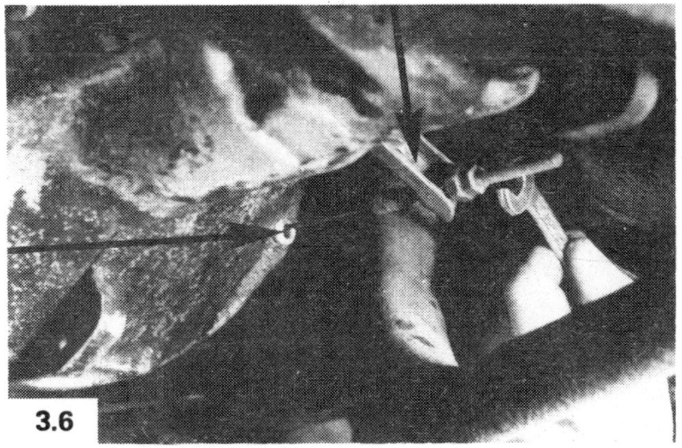

3.6

Clutch adjustment. Note cable through bellhousing and operating lever respectively (arrowed L and R)

5.2

Marks made on clutch cover and flywheel to ensure re-location in same place

6.3

Replacing the clutch assembly with the friction disc supported on one finger. Note the long bub section (arrowed) facing the flywheel

6.4

Clutch assembled with hub of friction plate centred with the cover

Fault Finding Chart – Clutch and Actuating Mechanism

Symptom	Reason/s	Remedy
Judder when taking up drive	Loose engine or gearbox mountings or over flexible mountings	Check and tighten all mounting bolts and replace any 'soft' or broken mountings.
	Badly worn friction surfaces or friction plate contaminated with oil carbon deposit	Remove clutch assembly and replace parts as required. Rectify any oil leakage points which may have caused contamination.
	Worn splines in the friction plate hub or on the gearbox input shaft	Renew friction plate and/or input shaft.
	Badly worn bush in flywheel centre for input shaft spigot	Renew bush in flywheel.
	Propeller shaft or rear axle mounting faults	Examine propeller shaft universal joints and rear axle to suspension arm attachment points
Clutch spin (or failure to disengage) so that gears cannot be engaged	Clutch actuating cable clearance from fork too great	Adjust clearance.
	Clutch friction disc sticking to pressure surface because of oil contamination (usually apparent after standing idle for some length of time)	As temporary remedy engage top gear, apply handbrake, depress clutch and start engine. (If very badly stuck engine will not turn). When running rev up engine and slip clutch until disengagement is normally possible. Renew friction plate at earliest opportunity.
	Damaged or misaligned pressure plate assembly	Replace pressure plate assembly
Clutch slip – (increase in engine speed does not result in increase in car speed - especially on hills).	Clutch actuating cable clearance from fork too little resulting in partially disengaged clutch at all times	Adjust clearance.
	Clutch friction surfaces worn out (beyond further adjustment of operating cable) or clutch surfaces oil soaked	Replace friction plate and remedy source of oil leakage.

Chapter 6 Gearbox

Contents

Specifications

General

Number of gears	4 forward, 1 reverse
Type	Helical constant mesh, with straight cut reverse gear Synchromesh on all forward speeds.
Oil capacity	**0.9 Imp pint (1.1 Imp pts for models with 0.375 in AF socket filler plug).**

Ratios:	Gearbox	Overall 3.89:1 axle	Overall 4.12:1 axle
First	3.765:1	14.64:1	15.53:1
Second	2.213:1	8.61:1	9.13:1
Third	1.404:1	5.46:1	5.79:1
Fourth	1:1	3.89:1	4.12:1
Reverse	3.707:1	14.42:1	15.29:1

* On 1256 cc engines 1st gear ratio is 3.46:1.

Mainshaft and Bearings

Mainshaft diameter - first gear	.8917 to .8923 in
First gear bore fit	.0014 to .0028 in clearance
Mainshaft diameter - second and third gears	1.0994 to 1.100 in
Second and third gears bore fit	.0014 to .0028 in clearance
Synchro hub: circlip thicknesses available	.061 in to .072 in in four stages - each clip covering .003 in
Speedo drive gear circlip: thicknesses available	.059 in to .076 in in six stages - each clip covering .003 in

Laygear and Layshaft

Overall length	6.021 in to 6.023 in
End float	.005 in to .017 in
Thrust washer thickness	.0615 in to .0625 in

Reverse Gear

Idler pinion shaft diameter	.5521 in to .5528 in
Idler pinion fit on shaft	.0022 in to .0039 in clearance
Clearance between rear face of reverse gear pinion and casing	.002 in to .012 in

Rear Extension Cover

Sliding sleeve diameter	1.1240 in to 1.1250 in
Sleeve clearance in bush	.0015 in to .004 in

Gear Lever

Clearance between selector shaft and change lever	.002 in to .012 in

Speedometer Driven Gears Available

Axle ratio	Tyre sizes	Number of teeth on gear
8:33	5.20 - 13	15
8:33	6.2S - 13	15
8:33	6.5S - 13	15
8:33	1.55SR - 13	15
9:35	As above	14

Automatic transmission

Gear ratios:	First	2.4:1
	Second	1.48:1
	Third	1:1
	Reverse	1.92:1
Oil capacity		9 pints
Stall speed		2100 - 2150 rpm

Approximate change speeds in 'D' range

	Change	Speed
Minimum throttle	1 - 2	10 mph
	2 - 3	12 mph
Closed throttle	3 - 2	10 mph
	2 - 1	8 mph
90% throttle	1 - 2	32 mph
	2 - 3	43 mph
Full throttle	1 - 2	35 mph
	2 - 3	58 mph
	3 - 2	52 mph
	3 - 1	32 mph

Torque Wrench Settings	lb ft
Flex plate to crankshaft bolts	25
Torque converter	42
Torque converter housing to transmission case bolts	25
Extension housing to transmission case bolts	20
Transmission sump to case bolts	7

1 General description

A four forward speed all synchromesh gearbox is fitted, and is the same on all HC series models. Any optional final drive ratios are made with a different rear axle ratio only.

The gearchange is a standard remote control on the floor. A single selector rod operates directly to two selector forks and a reverse striking lever, with a one piece collar device which locates and holds the forks through all gear changing operations.

All forward gears are helically cut, constant mesh, and gear engagement is by sliding hubs and cones engaging dogs on each mainshaft gear.

The laygear (supported at each end on needle roller bearings) runs on a stationary shaft. The reverse gear is straight cut and is part of the 1st/2nd speed synchro hub. The reverse idler gear runs on its own shaft without bushing or bearing.

All bearings and hub locations are made by circlips in grooves, and the stationary shafts by interference fit steel balls in depressions on the shafts and cut-outs in the casing. There are no set screws, grub screws, keys or pins used anywhere inside the gearbox.

The gearbox is basically the same as that used in all Viva models to date. The most recent modifications, introduced when the 1256 cc engine was fitted as standard are:

a) Fewer balls in ball bearing races

b) Two rows of needle roller bearings fitted at each end of the layshaft.

c) A slot introduced in one face of 1st gear to assist lubrication distribution.

d) Oil seal fitted in casing round the input shaft as part of shaft bearing. (Earlier versions had a scroll return groove).

e) The waisted portion of the mainshaft is now parallel.

For the foregoing reasons care should be exercised in assuming too much on the interchangeability of spares. There is no oil drain plug. Provision is made for level checking and topping up.

2 Gearbox - removal and replacement

1 Jack up the front of the car and support it properly on stands in the same way as for engine removal.

2 If the engine is not being removed together with the gearbox, jack up the rear of the car also, support it on stands and remove the propeller shaft as described in Chapter 7.

3 From inside the car unscrew the gear lever knob, and remove the rubber grommet round the lever. Another rubber cover will now be visible and this shrouds a metal cap (somewhat similar to the cap on an ordinary 1 gallon oil tin!) which is then unscrewed. This retaining cap is peened into a hole in the extension housing thread to prevent it unscrewing inadvertently. Do not puncture this cap in any way when removing or replacing it. The gear lever, cap, spring and retaining plate can then all be lifted from the selector rod extension tube. Remember to grease the assembly on replacement (Fig 6:1).

4 Undo the nut in the centre of the crossmember supporting the gearbox and remove it together with the washer (photo).

5 Disconnect the speedometer drive cable from the lower right hand side of the gearbox rear extension cover by undoing the knurled nut which holds it in place.

6 Disconnect the clutch actuating cable (for details see Chapter 5).

7 Support the gearbox just forward of the crossmember using a jack

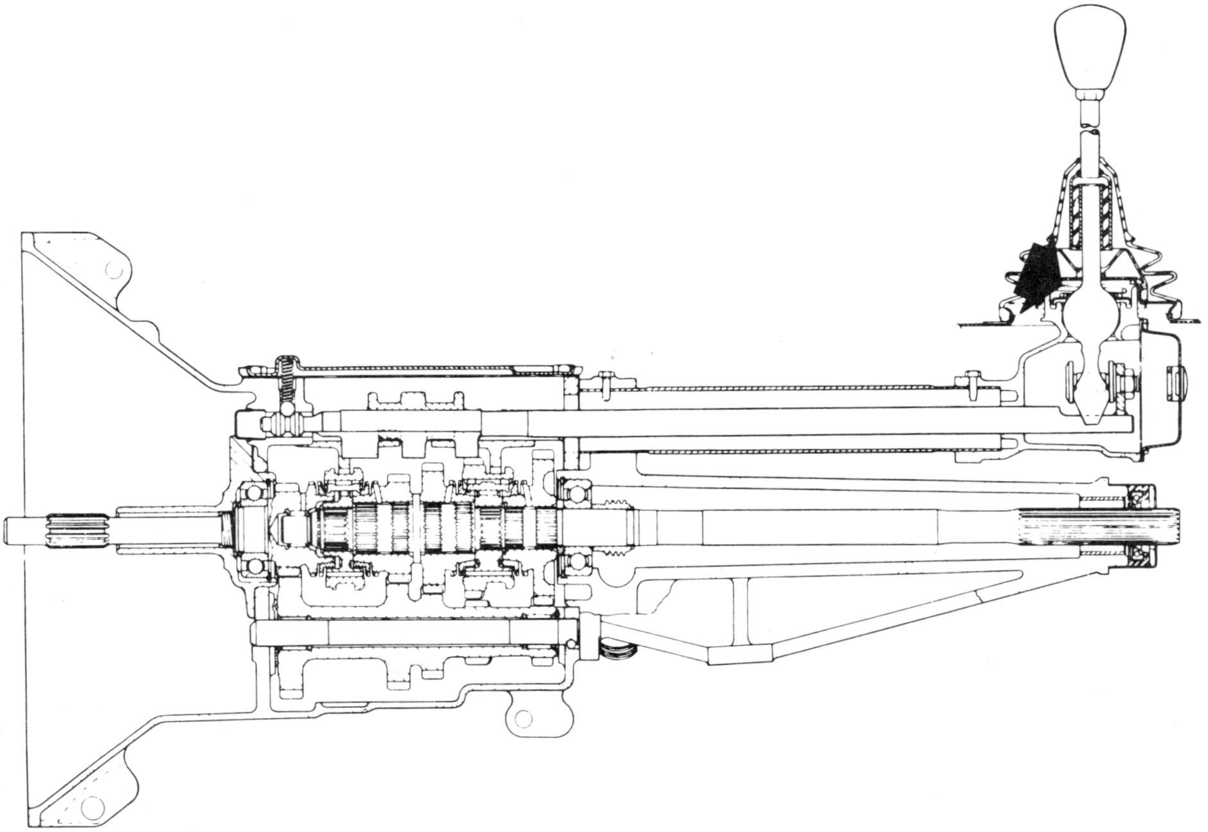

Fig 6:1 Gearbox - longitudinal cross section (arrow indicates point to pack with grease)

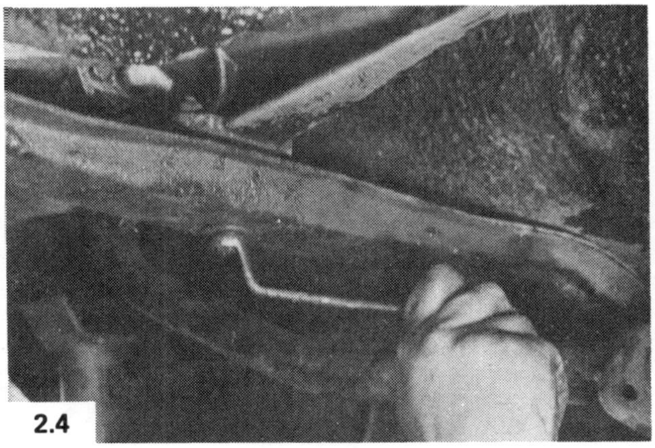

Removing the rear mounting nut followed by the gearbox support crossmember securing bolts

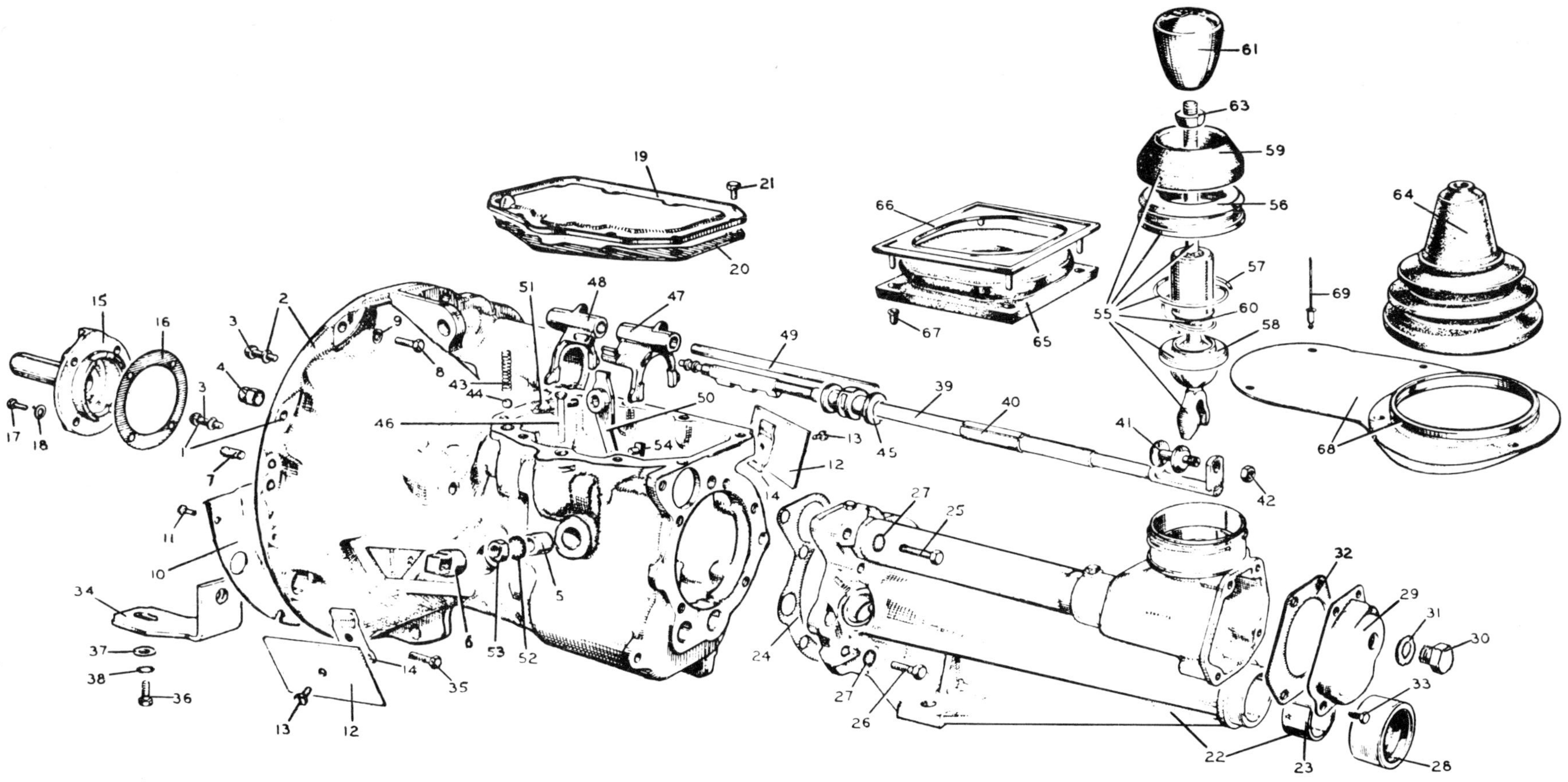

Fig 6:3 GEARBOX CASING AND SELECTOR MECHANISM

1 Gear casing and clutch lever pivot pin
2 Ditto
3 Ditto
4 Selector shaft cover
5 Bush
6 Level plug
7 Dowel pin - casing to engine
8 Bolt - casing to engine
9 Spring washer
10 Cover plate
11 Bolt
12 Cover plate
13 Screw
14 Locating plate
15 Bearing housing
16 Gasket
17 Screw
18 Washer
19 Cover plate
20 Gasket
21 Bolt
22 Rear extension housing and bush
23 Bush
24 Gasket
25 Bolt
26 Bolt
27 Lockwasher
28 Lockwasher
29 Cover plate
30 Reverse light blanking plug
31 Washer
32 Gasket
33 Screw
34 Brace bracket
35 Bolt
36 Bolt
37 Plain washer
38 Lockwasher
40 Sleeve
41 Selector lever pin
42 Locknut
43 Detent spring
44 Detent ball
45 Interlock collar
46 Interlock collar locking pin
47 Selector fork
48 Selector fork
49 Selector fork rail
50 Reverse striking lever
51 Eccentric pin
52 Lockwasher
53 Locknut
54 Reverse selector lug
55 Change lever assembly
56 Screw cover
57 Spring
58 Retaining cap
59 Grommet
60 Flexible coupling
61 Lever grip
63 Locknut
64 Boot
65 Sealing bolt
66 Boot retainer
67 Retaining pin
68 Lever cover plate
69 Pop rivet

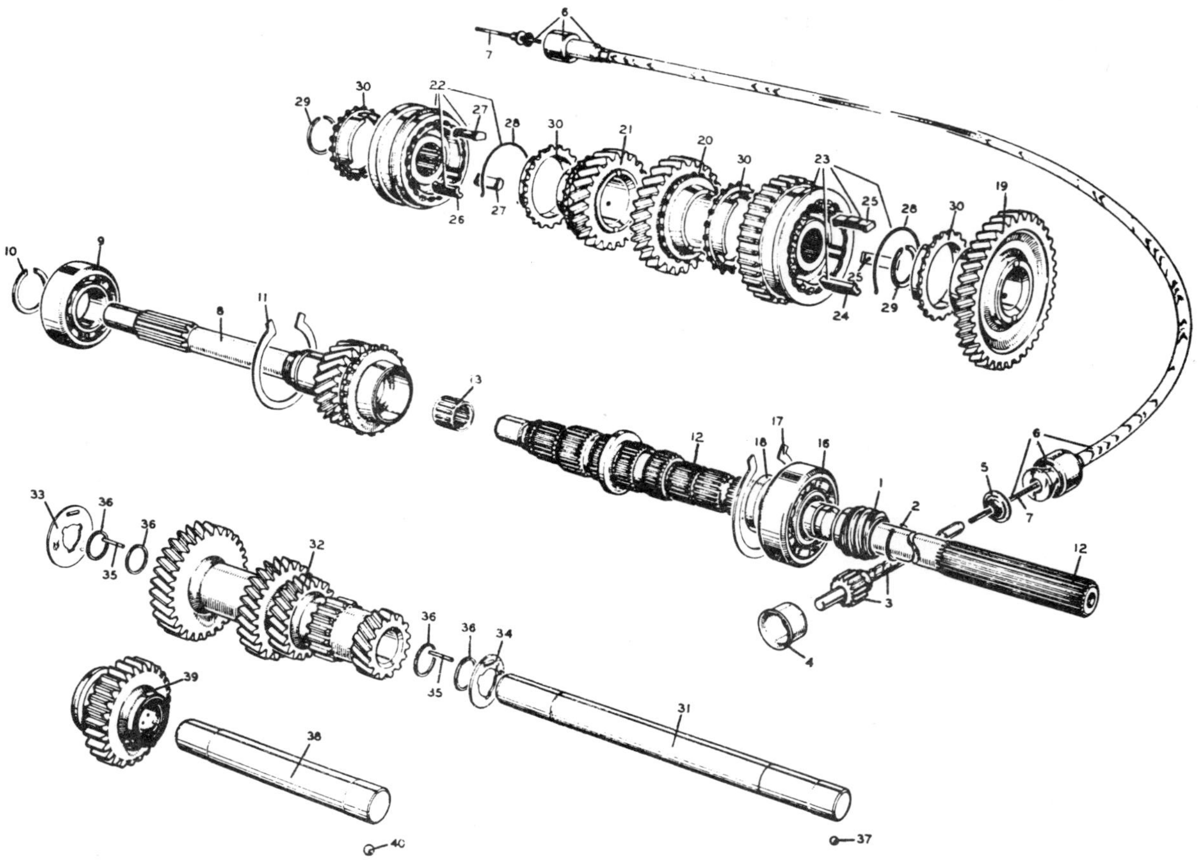

Fig 6:4 GEARBOX – INTERNAL COMPONENTS

1 Speedometer drive gear
2 Circlip
3 Speedometer cable assembly
4 Plug
5 Sealing disc
7 Cable inner
8 Input shaft
9 Input shaft bearing
10 Circlip - bearing to shaft
11 Circlip - bearing to housing
12 Main shaft
13 Needle roller bearing
16 Mainshaft bearing
17 Circlip - bearing to housing
18 Spacer
19 1st gear
20 2nd gear
21 3rd gear
22 3rd/4th synchro hub
23 1st/2nd gear synchro hub
24 Blocker bar (sliding key)
25 Blocker bar
26 Blocker bar
27 Blocker bar
28 Retaining spring
29 Circlip
30 Synchro ring
31 Layshaft
32 Laygear
33 Thrust washer
34 Thrust washer
35 Needle roller
36 Keep ring
37 Locating ball
38 Reverse idler pinion shaft
39 Reverse idler gear
40 Locating ball

Note: The mainshaft dividing flange comes between 2nd and 3rd gears (Items 20 and 21)

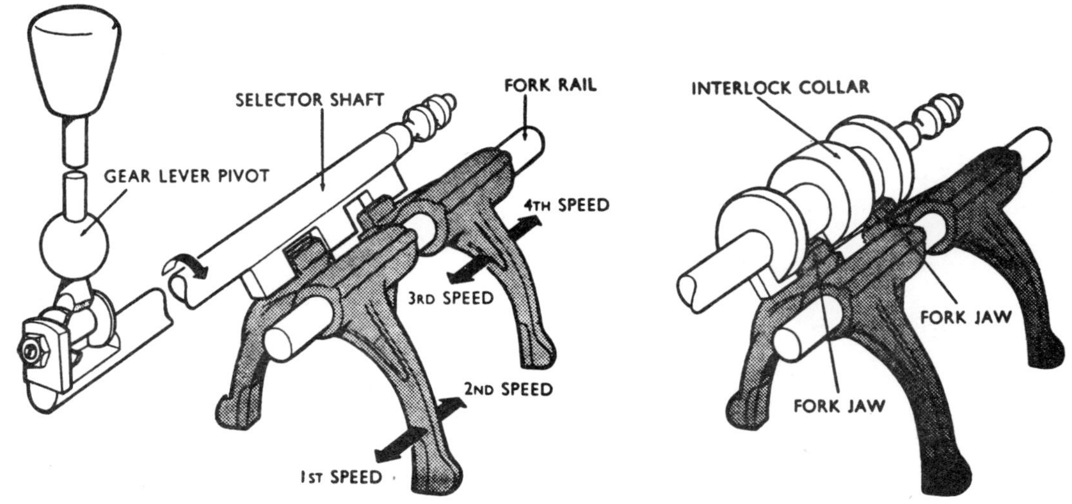

Fig 6:5 Gearbox - diagram of gearchange lever and selectors

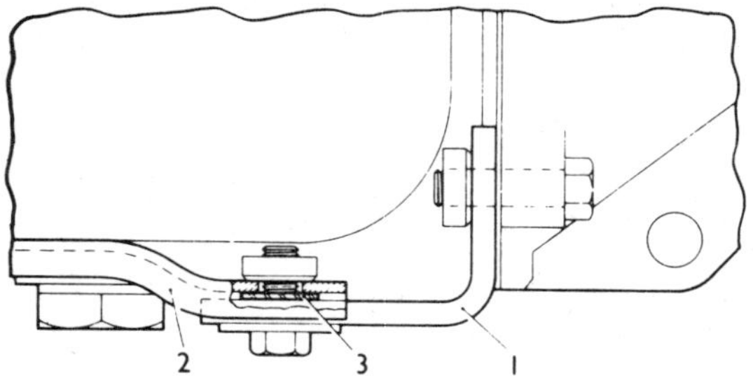

Fig 6:6 SUMP/BELLHOUSING BRACE PLATE ASSEMBLY
1 Brace plate
2 Sump flange
3 Serrated lockwasher

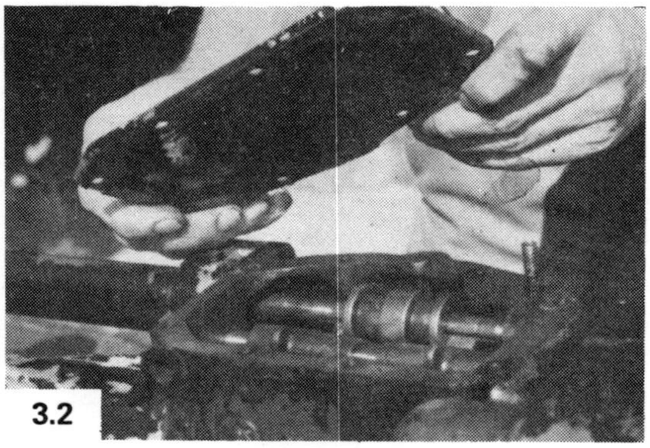

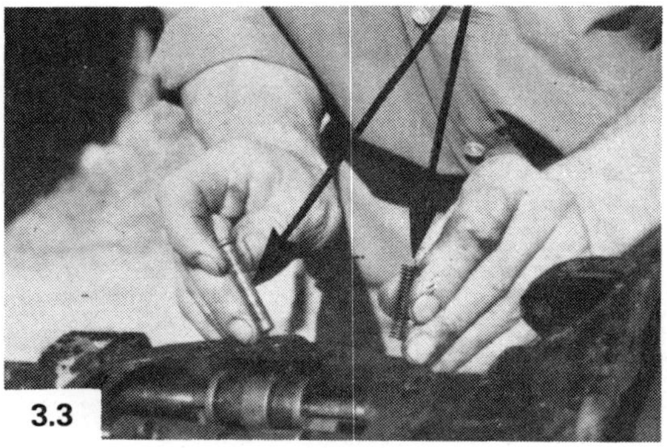

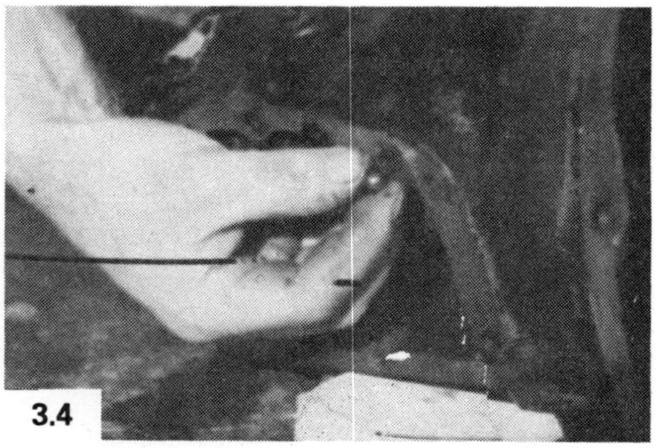

Removing the top cover, lifting out the interlock collar retaining pin and detent spring (arrowed) and tipping out the detent ball

and then remove the crossmember holding bolts, two at each end, and remove the crossmember (photo).
8 Remove all bolts holding the gearbox casing to the engine (still supporting the gearbox on the jack) as described in Chapter 1, 'Engine removal'.
9 Gently lower the jack, at the same time supporting the gearbox.
10 With very little effort it can now be drawn off the dowels on the engine block and, still being supported, the gearbox input shaft pinion disengaged from the clutch assembly.
11 It will be necessary to tilt the gearbox down at the tail end. At first the engine will tilt also on its mountings until the valve rocker cover comes up against the bulkhead. By this point the gearbox should be almost completely clear. It is very important that no strain in any direction is imparted to the input shaft as it, and the clutch assembly, could be damaged.
12 If the gearbox is to be taken out of the car together with the engine, it is necessary only to carry out the requirements of paragraphs 1, 3, 4, 5 and 6 in this Section and in addition to provide a support for the front end of the propeller shaft when it is released from the back of the gearbox, when the gearbox is taken away with the engine.
13 Remember to pause a minute or two after the propeller shaft is disengaged, and catch the oil which will drain out of the gearbox mainshaft rear cover as the front of the engine tilts up.
14 Replacement is a direct reversal of the removal procedure, having made sure that the clutch assembly is suitably lined up to accept the gearbox input shaft (see Chapter 5). Some juggling may be needed, but at all costs avoid forcing anything.
15 Once the gearbox is bolted up to the engine (which is firmly fixed to its forward mountings) there is no real need for support at the gearbox for the short time before the crossmember is replaced to support it.
16 Finally, see that the bracing plate is fitted securely between the lower edge of the bellhousing and onto the sump flange. This item is important to ensure complete rigidity of the engine gearbox assembly. So that there is no movement between the surfaces of the brace and the sump flange, an internal/external toothed lockwasher is fitted between the brace and the flange at the bolt holes (Fig 6:5).

3 Gearbox - dismantling

1 Place the complete unit on a firm bench or table and ensure that you have the following tools (in addition to the normal range of spanners etc) available:
a) Good quality circlip pliers, 2 pairs - 1 expanding and 1 contracting.
b) Copper headed mallet, at least 2 lb.
c) Drifts, steel $^{3}/_{8}$ in and brass $^{3}/_{8}$ in.
d) Small containers for needle rollers.
e) Engineer's vice mounted on firm bench.
f) Method of heating, such as blow lamp or butagas stove.

Any attempt to dismantle the gearbox without the foregoing is not necessarily impossible, but will certainly be very difficult and inconvenient, resulting in possible injury or damage.

Read the whole of this Section before starting work.

2 Remove the top cover holding bolts and lift off the cover (photo).
3 This will release the detent spring and expose the top of the interlock retaining pin which can be lifted out. There will probably be some oil left inside the casing so it is wise to invert the gearbox over a bowl(photo).
4 When this is done the ball under the detent spring will drop out. Retrieve this and keep in a jar (photo).
5 Whilst the oil is draining into the bowl, the clutch actuating fork can be removed, if not already done, by drawing the inner end off the fork ball and then drawing the whole assembly out of the bellhousing over the splined end of the input shaft (photo).
6 Remove the rubber cover from the bellhousing which shrouds the starter motor pinion.
7 Remove the end cover from the remote control selector rod hous-

ing by undoing the retaining bolts.
8 The selector rod can now be drawn out, twisting it to avoid obstructions at the gearbox end. A nylon bush in the housing supports the tail end of the selector rod. There is only just sufficient room at the end to draw the key part through, so do not try and free it through too quickly (photo).
9 The interlock collar can now be lifted straight out of the main gearbox casing (photo).
10 The mainshaft extension rear cover is now released (but not removed) by unscrewing the six bolts holding it in position. Three of these bolts are larger than the other three, but it is quite obvious where they go from the position of the shoulders in the casing.
11 The rear cover can now be rotated so that the small hole in its periphery can be lined up with the end of the selector fork mounting rod.
11 Using a brass drift, this rod can now be driven out from the front through the rear of the casing(photo).
13 The selector forks themselves can now be lifted out (photo).
14 Note that the selector forks are different (photo).
15 Remove the four front cover bolts (inside the bellhousing) and draw the cover and input shaft forward together as far as they will go. This will increase the clearance at the end of the mainshaft where it fits into the input shaft (photo).
16 Fourth speed should now be engaged by pushing the 3rd/4th speed clutch to the extremity of the mainshaft. Withdraw the mainshaft until the end spigot is clear of the input shaft. If there is a clatter of falling needle rollers at this juncture do not worry, they can be picked out of the casing later.
17 Now that the mainshaft is clear of the input shaft, it can be tipped so that the 3rd and 4th speed clutch clears the layshaft gears and the whole unit drawn out (photo).
18 Next the layshaft has to be driven out from the front of the casing through the rear. It is located at the rear end by a steel ball which locks it into the casing. Before driving the shaft right through with a drift, try and position the casing so that you can hold the laygear, and lower it rather than letting it drop to the bottom.

The laygear must be lowered before the input shaft can be removed.
19 The input shaft assembly can now be removed. This consists of the front cover, bearing and shaft and is held into the casing by a circlip on the inside. Remove this circlip and then tap the shaft with a soft metal hammer, and the cover, bearing and shaft will come out (photo).
20 By removing the other circlip the bearing can be taken off the shaft for renewal (photo).
21 The laygear is removed next by lifting it up and out through the rear of the casing. Keep it horizontal so that the needle rollers at each end can be collected rather than scattered all over the place. Each set of needle rollers also has a spacer ring. Note that on later models there are two rows of needle rollers at each end (photo).
22 Two thrust washers will have dropped down when the laygear is removed. These are of different sizes to suit each end of the laygear. They each have dimpled oil reservoirs on one side, (next to the laygear) and a locating lug on the other side which fits into a recess in the casing (photo).
23 The only remaining items are the reverse idler gear and shaft. Ideally, the reverse idler rod is drawn out with a special tool. However, it can be drifted out with a suitably long steel drift from inside the casing. It will need to be driven at an angle so the nose of the drift will need careful and firm location and a smart strike with a reasonably heavy mallet is necessary. Here again there is a locating ball at the rear of the shaft to prevent it from both revolving and moving axially (photo).
24 The reverse idler gear is then removed (photo).
25 With the reverse gear and shaft removed, the reverse selector arm can be lifted off its fulcrum pin.
26 The fulcrum pin is eccentric and held in position by a locknut. This can be slackened and the pin removed from the casing in which it is a push fit.

3.5
Pulling the clutch operating lever off the pivot pin

3.8
Lifting out the selector rod after removing the extension end cover

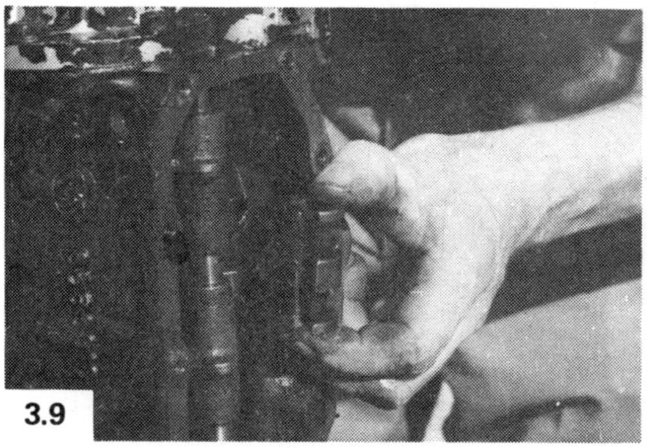
3.9
Lifting out the interlock collar

3.12
Driving out the selector fork rail after removing the rear cover bolts and turning the cover

3.13

3.14

Lifting out the selector forks and noting the difference between them left 3rd/4th right 1st/2nd

3.15

Pulling the input shaft forward to release it from the mainshaft

3.17

Removing the mainshaft assembly

27 The gearbox is now completely stripped out from its casing.
28 Clean out the interior thoroughly and check for dropped needle rollers.
29 The mainshaft is dismantled next. The assembly should first of all be held horizontally in a vice by the selector rod extension tube. This is made of sheet steel so be careful not to crush or distort its shape. Have the speedometer drive spindle facing towards you and, with a brass drift, drive it smartly through. There is a press fit cap opposite which will come out with it (photo).
30 Remove the paper gasket from the face of the rear cover and retain it, if undamaged. It may be refitted, although it would be as well to get a new one.
31 Now remove the circlip retaining the bearing into the cover. This is accessible via a cut-out in the edge of the cover flange (photo).
32 The mainshaft can next be drifted out of the cover. The bearing which retains it in place is an interference fit. Warm the housing with a gas burner, or similar, for a few minutes. Make sure that someone's hand supports the mainshaft as it comes out. Removal is easy if the housing has been warmed (photo).
33 Remove the cover from the vice and then grip the mainshaft in the vice at an angle, with the 3rd and 4th gear hub uppermost.
34 Remove the circlip around the nose of the shaft (photo).
35 The hub and third speed gear behind it should pull off quite easily. Sometimes, however, there can be a slight burr on the circlip groove which necessitates a few light mallet taps to assist their removal, or even a screwdriver behind the third gearwheel for leverage (photo).
36 Nothing more can be removed from this end of the shaft which should now be turned round in the vice, so that the longer end faces upwards, in preparation for the next steps.
37 The next item to remove is the speedometer driving gear from the end of the shaft. First remove the circlip around the shaft (photo).
38 More often than not this drive gear, and the bearing behind it is a very tight fit, and causes more difficulty than anything else on the mainshaft. Gather together two pieces of flat bar or two flat spanners, slightly thinner than the gap between the bearing and the 1st speed gear. The circlip between the bearing and the gear should be removed and drawn off over the bearing.
39 The two pieces of flat bar or spanners, one each side of the shaft behind the bearing, are now placed across the vice jaws so that the mainshaft is suspended below them clear of any obstruction. If the end of the shaft is now struck hard with a copper mallet some progress will be made, albeit slow. Two people are essential for this job as the flat bars across the vice need steadying. The man who does the striking must be confident with the mallet as half-hearted strikes will not serve. On no account support the shaft with the bars behind the 1st speed gear. Although hard, these gears are brittle and any shock force would probably shatter, or at least crack, them. If you feel that the speedo drive gear will never come off, then you would be well advised to get a local garage to do it for you. Any badly applied leverage, or blows, could be very expensive indeed (photo).
40 Assuming all went well with removal of the speedo gear, the thrust washer, 1st gear and synchroniser ring can be drawn off (photo).
41 Remove the final circlip from around the shaft (photo).
42 The 1st and 2nd gear hub and reverse gear, followed by second gear, can be drawn off, after a little tapping, if necessary (photo).
43 The mainshaft is now completely dismantled.
44 The oil seal in the rear cover around the propeller shaft can be removed by getting a bite with a sharp edge into the soft cover of the oil seal cap and driving it off (photo).
Note: Oil seals with or without metal casing may be encountered on the rear cover.

4 Gearbox - inspection

1 It is assumed that the gearbox has been dismantled for reasons of excessive noise, lack of synchromesh on certain gears, or for failure to stay in gear. If anything more drastic than this (eg total failure,

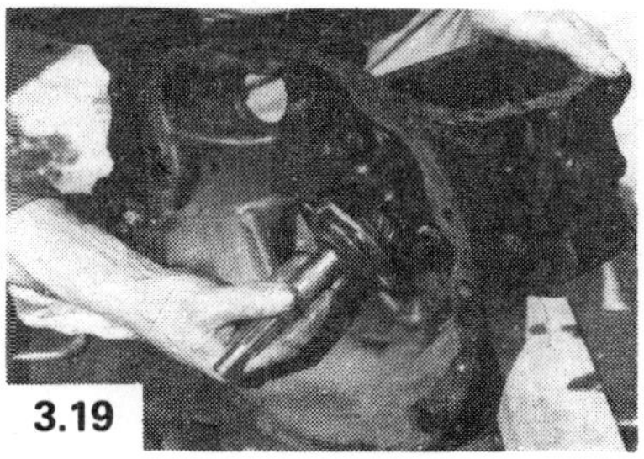
3.19

Removing the input shaft assembly after dropping the laygear into the bottom of casing

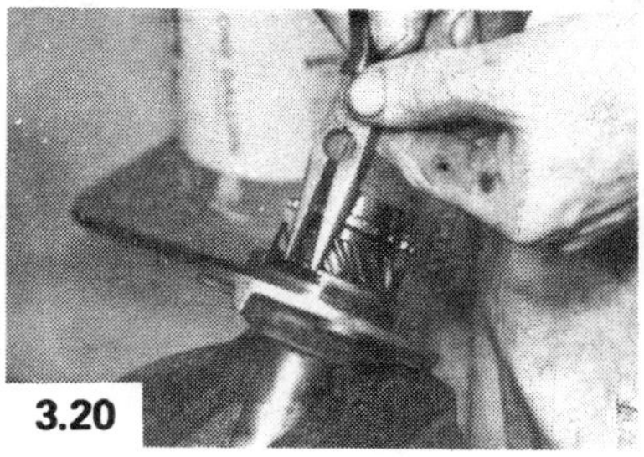
3.20

Removing the circlip retaining the input shaft bearing into the cover

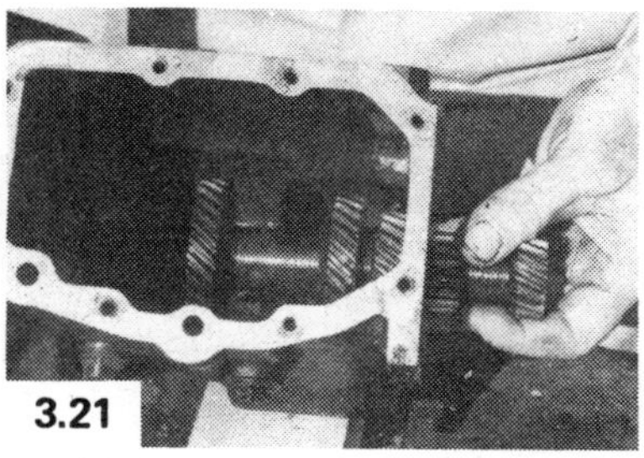
3.21

Lifting out the laygear

3.22

Laygear end thrust washers

3.23

Drifting out the reverse idler gear shaft

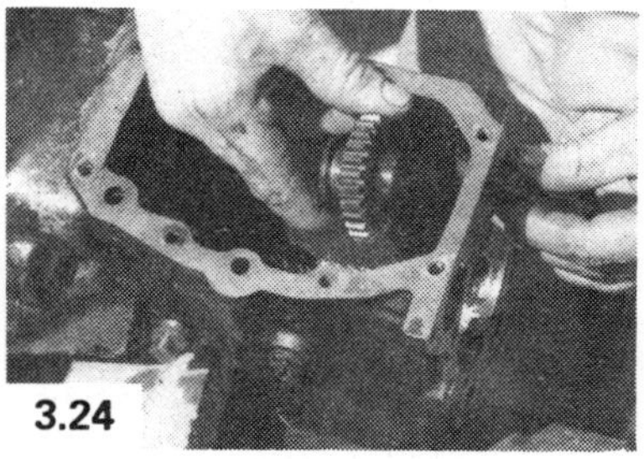
3.24

Lifting out the reverse idler gear

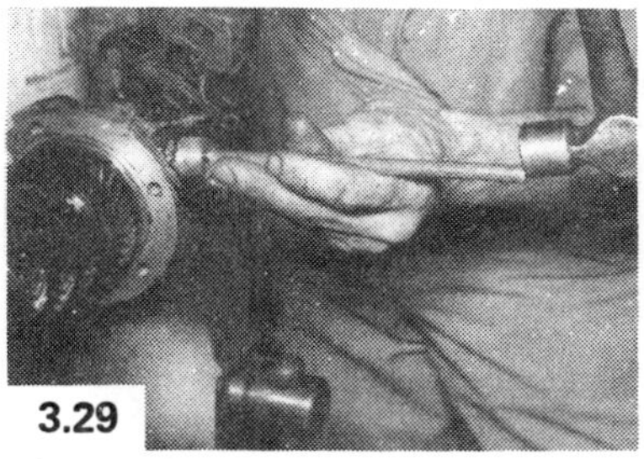
3.29

Driving out the speedometer drive gear

3.31

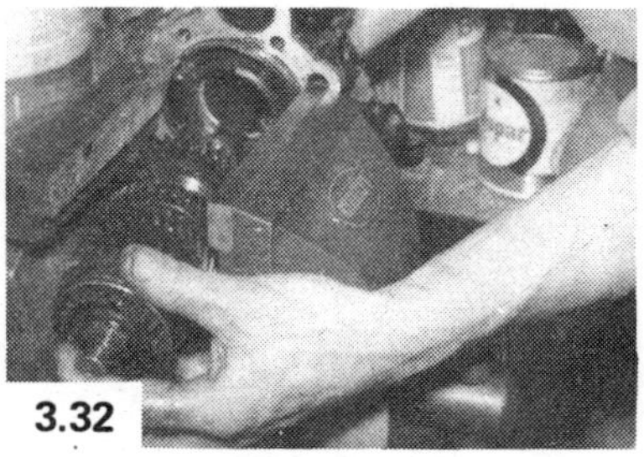
3.32

Releasing the circlip holding the mainshaft rear bearing into the cover and taking the shaft from the cover

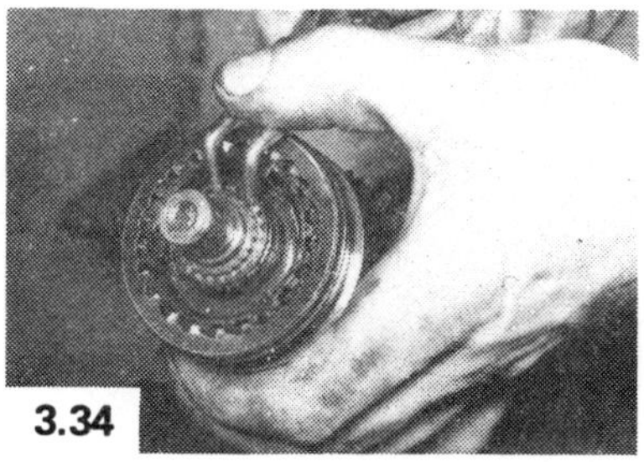
3.34

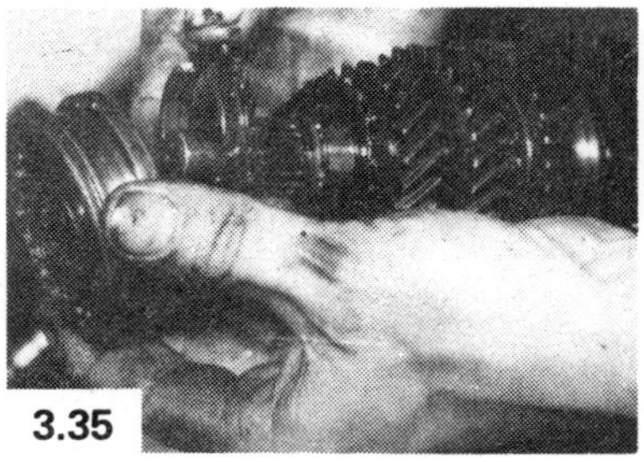
3.35

Removing the circlip from the nose of the mainshaft and taking off 3rd/4th speed synchro hub

3.37

Removing speedometer worm gear circlip

seizure, or gear case cracked) it would be better to leave well alone and look for a replacement, either secondhand or exchange unit.

2 Examine all gears for excessively worn, chipped or damaged teeth. Any such gears should be replaced.

3 Check all synchromesh cones for wear on the bearing surfaces, which normally have clearly machined oil reservoir lines in them. If these are smooth or obviously uneven, replacement is essential. Also, when the cones are put together - as they would be when in operation - there should be no rock. This would signify ovality, or lack of concentricity. One of the most satisfactory ways of checking is by comparing the fit of a new cone on the hub with the old one. If the teeth of the ring are obviously worn or damaged (causing engagement difficulties) the ring should be renewed.

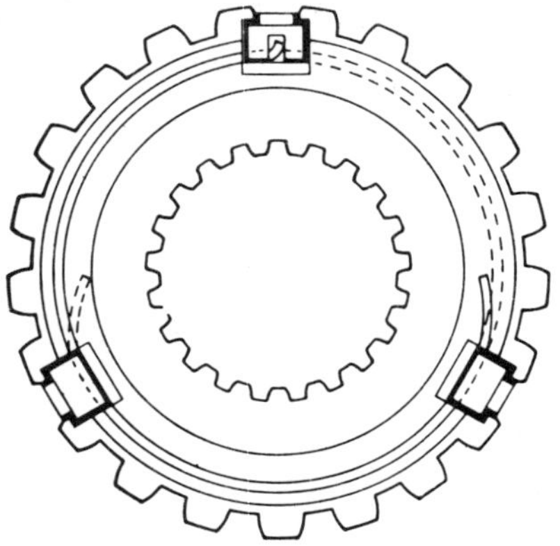

Fig 6:7 Synchro hub cross section showing position of blocker bars and retaining springs

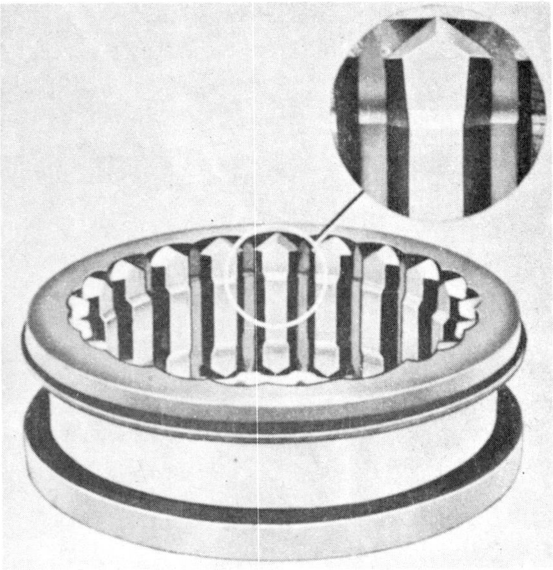

Fig 6:8 Synchro hub sleeve showing (inset) 'Keystone' tooth design

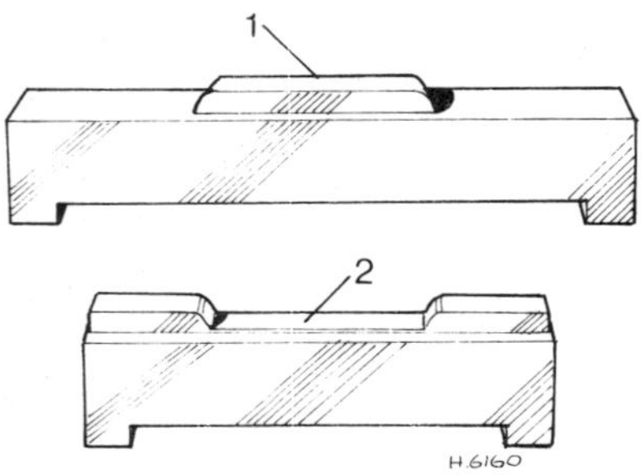

Fig 6:9 SYNCHRO HUB BLOCKER BARS
1 Pips on 1st/2nd gear hub 2 Grooves in 3rd/4th gear hub
Both subject to wear

4 All ball bearings should be checked for chatter. It is advisable to replace these anyway, even though they may not appear too badly worn.

5 Circlips which in this particular gearbox are all important in locating bearings, gears and hubs, should also be checked to ensure that they are undistorted and undamaged. In any case a selection of new circlips of varying thicknesses should be obtained to compensate for variations in new components fitted, or wear in old ones. The 'Specifications' at the beginning of this Chapter indicate what is available.

6 The thrust washers at the ends of the laygear should also be replaced, as they will almost certainly have worn if the gearbox is of any age.

7 Needle roller bearings between the input shaft and mainshaft and in the laygear are usually found in good order, but if in any doubt replace the needles as necessary.

8 The sliding hubs themselves are also subject to wear and where the fault has been failure of any gear to remain engaged, or actual difficulty in engagement, then the hub is one of the likely suspects. It is possible to examine the internal splines without dismantling. The ends of the splines are machined in such a way as to form a 'keystone' effect on engagement with the corresponding mainshaft gear. Do not confuse this with wear. Check also that the blocker bars (sliding keys) are not sloppy and move freely. If there is any rock or backlash between the inner and outer sections of the hub, the whole assembly should be renewed, particularly if there has been a complaint of jumping out of gear (Figs 6:7, 6:8, 6:9).

5 Gearbox - reassembly

1 Start by reassembling the mainshaft.

2 Place the 2nd gear wheel onto the longer end of the mainshaft with the helical teeth part towards the centre of the shaft (photo).

3 Next follows a synchro ring and the reverse gear/1st and 2nd synchro hub assembly (photo).

4 It is important not to let the hub and gear assembly come to pieces, so if it is a tight fit on the splines of the shaft, use a tube so that the centre of the hub may be tapped on to the shaft. It is obviously impossible to tap the outer gear part and keep the hub together (photo).

5 When the hub and synchro ring are up together, ensure that the sliding keys (blocker bars) mate with the corresponding cut-outs in the synchro ring (photo).

6 The retaining circlip is fitted next and must be selected from the range of thicknesses available so that the end float at the hub is eliminated (photo).

7 Fit another synchro ring, so that the cut-outs mate with the other ends of the blocker bars (sliding keys) in the hub and then follow with 1st gear, cone towards the synchro ring, of course. Place the main bearing circlip over the shaft (see paragraph 17) (photo).

8 Next fit the thrust washer and bearing over the shaft (photo).

9 To drive the bearing onto the shaft it is again necessary to select a piece of tube which will go round the shaft and onto the inner race of the bearing so that it may be driven on. Ensure that the end of the shaft rests on something soft to avoid damage (photo).

Note: The rear bearing used on later gearboxes has a nylon ball cage on one side. When installing the bearing make sure that the open side is towards the front of the gearbox.

10 Fit the speedometer drive gear over the shaft with the long boss towards the bearing. Then put the shaft between the soft jaws of a vice (two pieces of 'L' shaped lead are sufficiently soft) so that the jaws support the shoulder of the gear hub but do not grip the shaft. Then drive the shaft downwards with a soft headed mallet until the speedometer drive butts against the bearing (photo).

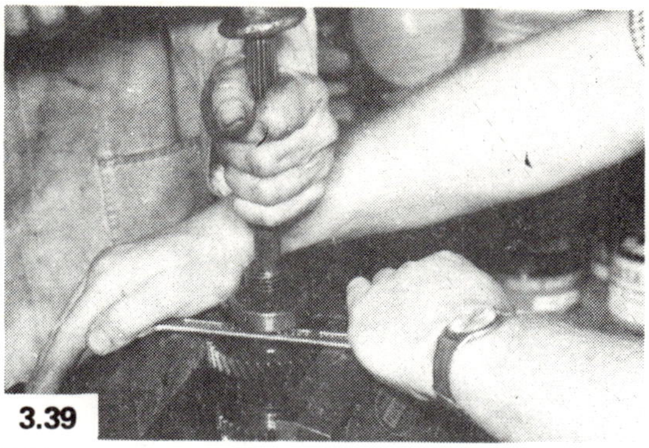

3.39

Driving out the mainshaft from the rear bearing and speedo worm drive gear

3.40

Removing 1st gear assembly

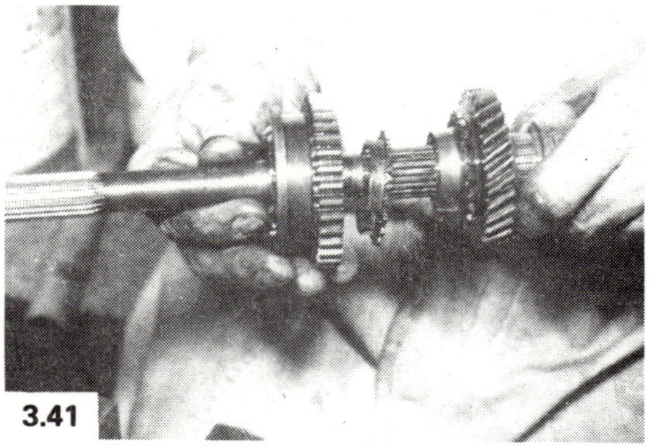

3.41

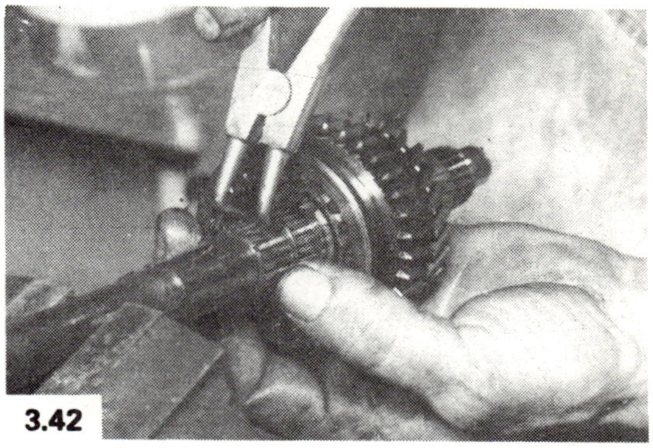

3.42

Removing, in order, circlip and 1st/2nd synchro hub assembly

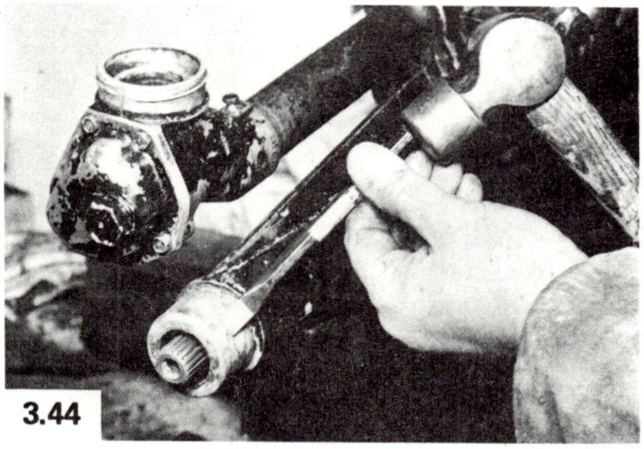

3.44

Removing rear extension oil seal

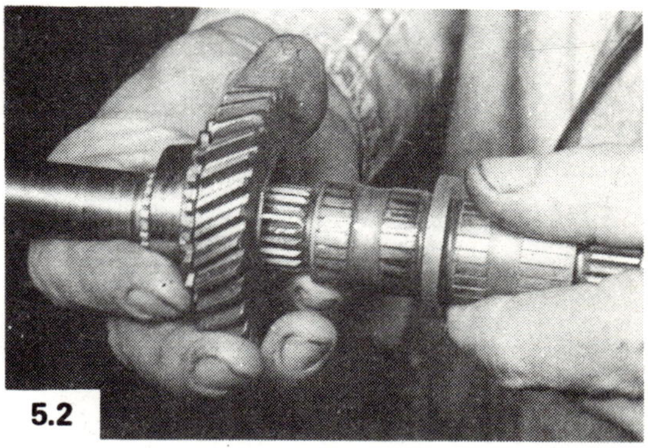

5.2

Replacing 2nd gear

5.3

5.4

Replacing synchro ring and 1st/2nd synchro hub and driving on the hub with a piece of tube

5.5

5.6

Lining up 2nd gear synchro ring cut-outs and fitting the circlip after the hub

5.7

5.8

Replacing synchro ring and first gear, thrust washer and bearing

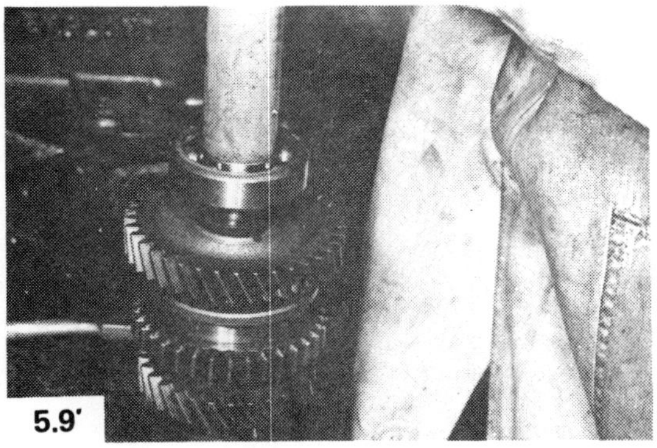

5.9

Driving on the bearing with a piece of tube

5.10

Driving the mainshaft assembly through the speedometer drive worm gear

5.11

Fitting a circlip of correct thickness

5.12

Replacing on the front of the mainshaft, third gear and synchro ring

5.13

Replacing 3rd/4th synchro hub

5.14

Selecting a circlip for the end of the mainshaft

5.15

Fitting a circlip to the end of the mainshaft

5.16

Tapping in a new rear extension oil seal using an old bearing round the shaft as a drift

5.17

Putting the rear bearing circlip behind the bearing prior to fitting the mainshaft into the rear cover

5.18

Holding the main bearing circlip with a piece of flattened tube

5.20

Guiding the main bearing circlip into the rear cover

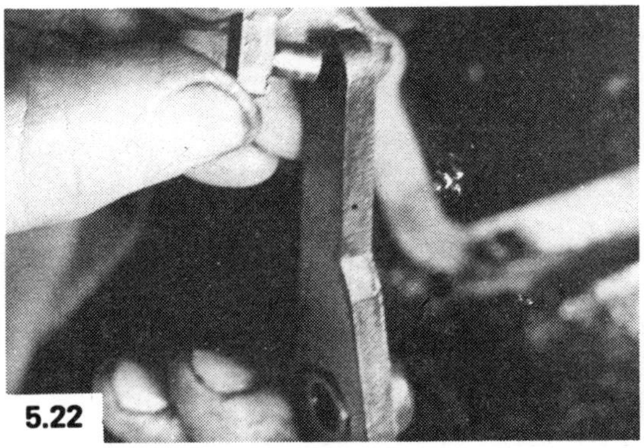

5.22

Fitting the reverse lever selector pin

5.23

Setting the reverse lever pivot pin

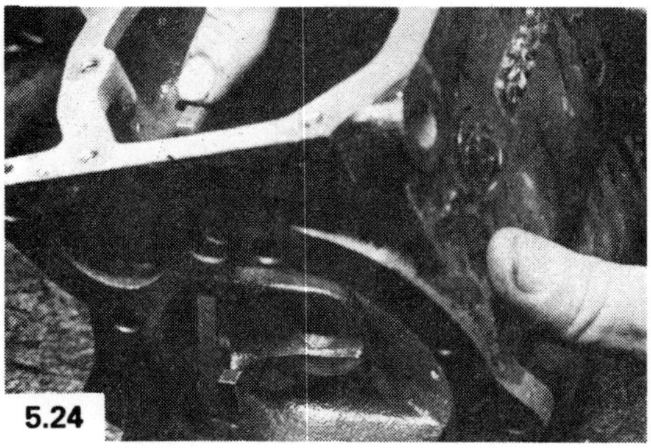

5.24

5.25

Putting the reverse lever on the pivot pin and replacing the idler gear and shaft

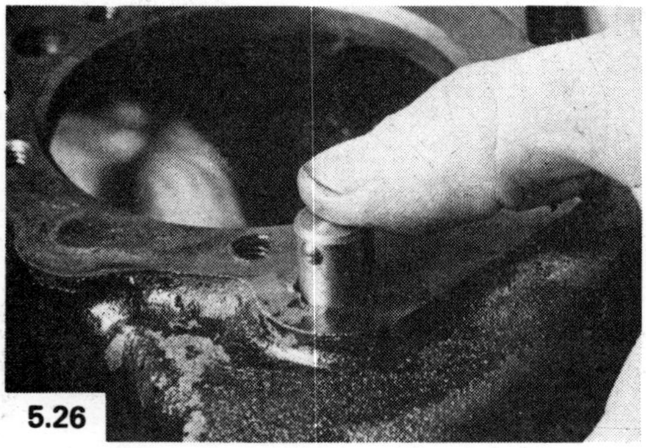

5.26

Lining up the locking ball to the groove for the idler shaft

11 Select and fit a circlip that will take up the clearance between the outer side of the groove and the gear (photo).

12 Turning to the other end of the shaft, fit 3rd gear and a synchro ring right up to the shoulder on the shaft (photo).

13 Fit the 3rd and 4th speed hub assembly ensuring it goes on the proper way round, which is with the larger hub of the boss towards the end of the shaft. The narrow groove in the outside also is towards the end of the shaft. Ensure that the blocker bars in the hub engage in the synchro ring cut-outs (photo).

14 Select a circlip which will take up the end float between hub and groove and try it first (photo).

15 When selected, fit the circlip in the groove. The mainshaft assembly is now complete (photo).

16 To fit a new extension cover oil seal, drive it in using something like an old bearing to spread the driving force. When nearly home it can be tapped with a soft headed mallet to finish off (photo).

17 The mainshaft is next installed in the rear housing and if you have forgotten to place the circlip over the shaft after 1st gear (paragraph 7) you should put it behind the bearing now, spreading it a little to get it over (photo).

18 To install the mainshaft in the cover is a tricky operation and the procedure detailed hereafter is the only way to do it. First of all find a suitable piece of soft metal pipe at least 6 inches long and approximately ¼–3/8 inch inside diameter and flatten one end in a vice in such a way that it will hold together the two ends of the retaining circlip (photo).

19 Mount the rear extension cover in a vice with the selector rod tube uppermost and start warming it up gently and generally with a blow lamp, butane gas lamp or similar. This is essential to ease the pressing of the bearing into the housing. Do NOT overheat, merely warm it up so that it is still just comfortable to touch.

20 When the casing is sufficiently warm, insert the mainshaft assembly and tap it in with a soft headed hammer until the circlip and its improvised holding pipe are up against the cover in the cut-away portion of the bearing housing (photo).

21 The circlip is then released so that it can be located in its groove after the bearing is driven fully home. Make sure that before it is released it will be held inside the bearing housing!

22 Returning to the main gearbox casing, fit the reverse selector lug into the reverse gear selector arm on the side opposite to the boss (photo).

23 Fit the reverse gear selector arm pivot pin (which is eccentric) so that the two punch marks are uppermost in relation to the gear casing and tighten the locknut (photo and Fig 6:10).

24 Place the reverse selector lever on the pivot pin so that the boss side goes on first, leaving the selector lug at the bottom, facing inwards (photo).

25 Put the reverse idler gear into the box so that the selector lug locates in the groove and the groove side of the gear is towards the front of the casing. Holding the gear in position, push the plain end of the reverse idler gear shaft through the housing and gear (photo).

26 Drive the shaft fully home so that the locating ball on the end of the shaft is in position in the cut-away part in the casing (photo).

27 In theory, one should now check the clearance between the idler gear and the rear of the casing, and adjust it. As this entails assembling the rear cover temporarily, it is recommended that the adjustment be made when the gearbox is assembled by actually operating the reverse gear in the normal manner.

28 Place a spacer in each end of the laygear and, using clean grease in the bore, assemble the needles for the laygear bearings, ensuring that there are twenty-five at each end. Place the outer spacers in after them with a little grease to hold them in place. On shafts fitted with two rows there will be double the number (photo and Fig 6:11).

29 Next put the laygear thrust washers in position by locating each one so that the small lug engages with the groove in the casing. The larger washer goes at the front of the gearbox. Hold them in position with a smear of grease (photo).

30 Take the laygear and, with care, lower it into the casing with the larger end towards the front of the casing. Be careful not to dislodge

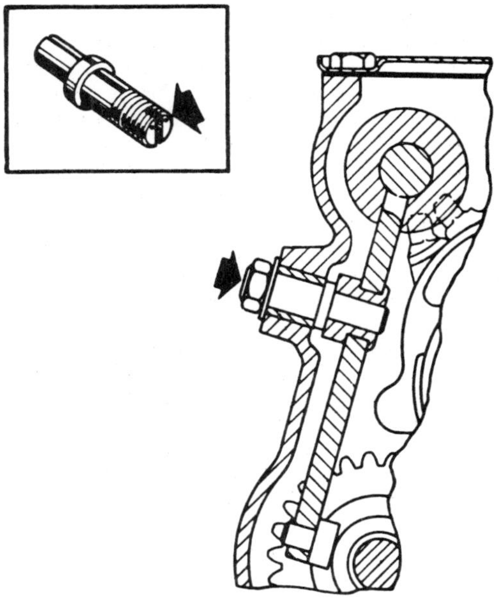

Fig 6:10 GEARBOX – REVERSE SELECTOR LEVER ECCENTRIC PIVOT PIN
Punch marks on pin (inset show high point of eccentric)

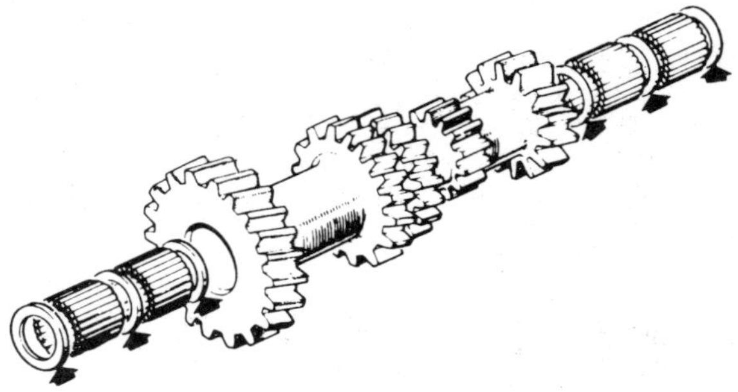

Fig 6:11 Laygear with double needle rollers in gearbox for 1256 cc engines

5.28

Replacing layshaft needle rollers into the laygear

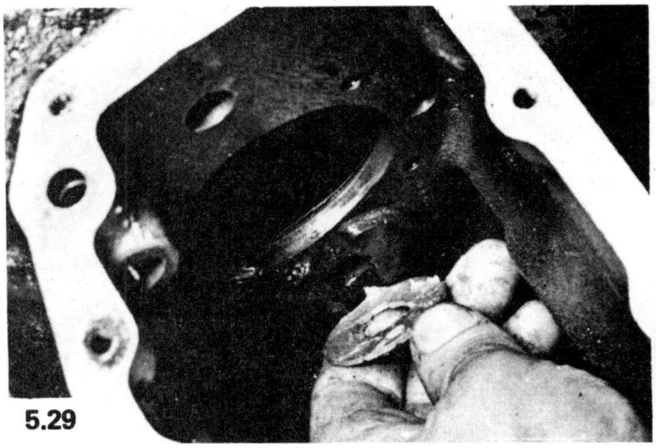

5.29

5.30

Positioning laygear thrust washers and replacing laygear

5.32

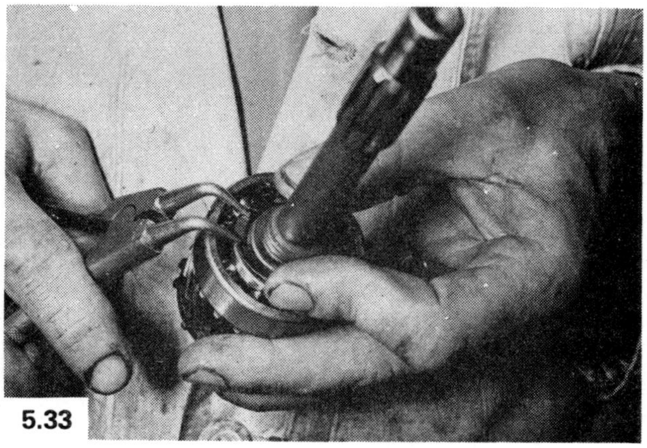

5.33

Driving bearing onto input shaft using a piece of tube and refitting the circlip retainer

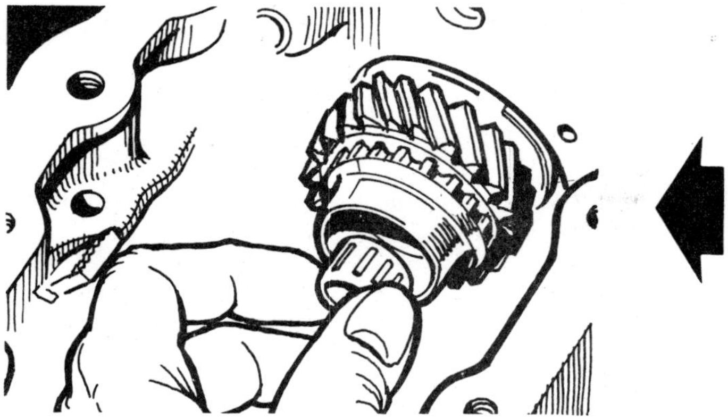

Fig 6:12 Fitting the caged needle roller bearing into the input shaft counterbore

:he thrust washers or the needle rollers and spacers in the laygear photo).

31 It is not possible to replace the layshaft yet because the input haft has to be put back first. Otherwise, with the layshaft in position, he fourth gear driving dogs will not get past the large laygear at the ront end.

32 Take the input shaft and, if a new bearing is to be fitted, drive it on with a piece of pipe around the shaft and on to the inner bearing ace. (photo).

33 Refit the circlip retaining the bearing (photo).

34 Place the circlip, to hold the bearing in the housing over the shaft nd put the shaft into the front housing (photo).

35 Hold the circlip with pliers, tap the bearing into the housing until he circlip can be released, and then ensure that it locates properly nto the groove in the bearing housing (photo).

6 Replace the needle roller bearing into the counter bore of the haft if not already done (Fig 6:12).

7 Fit a new gasket to the front cover flange (photo).

8 Put the input shaft assembly into the front of the casing, but do ot replace the bolts yet.

9 Next introduce the plain end of the layshaft into the hole in the ear of the casing and, with care, raise the laygear so that the layshaft nay be pushed into it without disturbing the bearings or thrust vashers. It should not be necessary to use a dummy layshaft as the ygear and thrust washers can be reached easily if necessary. When ne shaft is nearly home, ensure that the ball in the shaft is lined p with the recess in the casing and drive the shaft home with a soft ammer (photo).

) Fit a new gasket to the rear cover flange of the mainshaft. Now raw the input shaft assembly forwards (out) as far as it will go, (you ill see now why the layshaft could not go in first) and place a synnro ring over the end of the gear. Then carefully draw the 3rd/4th eed hub towards the forward end of the mainshaft as far as possle without pulling it right off. This is so that when you introduce the mainshaft assembly into the casing at an angle you will just be able to clear the 3rd speed laygear (photo).

41 Locate the spigot of the mainshaft into the needle roller bearings in the input shaft. Line up the slots in the synchro ring with the blocker bars (sliding keys) of the 3rd/4th speed hub and mate the two shafts together fully (photo).

42 Install the 3rd/4th speed selector fork so that it rests in the groove of the 3rd/4th speed hub (photo).

43 Install the 1st/2nd speed selector fork so that it sits in the groove of the 1st/2nd speed hub.

44 In order to fit the selector fork rod, rotate the rear cover until the hole in the cover lines up with the holes for the rod in the casing. The rod is fractionally larger (about 2 thou) at one end, so place the smaller end into the casing first. Run the rod through the mounting holes in the forks and drive it fully home with a brass drift (photo).

45 Match up the rear cover to the gearbox casing and replace and tighten all bolts evenly.

46 Next go to the input shaft bearing front cover and position it so that the small oil drain hole in the cover is at the bottom. Replace the bolts with new copper washers and tighten up (photo).

47 With both end covers now bolted up, check that all shafts rotate freely. If they do not, slacken off the front cover and then the rear cover until they do. Then take everything out and check all circlips and spacers for correct location. Line up the end of the reverse gear selector lever with the lug on the 1st/2nd speed selector fork (photo).

48 Replace the interlock collar so that the large peripheral grooves rest over the jaws on the selector forks and the longitudinal slot in the collar faces into the centre of the gearbox (photo).

49 Insert the selector rod, key end first, into the end of the remote lever extension tube (photo).

50 Pass it through the interlock collar so that the key flanges pass through the slot in the collar and the end of the rod locates in the hole in the front of the casing (photo).

51 Turn the interlock collar so that the groove in the centre section is

5.34

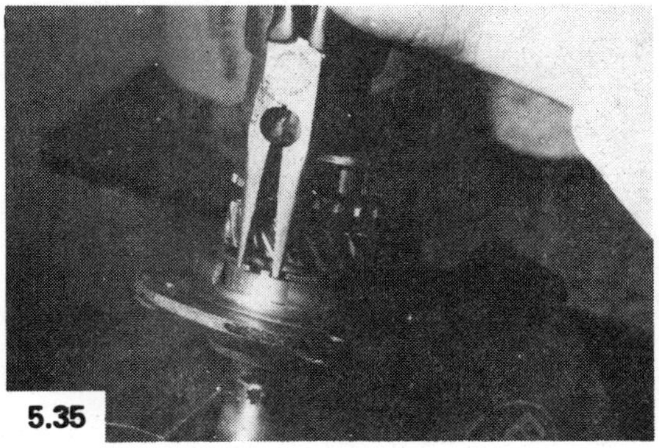
5.35

Fitting the input shaft into the front cover, locating the circlip to retain it in the cover

5.37

Placing a new gasket in position

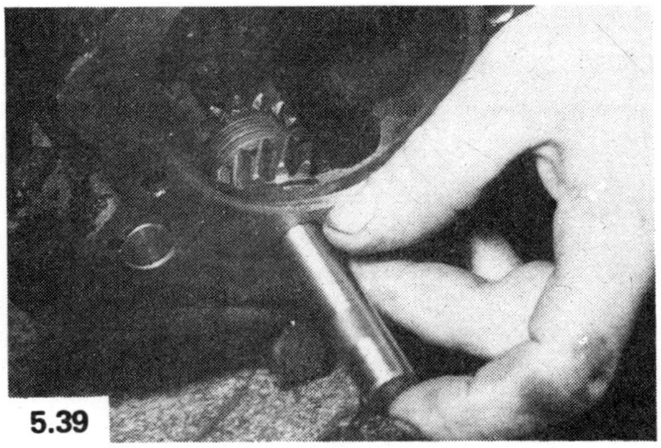
5.39

Fitting the layshaft through the laygear

5.40

5.41

Replacing the mainshaft into the casing and input shaft

5.42

Replacing 3rd/4th gear selector fork

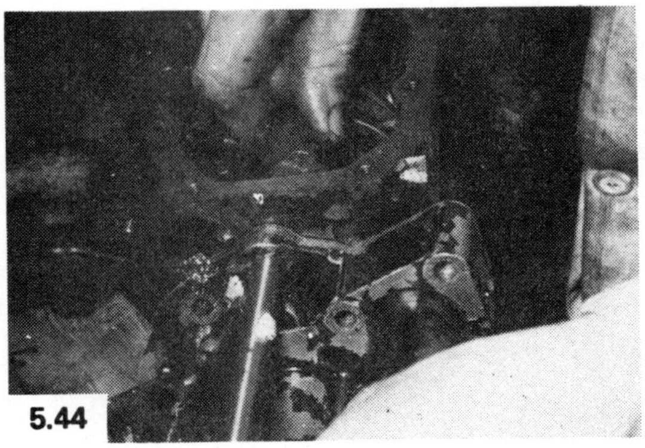
5.44

Refitting the selector fork rail

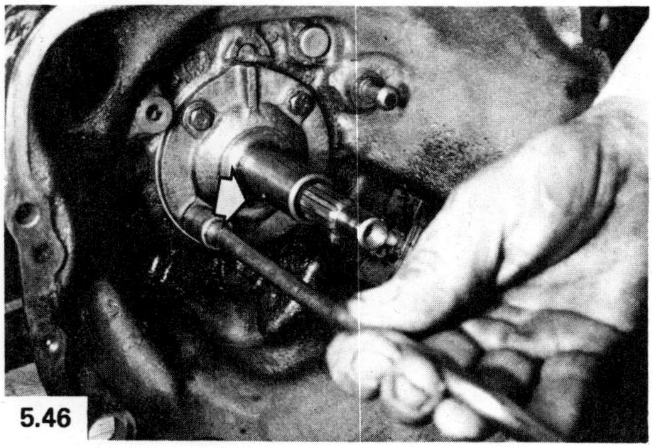
5.46

Tightening input shaft cover bolts. Note oil drain hole (arrowed) at bottom

5.47

5.48

Lining up the selector fork lugs prior to putting in the interlock collar

Fig. 6:12A Combined input shaft bearing/oil seal on later gearboxes

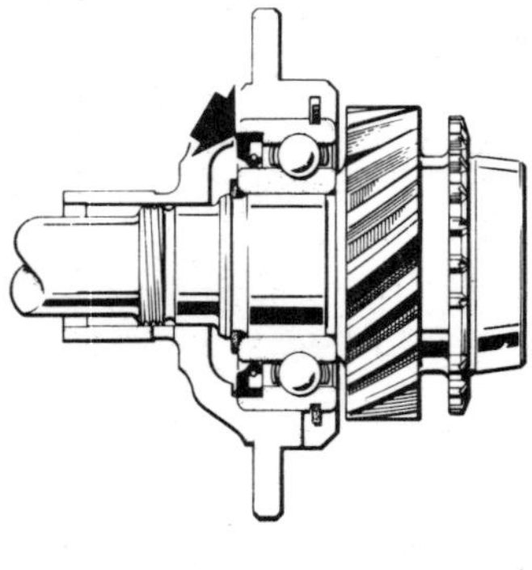

lined up to accept the collar retaining pin through the hole in the casing (photo).

52 Replace the detent ball and spring into the hole in the casing (photo).

53 Fit a new gasket and replace the top cover, ensuring that the detent spring locates properly in the dimple in the cover. Replace four bolts, one at each corner of the cover, for the moment (photo).

54 Refit the gear lever as it came out, with the longer lug on the lever fork to the left (photo).

55 Now engage and disengage all gears including reverse. Any difficulty with reverse may be due to the setting of the eccentric pivot pin mentioned in paragraph 23. Any difficulty with the selection and disengagement of the forward speeds could be due to the setting of the eccentric pin on the selector shaft. With 1st speed gear engaged, the gap between the reverse stop on the lever (ie the long lug) and the rod, should be between 2 and 12 thou. Adjust, if necessary, by unlocking the nut and turning the eccentric pin with a screwdriver (Fig 6:13).

56 Select neutral, remove the gear lever, and replace the extension tube end cover with a new gasket.

57 Before replacing the top cover bolts completely, it is a good idea to refill the gearbox with the specified quantity of oil - including a small amount in the rear cover extension (after replacing the speedo drive spindle and gear as described in the next paragraph).

58 Replace the speedometer drive by returning the spindle so that the gear enters the casing last on the left hand side. Replace the locating cup in the hole, having first smeared the sides and lip with sealing compound. Reassembly is now complete.

6 Speedometer drive gear

1 The speedometer drive gear and the associated oil seal disc can be removed with the gearbox mounted in the car. Proceed as described in Section 3, paragraph 29 for removal. A little oil may drain out.

After refitting the gear spindle, replace the sealing cup ensuring that it is properly sealed with compound. Do not forget to fit the oil seal disc round the spindle.

7 Oil seals - front and rear

1 The oil seal for the rear cover extension can be removed with the gearbox installed, once the propeller shaft has been removed as described in Chapter 7.

The procedure for removing the seal is as described in Section 3, paragraph 44 and replacement as in Section 5, paragraph 16.

2 The front oil seal is part of the bearing assembly. (Fig. 6.12A).

8 Automatic transmission - description

1 The GM automatic 3 speed transmission is fitted to the Viva as an optional extra. Normally one associates automatic transmission with larger and more powerful cars but, in recent years, development has enabled them to be fitted on engines of lower power output. Now that the Viva has a 1256 cc engine they are all the more attractive.

2 The automatic transmission replaces the conventional clutch and gearbox, and occupies the same space in the same way - being bolted onto the rear of the engine. It comprises two basic parts - the torque convertor and the three speed epicyclic gearbox.

The torque converter is a form of oil operated turbine which transmits the engine power from a multi-bladed rotor (the pump) directly connected to the crankshaft to another multi-bladed rotor (the turbine) directly connected to the input shaft of the transmission. At low engine revolution, the oil driven by the pump has little force imparted to it, so the turbine does not move. When the pump speed increases, so the forces of the oil are transferred to the turbine.

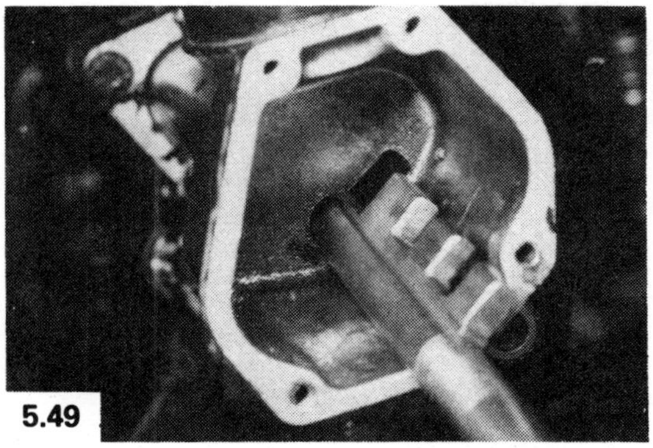
5.49

5.50

Inserting the change lever through the rear of the extension tube

5.51

Engaging the keys in the interlock collar and refitting the retaining pin

5.52

Replacing detent ball and spring

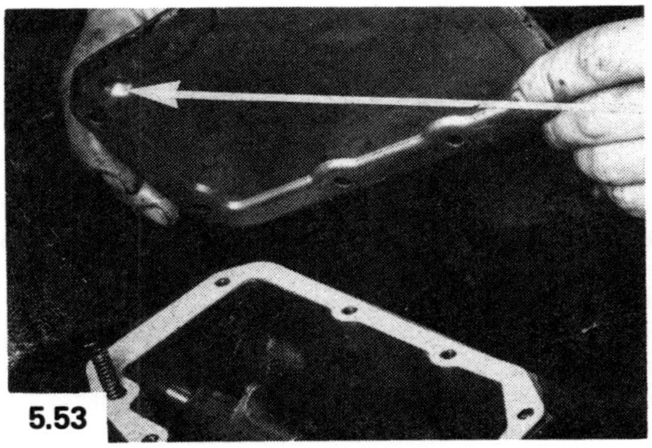
5.53

Replacing the top cover. Note recess in cover for detent spring

5.54

Fitting the gearchange lever into the extension housing

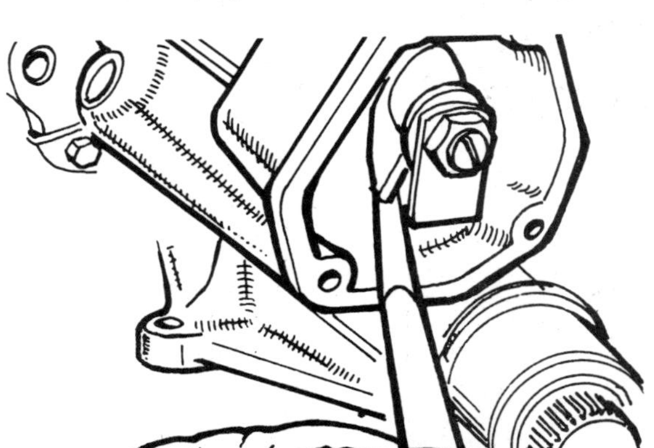

Fig 6:13 Checking the selector rod/reverse stop clearance with a feeler blade

Fig 6:14 Automatic transmission - fluid level dipstick

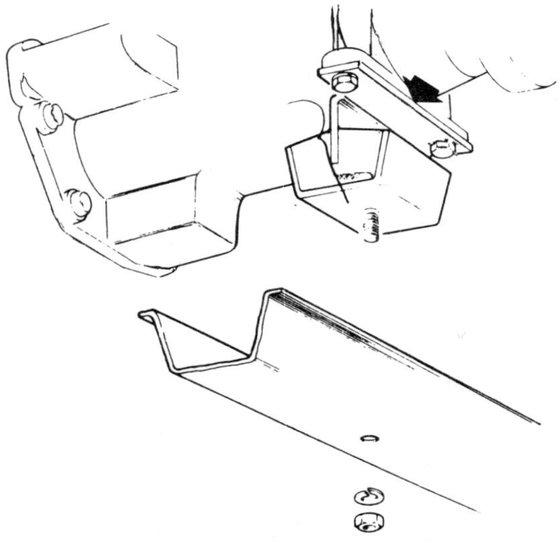

Fig 6:15 Automatic gearbox – rear mounting showing flange facing rearwards (arrowed)

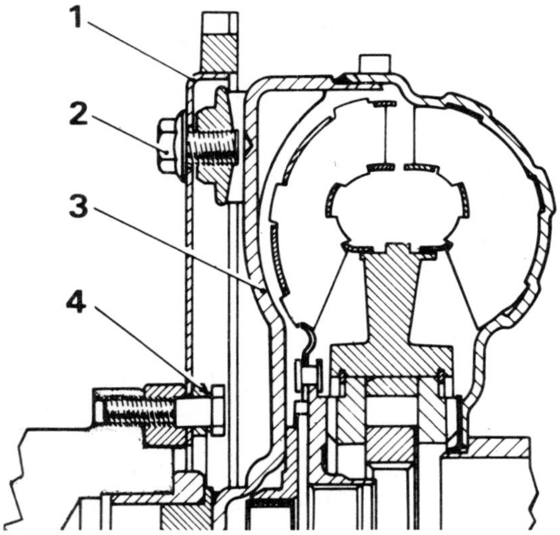

Fig 6:16 AUTOMATIC GEARBOX – COUPLING TO ENGINE FLEXPLATE
1 Flexplate
2 Flexplate to torque converter bolts
3 Torque converter
4 Distance plate for flexplate to crankshaft bolts

An intermediate multi-bladed rotor (the stator) regulates the flow of oil back to the pump after it has done its work through the turbine.

The gearbox consists of a ravignaux planetary gear set in constant mesh and the selection of the gears is by braking one or more of the components of this gear set.

This braking is effected by one of the three servo operated multi-plate clutches and a band - literally a brake band, which can be applied to the outer ring gear of the set. The automatic operation of three clutches and the low speed band is the complicated part, involving a servo/hydraulic pump system controlled by road speed, inlet manifold vacuum, and the position of the accelerator.

3 The capabilities of the automatic transmission are different from the manual system and in order that those unfamiliar with them may understand the difference, a full description of the functions at starting, parking and stopping, in all of the six selector positions is given below:

'P'. Park. In this position with the engine either stopped or running, no gears are 'engaged' and the gearbox output shaft is mechanically locked, which in effect means that the propeller shaft and rear axle are also locked. The car cannot be moved, therefore. The engine may be started in this position. In order to select the 'P' position, the selector lever button must be fully depressed. Do not select 'P' if the car is moving. Damage will result.

'N'. Neutral. The conditions for neutral are the same as for 'P', except that the gearbox output shaft is not mechanically locked. The car will, therefore, roll with the engine either running or stopped.

'R'. Reverse. The button on the selector lever must be partially depressed to engage 'R'. In this position reverse gear is 'engaged'. If the engine is not running, it cannot be started unless the selector lever is moved to 'P' or 'N'. With the brakes applied, the car will not move. With the brakes off, increase in engine speed will move the car backwards. When the engine speed is decreased, the engine will act as a brake through the transmission. If the car is standing with the brakes off, it may roll at low engine speed in either direction. Reverse should not normally be selected whilst the car is moving.

'D'. Drive. The selector position for normal driving requirements. In this position first gear is initially 'engaged' but, at low engine speed with the brakes off, the car may roll in either direction The engine cannot be started in this position. With the engine speed increased, the car will move forward in low gear.

When the speed and load conditions are right, the transmission will automatically move to second gear 'engagement' and then subsequently to top gear. When speed decreases, the gears will automatically shift back down as far as first, again according to speed and load situations.

The engine does not act as a brake or over-run in any of the three speeds in the 'D' position.

'I'. Intermediate. To select this position the lever button must be partially depressed. When selected, the automatic transmission will operate exactly as in 'D' except that it will not move up out of 2nd. It should not be used in excess of 60 mph.

It is possible to change to 'I' when the vehicle is moving. It will immediately put the vehicle in 2nd gear until speed or throttle position may cause it to change down to first. The intermediate range is normally used in traffic or on uphill sections where one would tend to get a lot of changing going on between 2nd and top if in the 'D' position. Although there is no over-run braking in 1st gear, there is on 2nd gear in the 'I' position.

'L'. Low. To select this position, the selector lever button is fully depressed. This position should not be selected above 35 mph. It would normally be used to provide engine braking on steep downhill sections of road, or to avoid unnecessary changing between 1st and

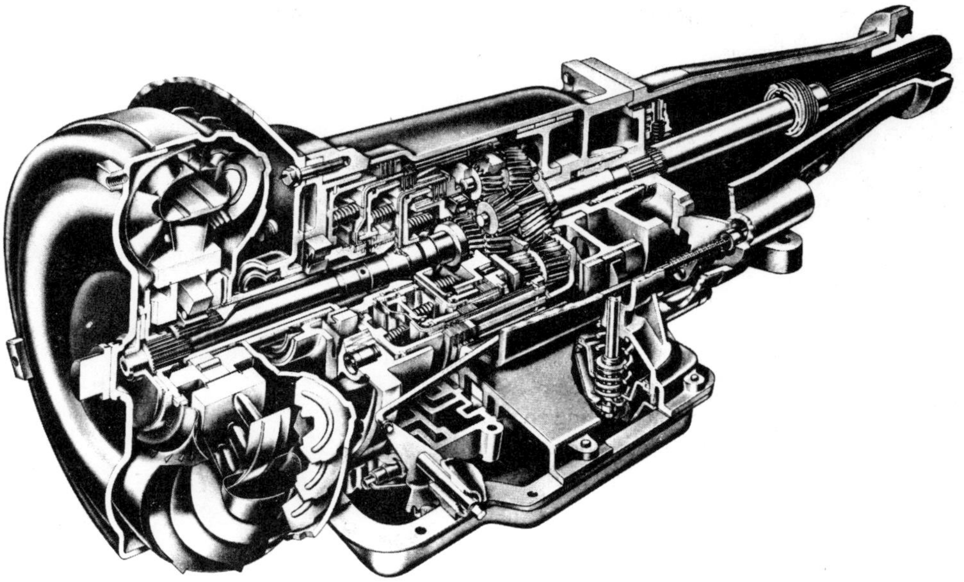

Fig 6:17 Automatic transmission - cutaway view of complete assembly

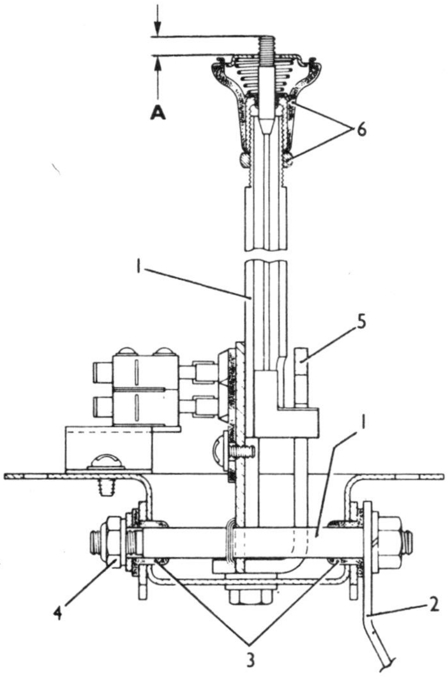

Fig 6:18 SELECTOR LEVER LATERAL CROSS SECTION
1 Selector lever and pivot shaft
2 Lower lever
3 Pivot shaft bushes
4 Pivot shaft locking nut
5 Selector plate
6 Locking ring and grip
Dimension 'A' = .24 inches with plunger up and selector lever in 'P' position

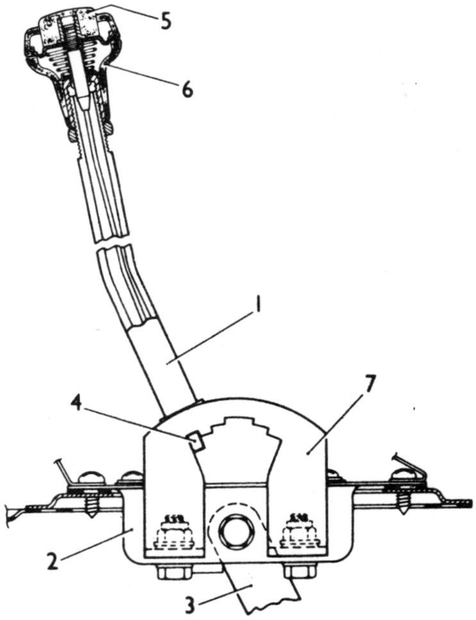

Fig 6:19 SELECTOR LEVER LONGITUDINAL CROSS SECTION
1 Selector lever
2 Housing
3 Lower lever
4 Plunger pawl
5 Push button
6 Grip
7 Selector plate

2nd in dense traffic or on continuous slow uphill climbs.

As implied, the engine acts as a brake on over-run in this range.

Some points to bear in mind in the operation of automatic transmission are:

a) It is possible to obtain a quick change down to provide instant acceleration by depressing the accelerator fully. This change will not take place, however, if the vehicle is already in excess of the maximum speed of the gear below.

b) Where continuous engine braking on over-run is wanted, 'L' or 'I' ranges must be selected. It follows, therefore, that when shifting into these ranges when on the move, engine braking will take place if the car speed is high. On slippery surfaces the possibility of skids occurring must, therefore, be considered due to the sudden braking effect on the rear wheels.

c) It is not possible to push to tow start the car.

d) If the car is to be towed for any reason, the speed must be kept below 30 mph and the selector be put in 'N'. Not more than 30 miles should be covered. If there is a suspected fault in the transmission, the car should not be towed at all unless the propeller shaft is disconnected or the driving wheels raised to prevent the transmission being turned.

e) Cars fitted with automatic transmission are also fitted with automatic chokes on the carburettor so that the engine speed is suitably governed until it is warmed up. There will be a tendency to a faster tickover and subsequent 'creep' when in any of the driving ranges until the engine is fully warm. Engine tuning and smooth running is much more significant where automatic transmission is fitted.

f) Transmission fluid normally heats up in use. Severe or abusive use, or failure to keep cooling areas clean, can cause overheating and damage.

9 Automatic transmission - fluid level

1 The total capacity of the system is 9 pints. A dipstick is provided in the filler pipe which is located on the right hand side and projects into the engine compartment at the rear of the engine (Fig 6:14).

2 To check the fluid level, the engine and transmission should be fully warmed up to normal working temperature. With the car stationary on level ground, engine ticking over at idling speed and the selector lever in 'P' or 'N', remove the dipstick, clean it off, replace and remove again to note the level. The level must be kept between the 'Full' and 'Add' marks. From 'Add' to 'Full' calls for 1 pint. Do not overfill or foaming and loss of fluid may occur. Use only the proper fluid for topping up the transmission (Castrol TQ Dexron 'R') and under no circumstances should additives of any kind be mixed with it.

3 It is generally best to check the level after a normal run. If otherwise, it is difficult to judge the correct working temperature. If starting from cold, then it will be necessary to select a drive range, apply brakes (driver in the driving seat for safety) and run the engine at a fast idle for no more than two minutes.

10 Automatic transmission - adjustments and attention

1 Automatic transmission systems are sophisticated and complicated, and require specialist tools, experience and skill if they are to be properly set up. As they tend to be the exception rather than the rule on anything other than larger vehicles, it follows that the availability of the tools and frequency of experienced mechanics is rare. Non-professional experience is rarer still. Consequently the owner is not advised to tamper with this unit himself.

2 A cross section of the selector lever mechanism and starter inhibitor switch is given so that adjustment can be made to ensure that the operation of the selector lever button and the safety start cut-out are correct. It should not be possible to start the engine when the selector lever is in the 'D', 'I', 'L' or 'R' positions. Similarly it should only be possible to select 'L' or 'P' when the selector button is fully depressed, and 'I' and 'R' when it is partially depressed.

3 Details are given in the next Section on how to remove the transmission unit but, it must be emphasised that, full testing can only be carried out when it is installed. Thus removal and replacement should only be carried out when it is known that the unit is beyond repair in its installed position.

4 The test which the owner may carry out, if he suspects that there is either slip or otherwise, is the stall test. However, it will be necessary for a tachometer to be fitted to the engine. With the transmission fully warmed up, apply the brakes fully (chock the wheels too for safety) engage a drive range and press the accelerator to the floor. The engine speed should settle at 2100-2150 rpm. Do not maintain this test for more than 10 seconds or overheating will result. If the engine rpm are too high then the torque converter oil supply should be suspect, and then the low band servo in the transmission itself. If the rpm are too low then the engine is not delivering full power or the torque converter unit is faulty.

5 The lower part of the torque converter housing is fitted with a perforated metal cover to permit cooling air into the housing. It is important to keep this clean as any restriction could result in overheating and loss of efficiency and damage.

11 Automatic transmission - removal and replacement

1 Before making any attempt to remove the transmission, make sure your reasons are valid. In other words get expert diagnosis first if transmission malfunctioning is the reason.

2 If you are removing the engine from a car with automatic transmission, the two should be separated at the flexplate which connects the crankshaft to the torque convertor. Do not try and separate the torque converter from the gearbox.

3 All the normal precautions for gearbox removal as described in Section 2 should be taken. It must be remembered that they are heavier than conventional gearboxes - approximately 110 lbs and, therefore, adequate support must be provided.

4 Proceed to remove the crossmember support as described in Section 2 after having first slackened the transmission brace bolts at the sump bracket, and moving the starter as far forward as possible to clear the starter teeth on the flexplate rim.

5 If an oil cooler is fitted (mounted under the air deflector panel above the radiator) it will be necessary to drain the oil out so that the cooler tubes may be disconnected from the transmission. Make sure the unions are perfectly clean first and seal the holes suitably to stop dirt entering. The combined filler/dipstick tube must be removed taking the same precautions.

6 The three bolts which hold the flexplate to the torque converter are accessible as soon as the semicircular sheet steel plate across the bottom half of the casing has been removed. These three bolts must be removed before the main housing bolts securing the transmission to the engine are undone. Otherwise, a strain could be put on the flexplate which would distort.

7 Once the flexplate bolts are removed, the casing bolts can come out with the whole unit properly supported. The transmission is then drawn a little to the rear and lowered in the normal way.

8 If the flexplate is to be renewed, it may be unbolted from the crankshaft flange. Seal the bolts on replacement as for the flywheel.

9 When replacing the transmission, proceed in the reverse order of removal. Line up the painted balance marks on torque converter and flexplate. When tightening the transmission brace, tighten the bolts on the torque converter housing first and then those on the sump bracket.

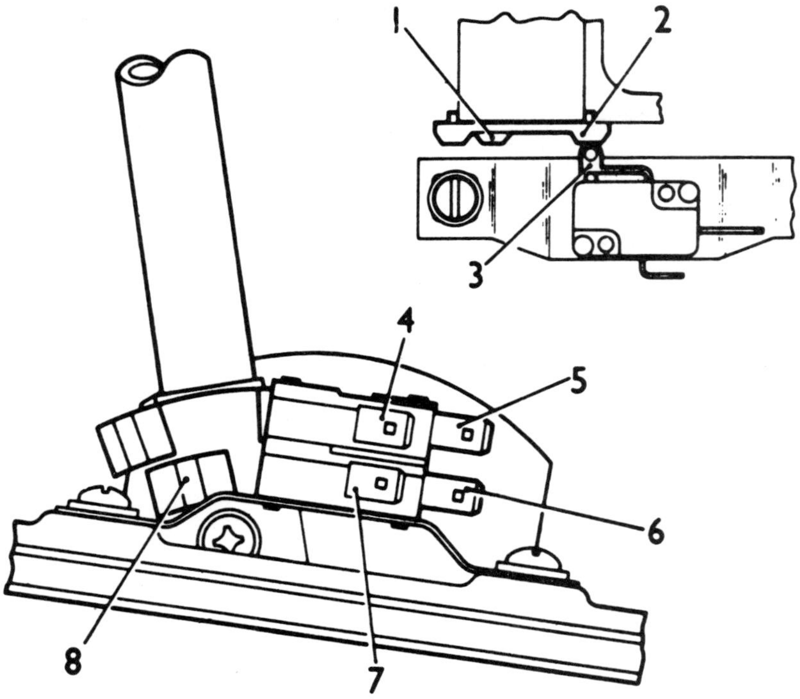

Fig 6:20 AUTOMATIC TRANSMISSION – STARTER INHIBITOR AND REVERSE LAMP SWITCH
1 Neutral cam 2 Park cam 3 Switch roller (upper) 4 Red/white lead connection (starter) 5 Yellow/white lead connection (starter) 6 Green/brown lead connection (reverse lamp) 7 Green connection (reverse lamp)
Position switch so that cams (1 and 2) operate switch lever in Neutral and Park positions

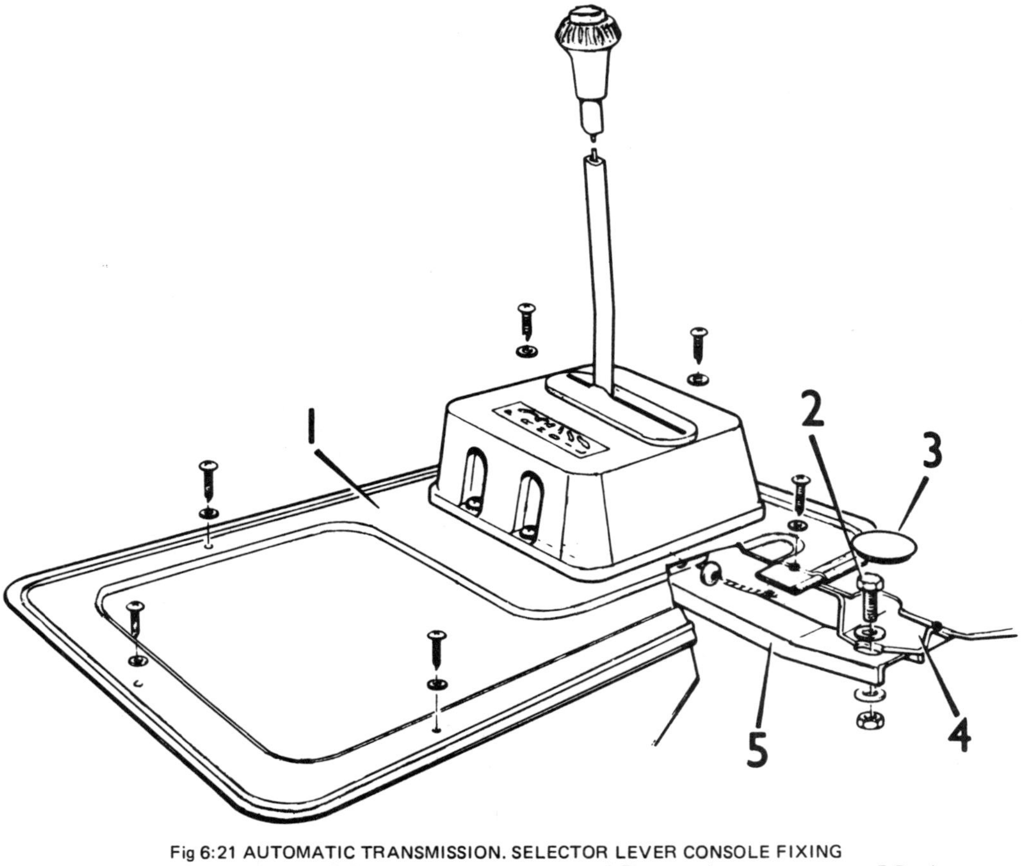

Fig 6:21 AUTOMATIC TRANSMISSION. SELECTOR LEVER CONSOLE FIXING
1 Housing assembly 2 Fixing bolt 3 Grommet 4 Floor member 5 Bracket

Fault Finding Chart – Manual Gearbox

Symptom	Reason/s	Remedy
Ineffective synchromesh	Worn baulk rings or synchro hubs	Dismantle and renew.
Jumps out of one or more gears (on drive or over-run)	Weak detent springs or worn selector forks or worn gears	Dismantle and renew.
Noisy, rough, whining and vibration	Worn bearings and/or laygear thrust washers (initially) resulting in extended wear generally due to play and backlash	Dismantle and renew.
Noisy and difficult engagement of gears	Clutch fault	Examine clutch operation.

NOTE: it is sometimes difficult to decide whether it is worthwhile removing and dismantling the gearbox for a fault which may be nothing more than a minor irritant. Gearboxes which howl, or where the synchromesh can be 'beaten' by a quick gear change, may continue to perform for a long time in this state. A worn gearbox usually needs a complete rebuild to eliminate noise because the various gears, if re-aligned on new bearings will continue to howl when different wearing surfaces are presented to each other.

The decision to overhaul, therefore, must be considered with regard to time and money available, relative to the degree of noise or malfunction that the driver has to suffer.

Chapter 7 Propeller shaft and universal joints

Contents

Specifications

Propeller Shaft

Make...	Hardy Spicer or BRD
Type...	**Tubular, single piece**
Sliding sleeve diameter	1.1240 to 1.1250 in

General description

The drive from the gearbox to the rear axle is via the propeller shaft which is, in fact, a tube. Due to the variety of angles caused by the up and down motion of the rear axle in relation to the gearbox, universal joints are fitted to each end of the shaft to convey the drive through the constantly varying angles. As the movement also increases and decreases the distance between the rear axle and the gearbox, the forward end of the propeller shaft is a splined sleeve which is a sliding fit over the rear of the gearbox splined mainshaft. The splined sleeve runs in an oil seal in the gearbox mainshaft rear cover, and is supported with the mainshaft on the gearbox rear bearing. The splines are lubricated by oil in the rear cover coming from the gearbox.

The universal joints each comprise a four way trunnion, or 'spider', each leg of which runs in a needle roller bearing race, pre-packed with grease and fitted into the bearing journal yokes of the sliding sleeve and propeller shaft and flange.

2 Propeller shaft - removal, inspection and replacement

1 Jack up the rear of the car and support it on stands.
2 The rear of the shaft is connected to the rear axle pinion by a flange held by four nuts and bolts. Mark the position of both flanges relative to each other, and then undo the bolts. An adaptor plate is mounted between the two flanges so make sure this is marked as well (photo).
3 Move the propeller shaft forward to disengage it from the pinion flange and then lower it to the ground.
4 Draw the other end of the propeller shaft, that is the splined sleeve, out of the rear of the gearbox extension cover, and the shaft is then clear for removal (photo).
5 Place a receptacle under the gearbox rear cover opening to catch any oil which will certainly come out if the gearbox is tilted.
6 If the propeller shaft is removed for inspection, first examine the bore and counterbore of the two flanges which mate at the rear. If they are damaged in any way, or a slack fit, it could mean that the propeller shaft is running off centre at the flange and causing vibration in the drive. If nothing obvious is wrong, and the universal joints are in good order, it is permissible to reconnect the flanges with one turned through 180° relative to the other. This may stop the vibration.
7 The replacement of the shaft is a reversal of the removal procedure. Ensure that the sliding sleeve is inserted into the gearbox end cover with care, and is perfectly clean, so as not to cause damage to, or failure of, the oil seal in the cover.
8 The flanges should be mated according to the position marks (unless a 180° turn is being done as mentioned in paragraph 6). Make sure that the adaptor plate is a snug fit on the spigot of the pinion shaft and correctly lined up as before. The plain side of the plate faces the pinion flange (Fig 7:2).
9 The four bolts should be fitted with the heads towards the universal joint.

3 Universal joints - inspection and renewal

1 Preliminary inspection of the universal joints can be carried out with the propeller shaft on the car.
2 Grasp each side of the universal joint, and with a twisting action determine whether there is any play or slackness in the joint.
3 If the universal joint (or joints) are obviously worn out, it is necessary, regrettably, to obtain a complete new propeller shaft. This is because the needle roller bearing assemblies are staked into the yokes rather than held in with circlips as before (Fig 7:3).

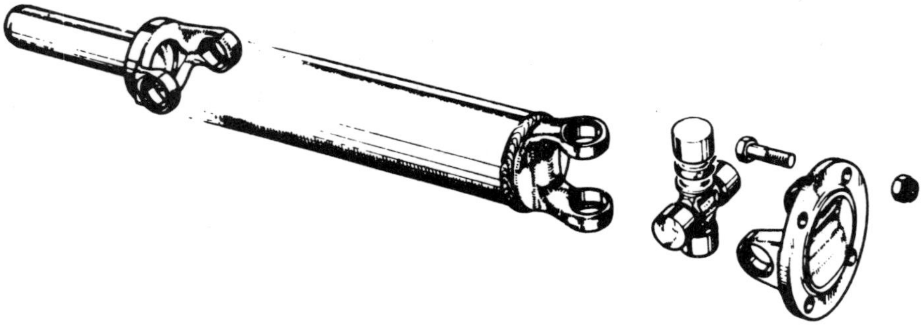

Fig 7:1 Exploded view of propeller shaft components

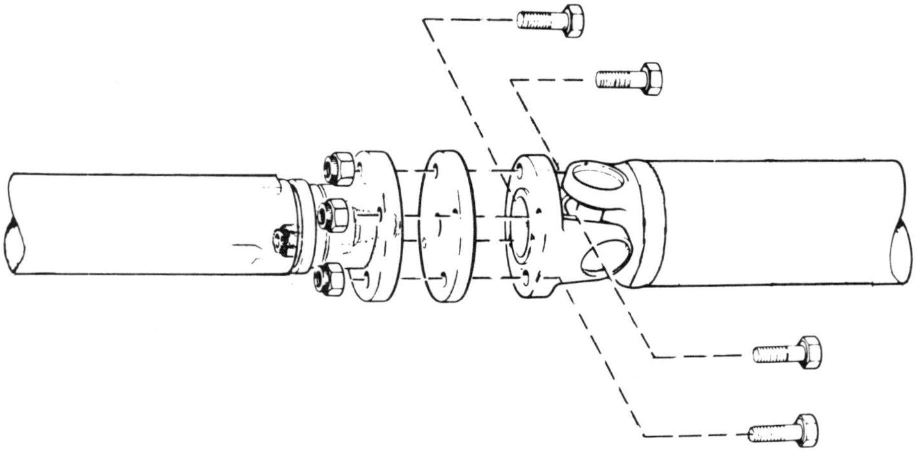

Fig 7:2 Assembly of adaptor plate between mating flanges

2.2

Disconnecting propeller shaft rear flange

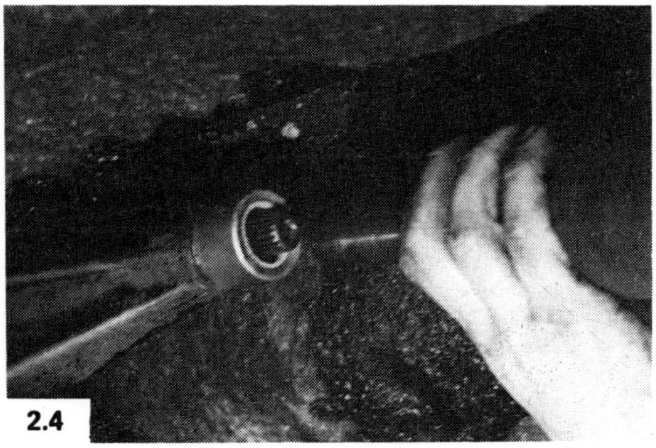

2.4

Withdrawing the sliding sleeve from the gearbox

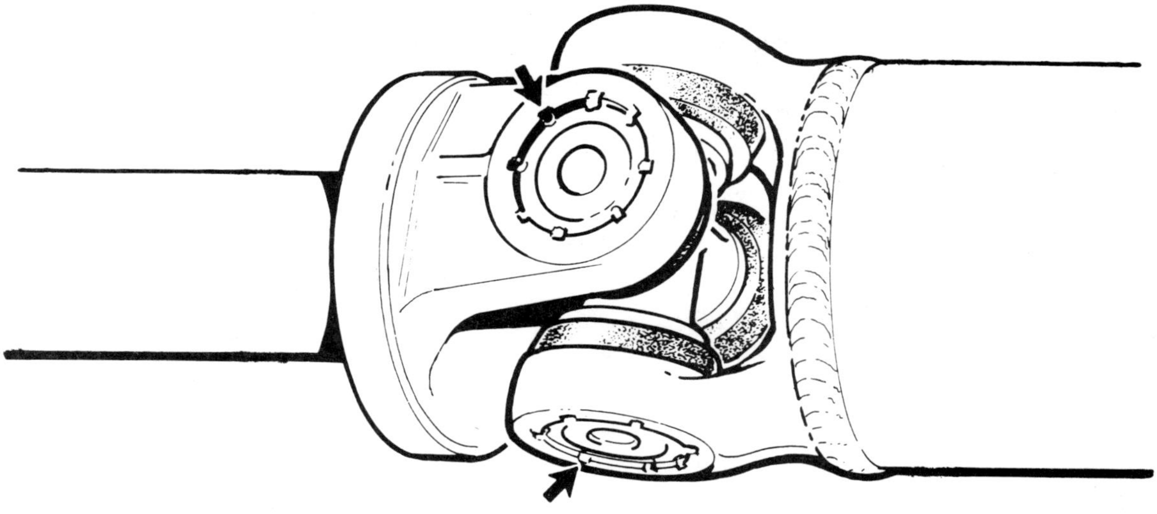

Fig 7:3 Universal joint showing bearings held in position by staking (arrowed)

Chapter 8 Rear axle

Contents

Specifications

Type	Hypoid - Semi-floating with overhung mounted pinion
Ratio	
1159 cc and 1256 cc Standard	8/33 (4.125:1)
1159 cc Extra performance	9/35 (3.89:1)
Oil Capacity	**1.85 pints (1976 models onwards)**
	1.25 pints (pre-1976 models)
Type	**SAE 90EP**
Half Shafts	
Permissible run-out on bearing land	.002 in maximum
Permissible run-out on flange face	.005 in maximum
Pinion	
Bearing pre-load - new	4–7 lb/in
- used	2–4 lb/in
Rear bearing - spacer thickness available	.059 in and .060 in
- shim thickness available	.003
Differential	
Side bearing pre-load - new	4 lb
- used	1¼ lb
Pinion fit on shaft	.0015 in to .0060 in clearance *(.0019 to .0056 in)
Pinion bore	.5528 in to .5567 in
Pinion thrust washer thickness	.016 in to .019 in
Pinion shaft diameter	.5507 in to .5513 in *(.6291 to .6297 in)
Side gear spacer thicknesses available	.019 in to .033 in in seven stages overlapping by .001
Differential case run-out permissible	.001 maximum
Side bearing shims available	.003 in
Side bearing spacers available	.122 and .123 in *(.050 and .051 in)

* 1256 cc engine models

Crown Wheel	
Run-out permissible	.002 in maximum
Backlash between pinion and crown wheel	.006 in to .008 in

Torque Wrench Settings	lb/ft
Differential side bearing cap bolts	24
Crown wheel to casing bolts	36
Half shaft bearing retainer plate nuts	13
Coupling flange to propeller shaft flange bolts	18

(ALL THREADS DRY)

1 General description

The rear axle is of the semi-floating type with a hypoid final drive. The pinion is overhung and contained in an extension to the axle housing

The rear axle casing assembly is located to the rear body members by means of four arms, two each side. Two of these are longitudinal and two diagonal, and their attachment points consist of steel bolts in rubber mounting bushes to hangers which are an integral part of the axle casing.

To control pinion to side gear backlash, thrust washers are used against spherical faces on the differential pinions and graded spacers are assembled to the side gears. The whole crown wheel and pinion unit is supported in the axle housing by two taper roller bearings secured by caps and bolts. The pinion runs in two pre-loaded taper roller bearings. The pinion is held in correct location to the crown wheel by spacers and shims between the front face of the rear bearing outer race and the abutment face in the axle housing.

An oil seal is pressed into the end of the pinion housing and operates directly on the pinion.

2 Rear axle - removal and replacement

1 Remove the hub caps and loosen the wheel nuts.
2 Raise and support the rear of the vehicle body, and remove the wheels. The easiest way to do this is to jack the car under the centre of the differential unit.
3 When it is sufficiently high, stands should be placed under the longitudinal lower suspension arms (Fig 8:4).
4 The front wheels should be chocked also to prevent any inadvertent movement. Under no circumstances should makeshift supports be used when doing work of this nature under the car.
5 Next, mark and disconnect the propeller shaft at the pinion flange as described in Chapter 7.
6 When the shaft is disconnected, tie it in position to one side rather than lower it, which could possibly impose a strain and cause damage to the gearbox extension housing bush and oil seal. If wished, the propeller shaft may be removed from the gearbox (see Chapter 7). Oil will not drain out of the gearbox provided the rear of the car is several inches higher than the front.
7 Disconnect the handbrake cable clevis pins from the brake shoe levers, having first slackened off the handbrake cable. For details refer to Chapter 9.
8 Disconnect the lower end of the flexible hydraulic fluid pipe where it is attached to a bracket on the rear axle.
9 Detach the lower end of each rear telescopic damper from the axle bracket by removing the two nuts and driving out the stud and bushes. For details see Chapter 11.
10 Remove the nuts and bolts from the longitudinal radius arm rear mountings on the axle, and then do the same with the diagonal arms where they locate to the brackets on the top of the differential casing.
11 The whole axle assembly is now free and, by pushing the hydraulic dampers out of the way, it can be drawn out straight back from the car.

3 Halfshafts - removal and replacement

1 Remove the hub caps, and loosen the wheel nuts.
2 Raise and support the side of the axle from which the halfshaft will be removed. Remove the wheel. NOTE: If both halfshafts are to be removed and the vehicle is raised level, oil may run out of the axle tubes. Precautions must be taken to prevent it running over the brake linings and the best way would be to remove the shoes as described in Chapter 9.
3 Release the handbrake. Undo the bolts which hold the brake drum to the half shaft flange and remove the drum.
4 Remove the nuts securing the half shaft bearing retaining plate. This must be done through one of the holes in the halfshaft flange. As each bolt is undone and removed, the flange must be rotated through 90° to the next bolt. Take care not to lose the four lock-washers.
5 The shaft can now be withdrawn and inspected. If the bearing is a very tight fit in the housing it will be necessary to borrow a slide hammer to fix to the flange to draw it out (Fig 8:5).
6 Carefully inspect the differential engagement splines for wear, and also the bearing and oil seal. If the oil seal shows any signs of failure it should be replaced. If the oil seal is to be replaced, the bearing will also have to be renewed because the oil seal is integral with the bearing.
7 Replacement of the halfshaft is a straightforward reversal of the removal procedure. See that the 'O' ring on the outside of the bearing is seated in its groove. However, before replacement, ensure that the oil drain hole in the brake backplate is clear and coat the halfshaft with oil. Also coat with oil the bearing bore in the axle halfshaft tube; this will allow easy replacement. The bearing retainer plate nuts should be tightened to 18 lb/ft with clean dry threads. Remember to adjust the brakes as described in Chapter 9.

4 Halfshaft bearings and oil seals - removal and replacement

If it is decided, after inspection, to replace the bearing and oil seal the procedure is as follows:
1 Slacken the bearing retainer ring by nicking it with a chisel. The retainer and bearing can be removed as one. As you will probably not have the correct Vauxhall pullers to remove the bearing, and the bearing is to be discarded, the following method of removal can be employed:
a) Clamp the bearing in a vice so that the halfshaft is parallel with the jaws.
b) Now using a hide hammer or mallet on the splined end of the halfshaft drive it back through the bearing and retainer. NOTE: A piece of wood MUST be interposed between the hammer and half-shaft.
2 Oil the new bearing journal and push the bearing down the halfshaft as far as it will go by hand. Ensure that the integral oil seal is facing towards the splines. Now drive the bearing right home against the shaft shoulder using a piece of steel tubing of a suitable length and diameter. Note that the tubing must only contact the bearing inner race, not the bearings, oil seal or outer race.
4 The bearing retainer can be driven home by the same method as the bearing, ensuring that the retaining ring collar faces the bearings.
5 Halfshaft replacement is described in Section 3.

5 Pinion oil seal - removal and replacement

If oil is leaking from the axle casing where the pinion emerges, it will mean that the oil seal needs renewal.
1 Raise the car, support it on stands and disconnect the propeller shaft as described in Section 2.
2 The pinion flange nut will now be exposed.
3 In order to hold the flange, when undoing the nut, it will be necessary to make up a piece of flat bar with two holes at one end which can be bolted to the flange. This can then be held firm while the socket wrench turns the nut which is, initially, a very tight fit.
4 It will be seen that the pinion nut is staked into a groove on the pinion flange and when the nut is removed this staking will break away. As the nut has to be retightened to exactly the same position afterwards, it will be necessary to grind another slot in the shaft to provide a staking position. If the nut is not tightened correctly the bearing pre-load will be upset (Fig 8:7).
5 Before pulling the flange off the pinion splines, mark the positions of flange to pinion so that it may be replaced in the same position.
6 The oil seal may then be literally dug out of the housing with a pointed punch and a hammer.
7 A new seal should be installed with the lip facing into the casing.

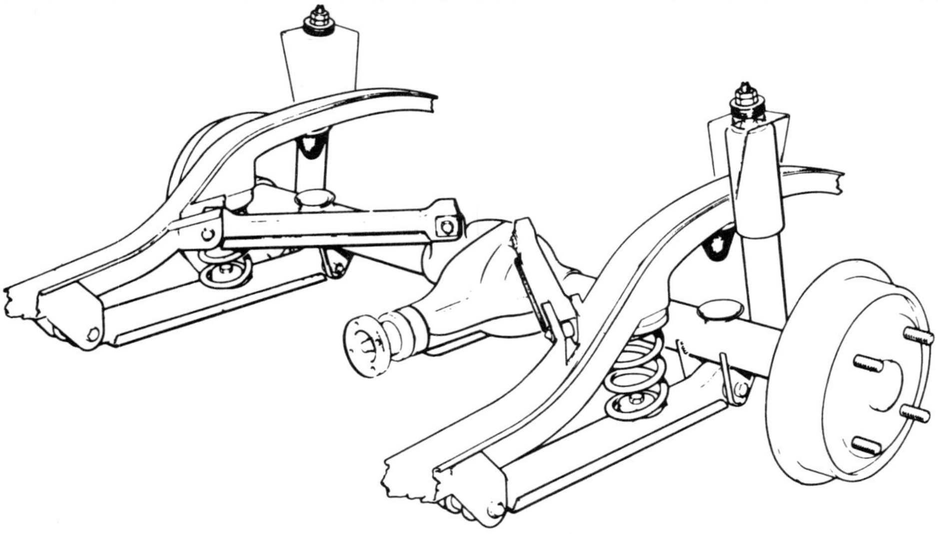

Fig 8:1 Rear axle and attachment points

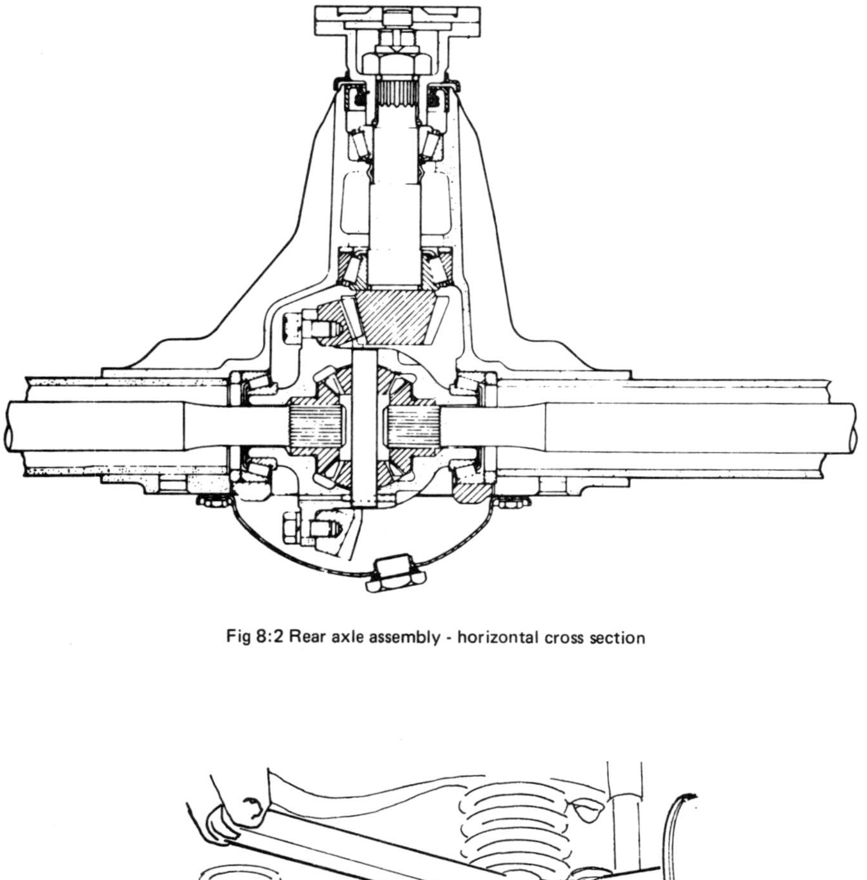

Fig 8:2 Rear axle assembly - horizontal cross section

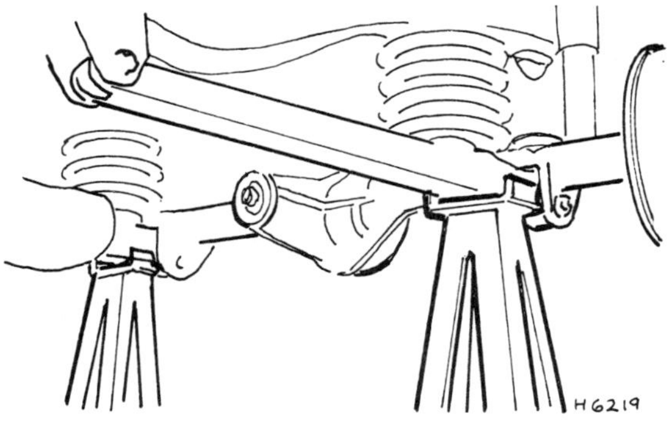

Fig 8:3 Supporting rear of car for axle removal

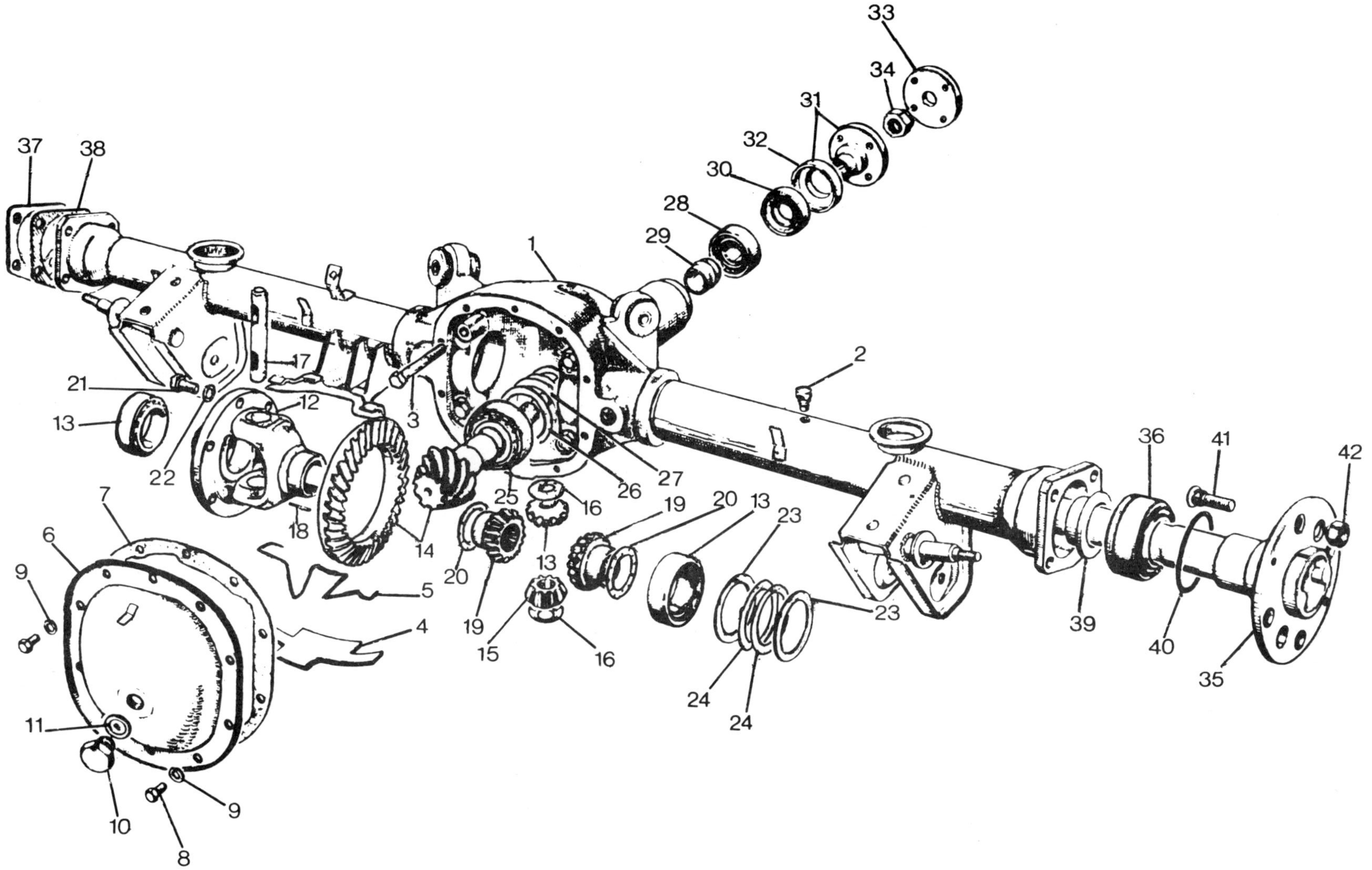

Fig 8:4 REAR AXLE COMPONENTS – EXPLODED VIEW

1 Axle housing
2 Vent plug
3 Bearing cap bolt
4 Filter pad
5 Retainer
6 Housing cover
7 Gasket
8 Cover bolt
9 Lockwasher
10 Filler plug
11 Gasket
12 Differential case
13 Side bearing
14 Crown wheel and pinion
15 Differential pinion
16 Pinion thrust washer
17 Pinion shaft
18 Shaft retaining pin
19 Side gear
20 Spacer
21 Crown wheel bolt
22 Lockwasher
23 Spacer
24 Shim
25 Pinion rear bearing
26 Pinion bearing spacer
27 Pinion bearing shim
28 Pinion front bearing
29 Collapsible spacer
30 Pinion oil seal
31 Flange
32 Oil slinger
33 Flange adaptor
34 Nut
35 Half shaft
36 Bearing
37 Bearing retainer plate
38 Gasket
39 Bearing retainer ring
40 Oil seal
41 Hub bolt
42 Wheel nut

Oil the seal lip and the pinion first and then carefully tap it home until it butts up to the bore face in the casing. It may assist in fitting the seal square if a piece of pipe, which will fit over the pinion, is used to drive it home.

9 Replace the pinion flange to the pinion ensuring the location marks made line up. Then fit the nut and tighten it to its previous position.

10 Finally, stake the nut rim into the new slot and reconnect the propeller shaft.

6 Pinion crownwheel and differential

This Chapter has so far shown how to replace bearings and oil eals for the halfshafts and the pinion oil seal, as these are considered o be within the average owner-driver's competence and facilities. We lo not recommend that owners go into the more complex problems of pinion to crownwheel settings, differential gear settings, differenial side bearing replacement, or pinion bearing replacement.

If, however, an owner feels he has the requisite tools and competnce we give the procedures to be followed. We shall refer to the pecial tools needed as it is not considered either sensible to attempt, r feasible to carry out, this work without them.

Remove the rear axle from the car and the halfshafts from the xle as described in Sections 2 and 3.

Remove the rear cover plate from the axle casing and drain the oil.

Remove the bolts holding the differential side bearing caps and, /hen removing the caps, note the marks on the right hand cap and ne casting to prevent inadvertent mixing up on replacement.

Place a bar under one of the differential case bolts and lever the asing assembly and crownwheel out of the housing.

Carefully remove the spacers and shims from each side of the ousing and keep the bearing inner races on the bench noting which de they are from.

Remove the pinion flange (see Section 5).

Tap out the pinion and remove the front inner bearing, bearing 'asher and compressible spacer.

0 Press the outer races of the pinion bearings from the housing using ne removers Z8545 and Z8546.

1 Press the pinion rear bearing inner race from the shaft.

2 Dismantle the hypoid gear and differential by first removing the ypoid gear from the differential casing, by removing the bolts. Tap ne gear off from alternate sides using a soft mallet.

3 Punch out the pin which locates the differential pinion shaft, and move the shaft.

4 The pinion can be lifted out with the hemispherical shaped thrust ashers followed by the side gears and spacers.

5 If the side bearings are to be replaced, draw them off the differenal case using a puller.

6 When new races are fitted ensure that they are pressed fully home the shoulder on the casing (using installer Z8552).

7 Fit a new filter pad in the axle housing.

8 Examine the pinion shaft for wear at the oil seal land. If badly orn, a new pinion (and therefore crownwheel as well) will be reuired. If only lightly scored clean up the shaft with very fine emery oth.

9 To reset the differential pinions and side gears in the differential sing, first dip all the components in oil to ensure initial lubrication hen first put back into use.

Replace the pinions together with new thrust washers and the side ars, each with a selected spacer.

Replace the pinion gear shaft, lining up the locating pin hole in the sing, but do not replace the retaining pin yet.

With a feeler gauge behind opposite sides of the sidegear spacers, eck the clearance between each of them and the case, which should .006 inch with no gear backlash evident.

Increase or decrease the spacer thickness accordingly. Spacers are ailable in seven thicknesses with a total range from .019—.033 inch clusive. It is permissible for the spacers used on each side to be of fferent thicknesses if necessary. When tolerances are correct, fit a new shaft retaining pin and punch the end flush with the casing. As a final check, the gears should turn by hand with the halfshafts inserted in the side gears.

24 Install the hypoid gear to the differential case by first warming the gear evenly (on a hot plate). It is best also to make up two guide studs that can be screwed into two opposite bolt holes in the gear. These will ensure that the register on the case fits into the gear, first time, squarely. Draw the gear on with the mounting bolts and new lockwashers and tighten to specified torque.

25 The lateral location of the differential case and hypoid gear assembly in the axle casing is controlled by spacers and shims. These also determine the side bearing pre-load and are available in two thicknesses of spacer, viz .050 inch and .051 inch and shims of .003 inch. These cover from .100 inch upwards, therefore, in steps of .001 inch, using two spacers for each bearing and the requisite shims (Fig 8:8).

26 Replace the differential case in the housing complete with side bearings, and then select four spacers and the appropriate number of shims which will remove all the end float between the bearing outer races and the ends of the axle housing tubes.

27 Remove the differential casing once more and then divide the spacers and shims equally into two lots, ie two spacers for each side with the necessary shims. Fit one lot on one side of the housing, ensuring that any shims are sandwiched between the spacers, and that the spacer chamfers are facing outwards.

28 Add one extra shim of .003 inch to the second lot, arranging them in the same way, and placing these at the other end of the axle housing, fit the casing back into the housing once more. Some pressure will be needed to force the assembly in and care must be taken to ensure the side bearings do not tilt and jam.

29 Tap the axle housing lightly near the bearings and rotate the assembly so that the bearings will settle properly. Then replace the bearing caps in the correct sides and tighten the bolts to the specified torque.

30 The pre-load on the bearings can now be checked by measuring the torque resistance at the periphery of the crownwheel. This is simply done by tying a piece of string around the crownwheel and measuring the turning resistance with a spring balance. If the reading is outside the 'Specifications' given at the beginning of this Chapter, then the unit must be removed and the shim thicknesses adjusted accordingly (Fig 8:9).

31 Check that the run-out of the hypoid gear rear face does not exceed .002 inch on a clock gauge micrometer. If it does, it indicates that dirt or burrs may have affected it when being reassembled to the differential case.

32 Once the differential has been satisfactorily fitted and checked, remove it once more, keeping the shims and spacers carefully for final assembly on their respective sides.

33 The pinion fitting is somewhat more complex and there are four factors which determine the initial selection of pinion spacers and shims to control the pinion/crownwheel mesh. These are:

a) The pinion bearing correction - being the variance from the maximum of .8445 inch of the thickness of the pinion rear bearing - is determined by measuring the actual thickness of the rear pinion bearing on a special jig plate, with a micrometer.

b) The pinion meshing correction - being the variance from standard required in production assembly - is stamped on the nose of the pinion in single figures representing thousandths of an inch.

c) The designed basic spacer thickness - being the standard spacer used in production assembly of .1230 inch.

d) The axle housing correction - being the deviation from the nominal depth of the bearing abutment face in relation to the centre line of the crownwheel axis. This figure is stamped on the axle housing rear face at the top - the single figure representing thousandths of an inch.

34 The calculation of the pinion spacer/shims required is best indicated using an example:

Pinion rear bearing maximum thickness	.8445 inch
Pinion rear bearing actual thickness	.8410 inch
Difference (pinion bearing correction)	.0035 inch

Fig 8:5 Half shaft removal using a slide hammer

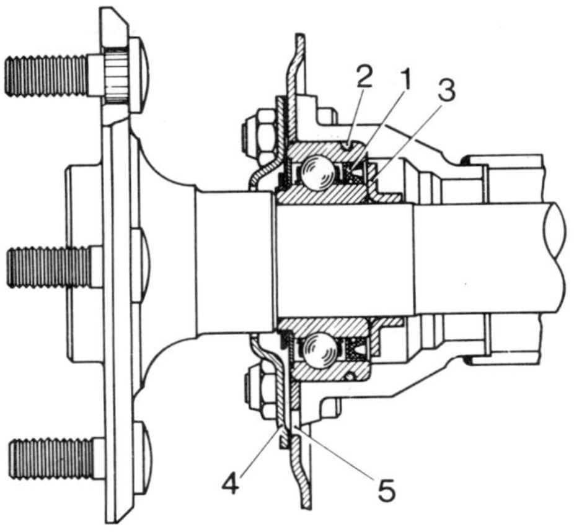

Fig 8:6 HALF SHAFT BEARING CROSS SECTION
1 Bearing seal
2 Bearing bore sealing ring
3 Bearing retainer ring
4 Retainer plate
5 Oil drain hole

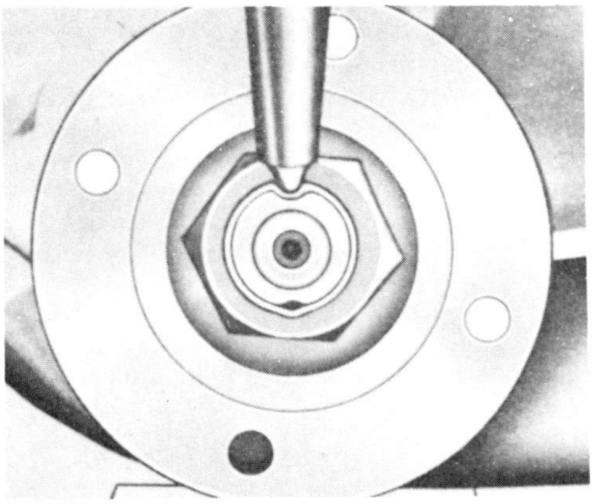

Fig 8:7 Staking a new part of the pinion flange nut into a new groove. Note how nut is aligned on the old groove and staking

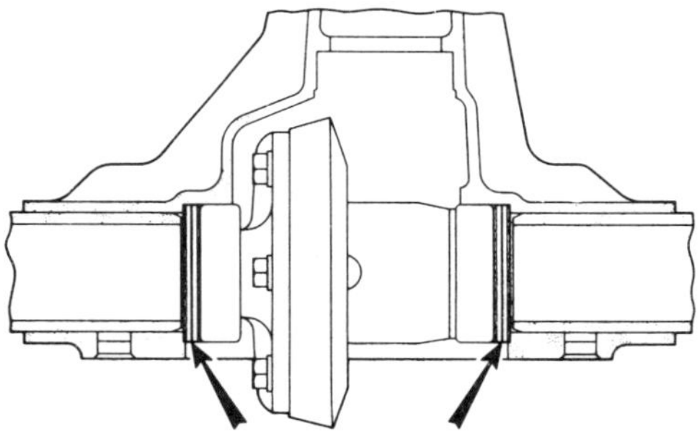

Fig 8:8 Location of side bearing pre-load shims

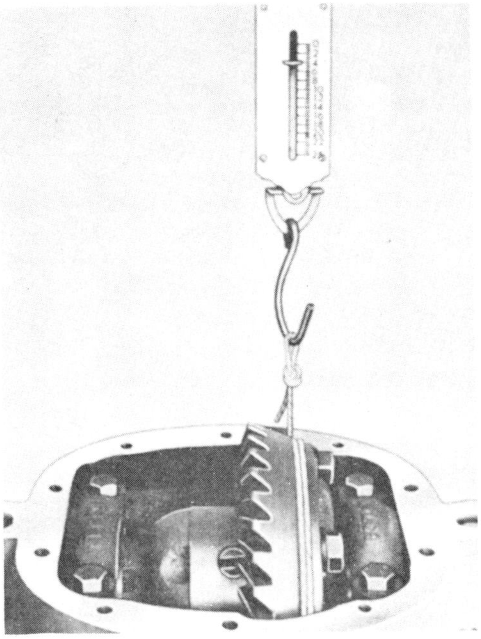

Fig 8:9 Checking differential side bearing pre-load with a spring balance

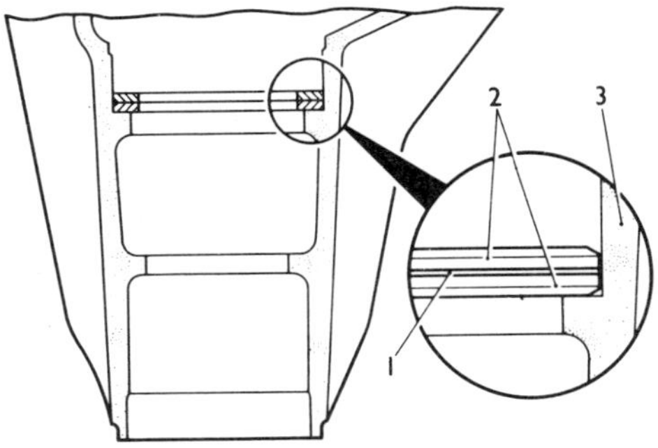

Fig 8:10 LOCATION OF PINION SHIMS BETWEEN INNER BEARING OUTER RACE AND CASING
1 Shims
2 Spacers
3 Casing

Axle housing correction '3'	.0030 inch
Pinion meshing correction '5'	.0050 inch
Basic spacer thickness	.1230 inch
Total spacer thickness required	.1345 inch

Spacers are available in thicknesses of .059 inch and .060 inch and shims in a thickness of .003 inch. In this case, therefore, one could select:

2 spacers (.060)	.1200	OR	1 spacer	.0600
5 shims (.003)	.0150		1 spacer	.0590
			5 shims (.003)	.0150
Total	.1350			.1340

It is better to select the larger thickness as this will reduce backlash fractionally, rather than increase it.

35 Having checked each shim and spacer with a micrometer, the shims should be sandwiched between the spacers as before and placed on the pinion rear bearing abutment face in the axle housing.

36 The pinion rear bearing outer race should then be pressed firmly into place in the housing using installer No Z8544 followed by the front bearing outer race using installer Z8543.

37 Having fitted the pinion rear bearing inner race to the shaft with installer Z8553, then fit a new compressible spacer and bearing washer over the shaft. Place the shaft into the axle housing.

38 Next place the front bearing inner race over the shaft and, whilst supporting the end of the pinion, tap it home on the shaft using installer Z8537.

39 Now replace the pinion flange and fit a new pinion nut lubricated with rear axle oil. Tighten it until a positive resistance is felt, indicating that all end float is taken up between the front bearing inner race, bearing washer and compressible spacer.

40 Using a torque pre-load gauge fitted to the pinion shaft, the pinion nut should now be tightened gradually, (compressing the spacer as it does so) until the pre-load reading is within specification.

41 If the pre-load is exceeded it will be necessary to fit a new compressible spacer and start again.

42 The differential case/crownwheel assembly, together with the selected shims, should now be installed in the axle housing as previously described. If there is an odd number of shims, the greater number should be put between the right hand pair of spacers.

43 Check the backlash between crownwheel and pinion. If incorrect, then the shims can be moved from one side to the other (fitting them always between the spacers), until it is correct. On no account alter the total number of shims and spacers originally selected, or the side bearing pre-load will be altered.

44 With the backlash correct, according to specification, the bearing caps can be replaced and tightened down to specified torque, after which backlash needs to be rechecked.

45 A final test for correct pinion/crownwheel location can be carried out by applying a load to the crownwheel, and driving it by turning the pinion so that marks will be made on the teeth. Use engineer's blue to emphasise the marks if necessary (Fig 8:14).

46 If the pinion is too far out (ie too far away from the crownwheel) the marks will be on the peaks of the crownwheel teeth, whereas if the pinion is too far in, they will be in the valleys.

47 Carefully mark the position of the pinion nut and take if off once more and remove the flange. Fit the new oil seal in the casing. Replace the flange and nut, tightening up to the same position in order to get the same bearing pre-load. The nut may then be staked into the slot in the pinion.

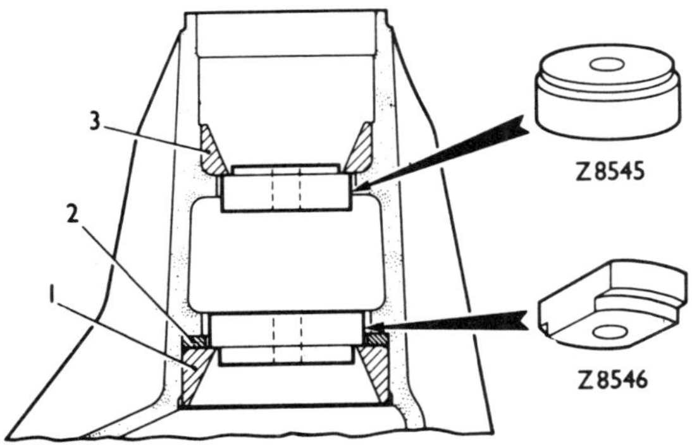

Fig 8:11 Special tools for removing pinion bearing outer races

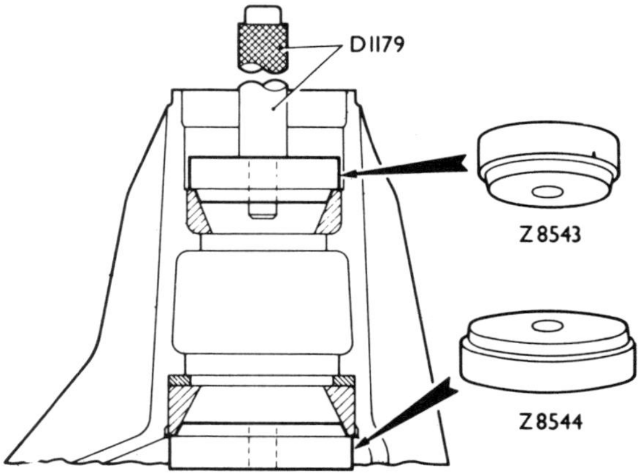

Fig 8:12 Special tools for installing pinion bearing outer races

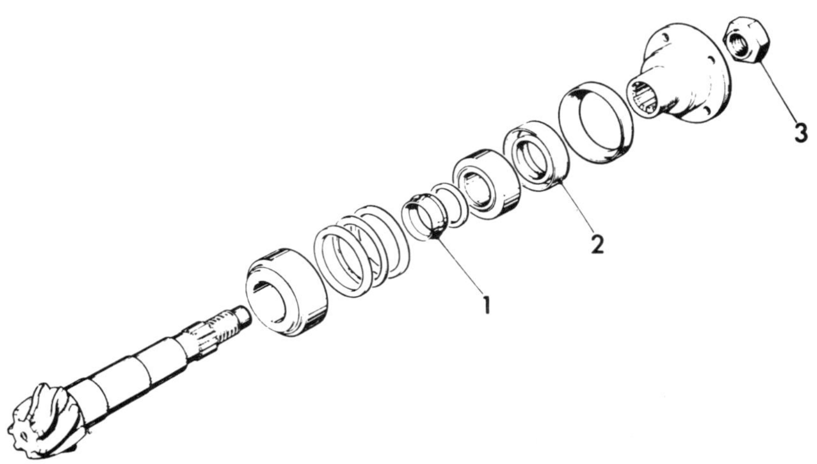

Fig 8:13 PINION ASSEMBLY
1 Compressible spacer * 2 Oil seal * 3 Pinion flange nut *
* Must be renewed on reassembly

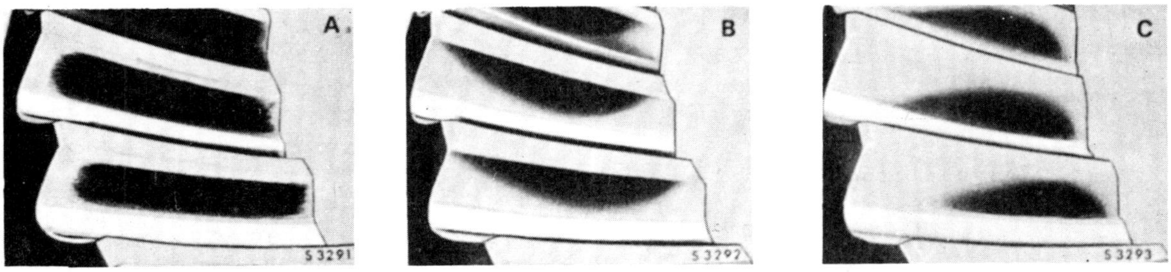

Fig 8:14 CROWN WHEEL TEETH MESHING MARKS
A Correct B Pinion too far out (increase pinion shims) C Pinion too far in (decrease pinion shims)

Chapter 9 Braking system

Contents

Specifications

Type	
Girling or Lockheed	Hydraulically operated drums all round or servo-assisted system with discs at the front. Cable operated handbrake to rear wheels.
Brake Drums	
Internal diameter	8.000 in to 8.005 in
Maximum internal diameter	8.06 in
Maximum run-out of front drums on hubs	.002 in
Maximum run-out of rear drums on halfshafts	.004 in
Disc Brakes	
Disc thickness	.375 in to .380 in
Maximum run-out of discs	.004 in
Friction pad minimum thickness	.06 ins (1½ mm)
Torque Wrench Settings	lb/ft
Front brake flange plate to steering knuckle nuts	25
Rear brake flange plate to rear axle nuts	13
Disc caliper to steering knuckle bolts	33
Lockheed caliper bolts	37
*Disc to hub bolts	18

*Use Loctite on bolt threads - all others clean and dry.

1 General description

The standard models of the Viva HC are fitted with hydraulically operated, internally expanding drum brakes on all wheels. The drums are 8 inches in diameter and equipped with two leading shoes on the front wheels, and one leading and one trailing shoe on the rear wheels. The pedal operated hydraulic master cylinder is located behind the brake pedal. It has an integral fluid reservoir which protrudes through the bulkhead into the engine compartment. A dual line type is fitted ensuring that a break in the hydraulic system does not make all brakes automatically useless. Each of the front wheels has two fixed single ended cylinders, one operating each shoe. Each rear wheel has one hydraulic cylinder, single ended but floating, operating the trailing and leading end of both shoes in each drum.

On four-door de luxe and SL models, disc brakes are fitted to the front wheels, and the hydraulic pressure to all wheels is boosted by a vacuum-servo unit interposed in the system between the master cylinder and the wheel cylinders. Disc brakes are optional extras on the standard models.

The parking or emergency brakes are mechanically operated to the rear wheels only, by a hand operated lever mounted on the floor between the front seats. The lever is connected to the rear wheel brake shoes by a stranded steel cable, which can be adjusted, and operating links in the rear wheel drums. All drum brakes are equipped with adjusters to the shoes which bring them nearer to the drums when the linings of the shoes start to wear down. On front wheels with drum brakes, these adjusters are the cam type, one to each shoe, and the rear wheels have the conical ended screw type, one adjuster serving both shoes in each rear drum. None of the brake shoes is pivoted on a fixed arc, each end being located in a groove in the cylinders and adjuster assemblies. By this means all shoes are automatically self-centring on the drums when applied. Disc brake pads are automatically adjusted to the discs by the hydraulic systems.

The basic principles of operation of the hydraulic brake system are as follows:

The brake pedal, when depressed, operates the plunger of a pump containing hydraulic fluid (the master cylinder) and forces this fluid along small diameter pipes, both rigid and flexible, to a series of small cylinders located at each wheel. The hydraulic fluid pressure at these cylinders (wheel cylinders) forces their pistons outwards. These pistons are connected to the ends of the brake shoes which are then forced against the drums, thereby applying the brakes. When the brake pedal is released, the shoes are drawn off the drums by springs which link the pairs of shoes together inside each wheel drum.

With disc brakes the conventional drum is replaced by a disc, against each side of which pads of friction material are forced by hydraulic pressure from a caliper - rather like gripping a gramophone record between a thumb and forefinger.

As the pads wear, so the hydraulic piston which forces them against the disc advances further towards the disc and obviates the need for adjustment. There is no return spring of any sort for the brake pads, so that when the hydraulic pressure is relieved, the fractional reverse movement of the pistons relieves the pads sufficiently to clear the disc surfaces. This automatic adjustment is the reason why an additional reservoir of hydraulic fluid is required. The rear wheel brakes and handbrake on models fitted with front disc brakes, are identical to those fitted on standard models.

In addition to the disc brakes where fitted, a vacuum-servo unit is installed. This unit uses the vacuum of the inlet manifold of the engine to operate what is, in effect, another pump to apply pressure to the hydraulic system. This reduces the pressure required on the conventional brake pedal when operating the brakes.

Although brakes fitted are made by either Girling or Lockheed, the basic design principles are the same. It is necessary to know what type they are in order to obtain the correct spare parts.

2 Adjustment - drum brakes

1 If the pedal travel becomes noticeably excessive before the brakes operate, and presuming that the pedal pressure is still firm and hard when pressure is applied, then the brake shoes need adjustment. This will be necessary on average about every 2–3000 miles.

2 Adjust the front wheels first. Jack up the car so that one front wheel is just clear of the ground and spins freely.

3 Behind the brake backplate are two square headed adjusters, one at the top and the other at the bottom of the backplate. With Girling brakes, turn the top one clockwise (using a square headed ring spanner preferably, to prevent burring of the screw head), until the shoe is locked tight on the drum. Then release the adjuster in the opposite direction for two notches (which can be felt when turning it). Spin the wheel to ensure the shoe is not binding on the drum. Repeat this process with the lower adjusting screw. With Lockheed brakes the process is the same except that the adjuster must be rotated in the direction of wheel rotation. This means, therefore, that on the right front wheel they are turned anticlockwise.

4 Lower the wheel to the ground and repeat the full adjustment process for the other front wheel.

5 On rear brakes the adjusters are turned clockwise on both sides. The difference is that the Girling one is at the top of the backplate, and for Lockheed it is at the bottom.

Release the handbrake, block the front wheels and jack up the rear wheels in turn. The single adjuster also has a square head and is located at the bottom of the brake flange and towards the rear of the car. Turn the adjuster clockwise until the brake shoes lock the wheel, then reverse it two notches as for the front wheel adjusters. Turn the wheel to ensure that the shoes are not binding on the drum. The rear wheels will not spin quite so freely as the front ones because the differential gear and propeller shaft will be revolving with them, so do not confuse this turning resistance with brake drag. Repeat the adjustment process with the other rear wheel.

6 It is often possible that a little shoe rubbing can be detected even after the adjusters have been slackened off the required two notches. Provided the degree of drag in such instances is negligible, ignore it. The shoes will bed down into their new positions on the drums after a mile or two. If there is serious binding after the adjusters have been slackened off two or more notches it will be necessary to remove the drum and examine the shoes and drums further.

3 Adjustment - disc brakes

1 Disc brakes are fitted to the front wheels only and do not require adjustment.

2 Drum brakes are fitted to the rear wheels of all cars fitted with front disc brakes and adjustment should be carried out in accordance with Section 2.

4 Drum brakes - removal, inspection and replacement of drums and shoes

1 If the brakes are inefficient, or the pedal travel excessive and the hydraulic system is showing no signs of leaks, first try adjusting the brakes. If little or no improvement results it will be necessary to examine the drums and shoes.

2 It is possible to attend to each wheel individually, so start with the front and remove the hub cap and slacken the wheel nuts. Then jack up the car and remove the nuts and wheel.

3 Next undo the two locating bolts which position the drum on the front hub (the purpose of these bolts is to hold the drum tight in position when the wheel is off. The full load of the braking force applied to the drum is carried by the four wheel studs and the drum is only fully clamped when the wheel is on) (photo).

4 Now draw the drum off the studs. A light tap around the periphery with a soft mallet will help to start it moving if it tends to stick at the roots of the studs.

5 It is also possible that the drum, although loose on the studs, is restricted by the shoes inside from coming off. In this case slacken off the adjuster screws (Section 2), until the drum can be removed (photo).

6 Examine the friction surface on the interior of the drum. Normally this should be completely smooth and bright. Remove any dust with a dry cloth and examine the surface for any score marks or blemishes. Very light hairline scores running around the surface area are not serious but indicate that the shoes may be wearing out, or heavy grit and dirt have got into the drum at some time. If there are signs of deep scoring the drum needs reconditioning or replacement. As reconditioning will probably cost as much as a new drum, and certainly more than a good second hand one (obtained from car breakers), it is not recommended. In theory a drum should not be renewed without replacing the front hub assembly also, but in practice the variations of concentricity which may occur with matched drums do not significantly affect the braking efficiency unless you are particularly unfortunate and have hub and drum tolerances at the extreme limits.

7 Examine the brake shoes for signs of oil contamination, deep scoring or overall wear of the friction material. Deep scoring will be immediately apparent and will relate to any scoring in the drum. Oil contamination is evident where there are hard black shiny patches on the linings caused by the heat generated in braking which carbonises any oil that may have reached them. As a temporary measure, these areas can be rasped down but it is far better to replace the shoes. Normal wear can be judged by the depth of the rivet heads from the surface of the linings. If this is .025 in (.65 mm) or less, the shoes should be renewed.

8 To remove the brake shoes, first of all slacken the two cam adjusters (front brakes) or cone adjuster (rear brakes) anticlockwise until the contracting movement of the shoes (pulled inwards by the

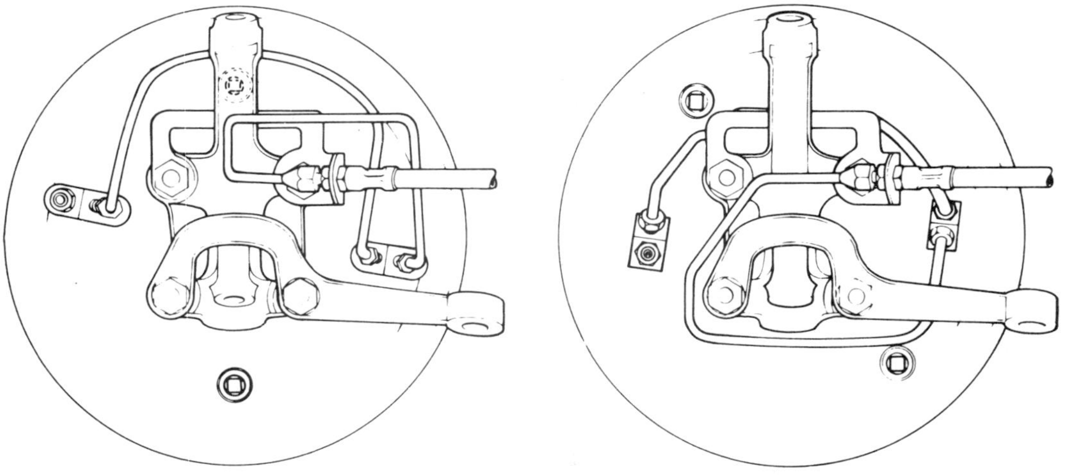

GIRLING LOCKHEED

Fig 9:1 Front brake backplates showing difference between Lockheed and Girling. Note positions of square-headed shoe adjusters

Fig 9:2 FRONT WHEEL BRAKE SHOE ASSEMBLIES
Left: Girling - Note springs hooked into backplate Right: Lockheed

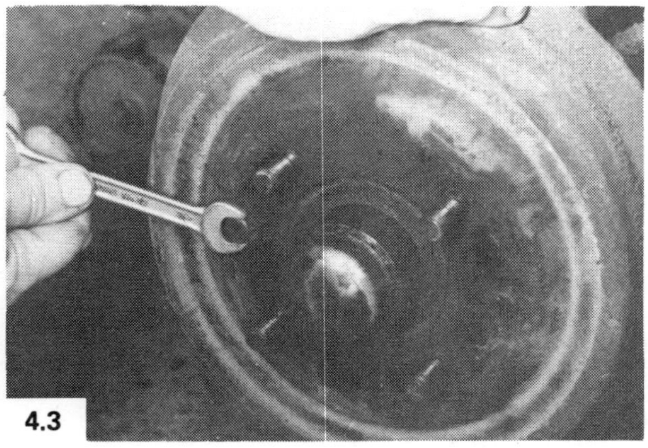

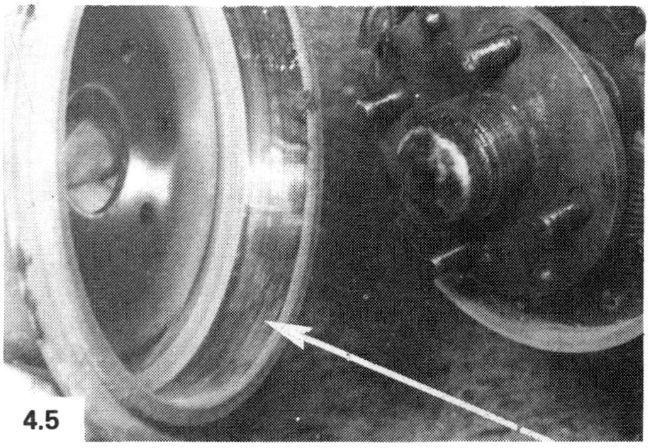

Removal of front brake drums and examination of friction surface (arrowed) for scoring

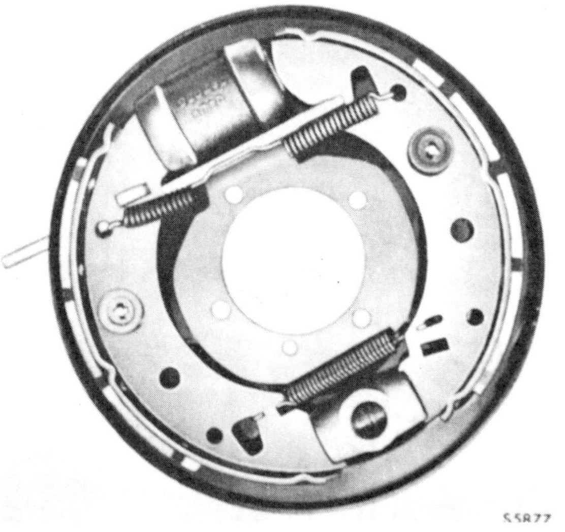

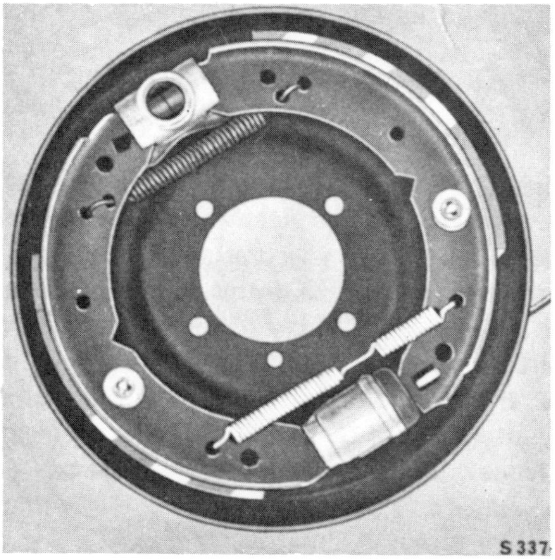

Fig 9:3 REAR WHEEL BRAKE SHOE ASSEMBLIES
Left: Lockheed Right: Girling
Note difference in position of pistons and adjusters

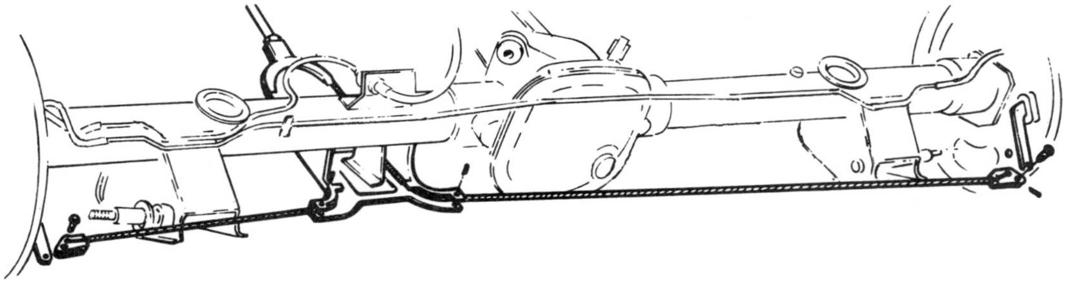

Fig 9:4 Handbrake cable loop and connections

Fig 9:5 Handbrake cable connections to rear wheel shoe levers

springs) ceases. On the rear brakes only, each shoe is held by a locating pin and spring in the centre of each shoe. With a pair of pliers turn the slotted washer on the pin so that it comes off the end of the pin. The spring and pin can then be removed.

9 Continuing with the rear brakes, release one of the shoes from the adjuster by levering it away with a screwdriver. Once this spring tension is released it will be quite easy to remove the shoes and springs.

10 With Girling front brakes, it is much simpler to remove the front wheel hubs before trying to remove the shoes, as the return springs are hooked into the backplate. Shoe replacement is virtually impossible with the hubs fitted, so, much time will be saved in the long run by turning to Chapter 11 and removing the hubs as detailed there. (It is easier with Lockheed brakes. On these the retractor springs are between the shoes rather than from the shoes to the backplate. It is possible to remove and replace these without removing the hub).

11 With the hub removed, release the trailing end of a shoe from a cylinder slot with a screwdriver. Both shoes can then be lifted off. As soon as any shoes are removed, make sure that the hydraulic cylinder pistons are prevented from coming out of the cylinders. This can be done by tying wire or even string around the slots to hold the pistons in.

12 Replacement of the shoes is a direct reversal of the removal procedure paying special attention to the following:

a) Ensure that the shoes and springs for the front brakes are reassembled exactly as shown in the illustrations - according to whether they are Lockheed or Girling.

b) Reassemble the rear brake shoes exactly as indicated in Fig 9:3 (Girling). The spring next to the hydraulic cylinder should have the smallest number of coils towards the piston end of the cylinder. Ensure the cylinder slides freely in its slot in the backplate. Lockheed rear brakes should be reassembled as shown in Fig 9:3. Note that Lockheed rear brake shoes are actuated by a double ended cylinder with two pistons. The shorter coil of the longer spring should be hooked into the hole near the parking brake lever.

c) Handle the brake shoes with clean hands. Even a small oil or grease deposit could affect their performance.

d) Apply a thin film of grease (Duckhams Keenol) to the backplates where the edges of the shoes rub against them.

e) If any shoe requires replacement it means that all shoes at the front or all shoes at the rear (ie a minimum of four shoes) be replaced together. Anything less can only lead to dangerous braking characteristics and uneconomical wear. Front and rear shoes are not interchangeable, except on those cars fitted with Lockheed brakes.

13 Replace the drums to the wheels from which they came and screw up the locating bolts.

14 Adjust the brakes as described in Section 2, and road test as soon as possible.

5 Handbrake - adjustment

1 Assuming that the rear brake shoes have been adjusted in accordance with Section 2, the handbrake should be fully on when five or six clicks of the ratchet have been pulled on. If more than this, then the cable adjuster locknut should be slackened and the adjuster screwed up until the desired condition is achieved. Tighten the locknut. There is only one adjuster, as the cable is a single loop with the ends at each wheel and the point of the loop on a sliding bridle at the handbrake lever. If the cable has stretched so much that the adjustment is fully taken up, it is possible to take up more by selecting an alternative pin hole at the brake lever clevis.

6 Disc brakes - removal, inspection and replacement of pads

1 The thickness of the pads can be visually checked by jacking up the car and removing each front wheel, when the pads can be seen between the disc and caliper body. If the thickness of the friction material is less than .06 inch (1½ mm) they must be replaced. Sometimes the pads wear unevenly, but if one of a pair is under specified thickness, the pair should be renewed. As a general rule pads on both front wheels should be renewed even if only one set needs it.

2 Remove the fluid reservoir cap and syphon out some fluid - say ¼ inch down. This will prevent the fluid overflowing when the level rises as the pistons are pushed back for fitting new pads.

3 To remove the pads, first pull the clips off the retaining pins and withdraw the pins. On Lockheed brakes the pins are split and spread to retain them in position. Also there are retaining spring plates behind them (photos).

4 Apply pressure to the faces of the old pads with the fingers, so pressing the pistons behind them back into the caliper. Then lift out the pads and shims from the caliper body (photos).

5 Refit new pads and new shims also if the old ones show signs of distortion or deterioration. On Girling brakes, make sure the shims are fitted with the arrow shaped hole in the top edge pointing in the direction of forward disc rotation. On Lockheed brakes, the shims are symmetrical and it does not matter (Fig 9:7).

6 Replace the locating pins and clips.

7 Depress the brake pedal two or three times to position the pistons and pads once more, and top up the fluid reservoir to the specified level.

7 Drum and disc brakes - hydraulic pipes, rigid and flexible - inspection, removal and replacement

1 Periodically, normally at safety check services, all brake pipes, pipe connections and unions should be completely and carefully examined.

2 First examine for signs of leakage where the pipe unions occur. Then examine the flexible hoses for signs of chafing and fraying and, of course, leakage. This is only a preliminary part of the flexible hose inspection, as exterior condition does not necessarily indicate the interior condition, which will be considered later.

3 The steel pipes must be examined equally carefully. They must be cleaned off and examined for any signs of dents, or other percussive damage and rust and corrosion. Rust and corrosion should be scraped off and, if the depth of pitting in the pipes is significant, they will need replacement. This is particularly likely in those areas underneath the car body and along the rear axle where the pipes are exposed to full force of road and weather conditions.

4 If any section of pipe is to be taken off, first of all remove the fluid reservoir cap and line it with a piece of polythene film to make it air tight, and replace it. This will minimise the amount of fluid dripping out of the system, when pipes are removed.

5 Rigid pipe removal is usually quite straightforward. The unions at each end are undone, the pipe and union pulled out, and the centre sections of the pipe removed from the body clips where necessary. Underneath the car, exposed unions can sometimes be very tight. As one can use only an open ended spanner and the unions are not large burring of the flats is not uncommon when attempting to undo them. For this reason a self-locking grip wrench (Mole) is often the only way to remove a stubborn union.

6 Flexible hoses are always mounted at both ends in a rigid bracket attached to the body or a sub-assembly. To remove them it is necessary first of all to unscrew the pipe unions of the rigid pipes which go into them. Then with a spanner on the hexagonal end of the flexible pipe union, the locknut and washer on the other side of the mounting bracket need to be removed. Here again exposure to the elements often tends to seize the locknut and in this case the use of penetrating oil or 'Plus-gas' is necessary. The mounting brackets, particularly on the bodyframe, are not very heavy gauge and care must be taken not to wrench them off. A self-grip wrench is often of use here as well. Use it on the pipe union in this instance, as one is able to get a ring spanner on the locknut.

7 With the flexible hose removed, examine the internal bore. If it is blown through first, it should be possible to see through it. Any specks of rubber which come out, or signs of restriction in the bore,

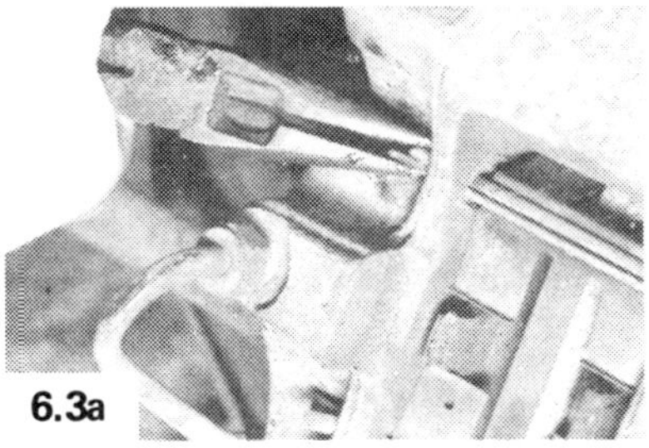
6.3a

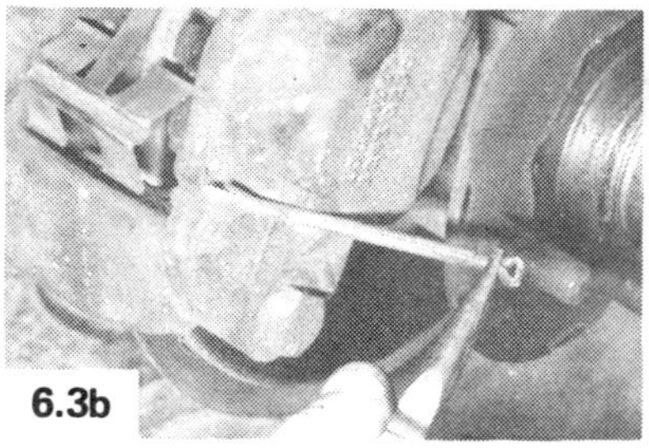
6.3b

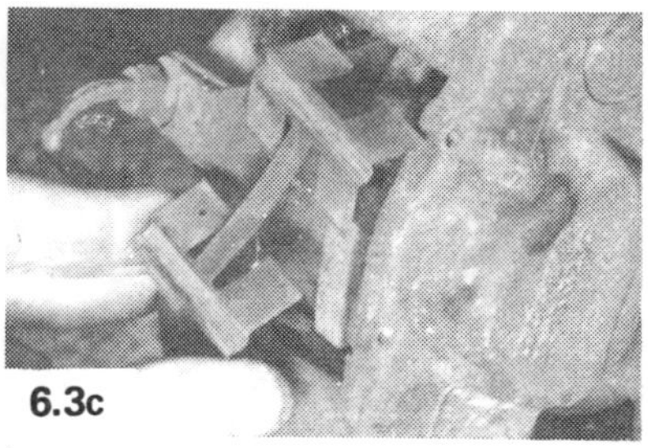
6.3c

Lockheed disc brakes: Removal of pad retaining pins and springs

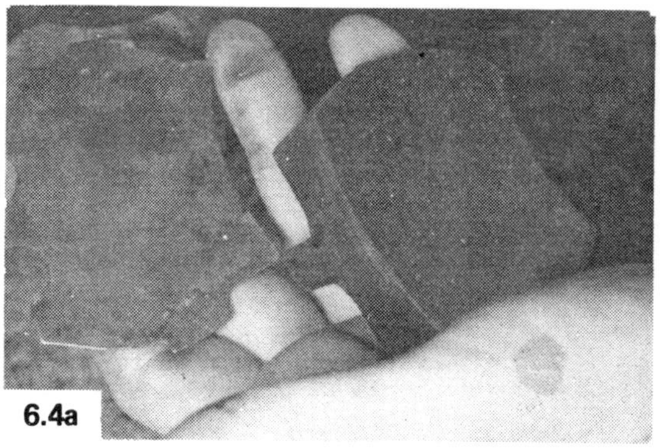
6.4a

6.4b

Lockheed disc brakes: Removal of pads and anti-squeal shims

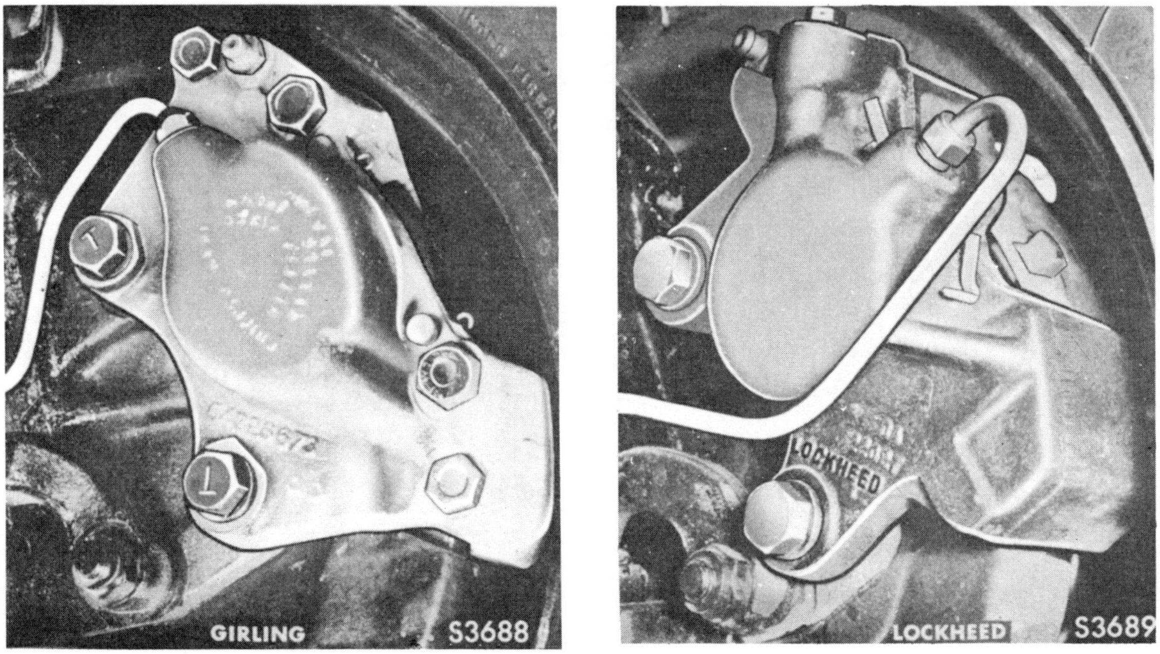

Fig 9:6 Disc brakes. Showing difference between two types fitted

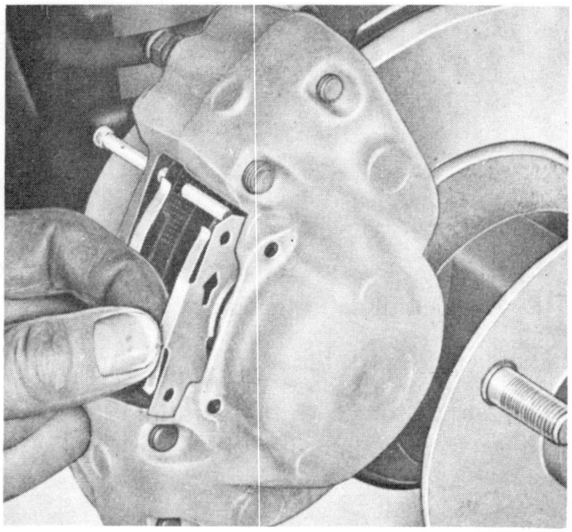

Fig 9:7 Disc brakes - Girling. Correct fitment of anti-squeal shims

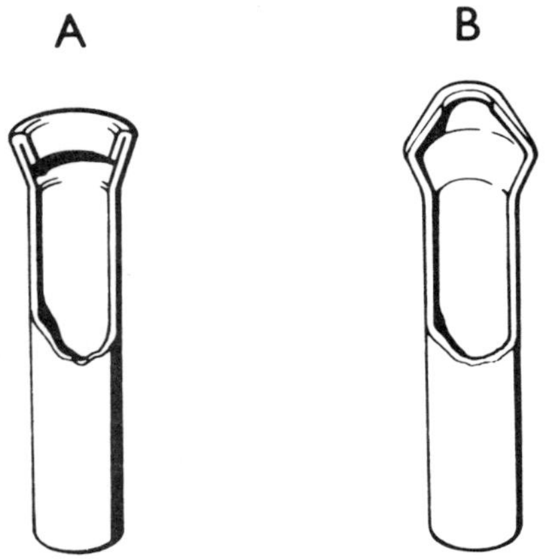

Fig 9:8 TYPES OF HYDRAULIC PIPE END FLARES
A Concave B Convex

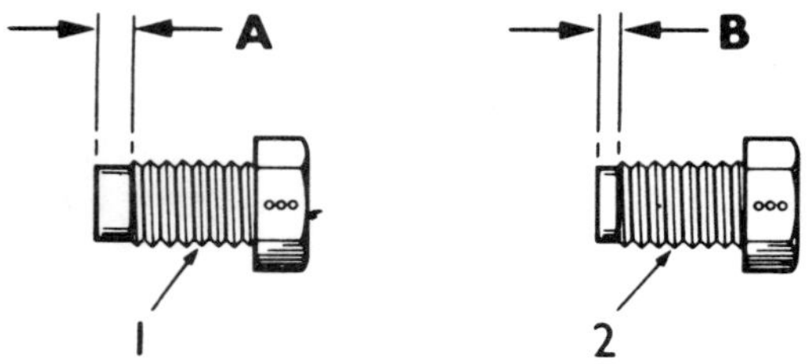

Fig 9:9 REAR BRAKES HYDRAULIC TRANSFER PIPE UNIONS
1 RH A = .012 inch 2 LH B = .07 inch

mean that the inner lining is breaking up and the pipe must be replaced.

8 Rigid pipes which need replacement can usually be purchased at any local garage where they have the pipe, unions and special tools to make them up. All they need to know is the total length of the pipe, and the type of flare used at each end with the union. This is very important as it is possible to have a convex flare at one end and a concave flare at the other (Fig 9:8).

Note also that on Lockheed brakes, the pipe linking the rear wheel cylinders has a different union at each end. This difference is in the length of the unthreaded portion as shown in Fig 9:9.

9 Replacement of pipes is a straightforward reversal of the removal procedure. If the rigid pipes have been made up it is best to get all the sets (bends) in them before trying to install them. Also if there are any acute bends, ask your supplier to put these in for you on a tube bender. Otherwise you may kink the pipe and thereby restrict the bore area and fluid flow.

10 With the pipes replaced, remove the polythene film from the reservoir cap and bleed the system as described in Section 14. It is not necessary always to bleed at all four wheels. It depends which pipe has been removed. Obviously if the main one from the master cylinder is removed, air could have reached any line from the later distribution of pipes. If, however, a flexible hose at a front wheel is replaced, only that wheel needs to be bled.

8 Drum brakes - hydraulic wheel cylinders, inspection and repair

1 If it is suspected that one or more wheel cylinders is malfunctioning, jack up the suspect wheel and remove the brake drum.

2 Inspect for signs of fluid leakage around the wheel cylinder, and if there are any, proceed with instructions at paragraph 4.

3 Next get someone to very gently press the brake pedal a small distance. On rear brakes, watch the wheel cylinder to see that the piston moves out a little, and on Lockheed brakes, check that the piston at each end of the cylinder moves out. On no account allow it to come right out or you will have to reassemble it and bleed the system. Then release the pedal and ensure that the shoe springs force the piston back. On front brakes, block the shoe on each cylinder in turn with a piece of wood and see that the other one moves in and out as the pedal is depressed and released. Do not let the piston move too far out.

4 A wheel cylinder where there is leaking fluid or which does not move at all (ie with a seized piston) will have to be fitted with new seals at least.

5 Remove the brake shoes as described in Section 4 and seal the fluid reservoir cap.

6 Remove the rubber dust cover and clip from the end of the cylinder, and draw out the piston (or pistons - Lockheed rear brakes), on the inner end of which a seal will be fitted. On front brake cylinders only draw out the spring behind the piston.

7 If the piston is seized in the cylinder it may be very difficult to remove, in which case it may be quicker in the long run to remove the cylinder from the wheel (see paragraphs 14–20).

8 Examine the bores of the wheel cylinders. Any sign of scoring or ridging in the walls where the piston seal travels means that the cylinder should be replaced.

9 If the cylinder is in good condition it will be necessary to replace only the seal on the piston. Pull the old one off and carefully fit a new one over the long boss of the piston, engaging it over the raised rim. The lip of the seal must face away from the centre of the piston.

10 Clean out the interior of the cylinder with a dry cloth and ensure the piston is quite clean. Then use a little rubber grease or brake fluid and lubricate the piston seal before replacing the spring (front only) and piston in the cylinder. Be careful not to damage or turn over the seal lip on replacement.

11 If the old seal shows signs of swelling and deterioration, rather than just wear on the lip, it indicates that the hydraulic fluid in the system may have been contaminated. In such cases all the fluid must be removed from the system and all seals replaced including those in

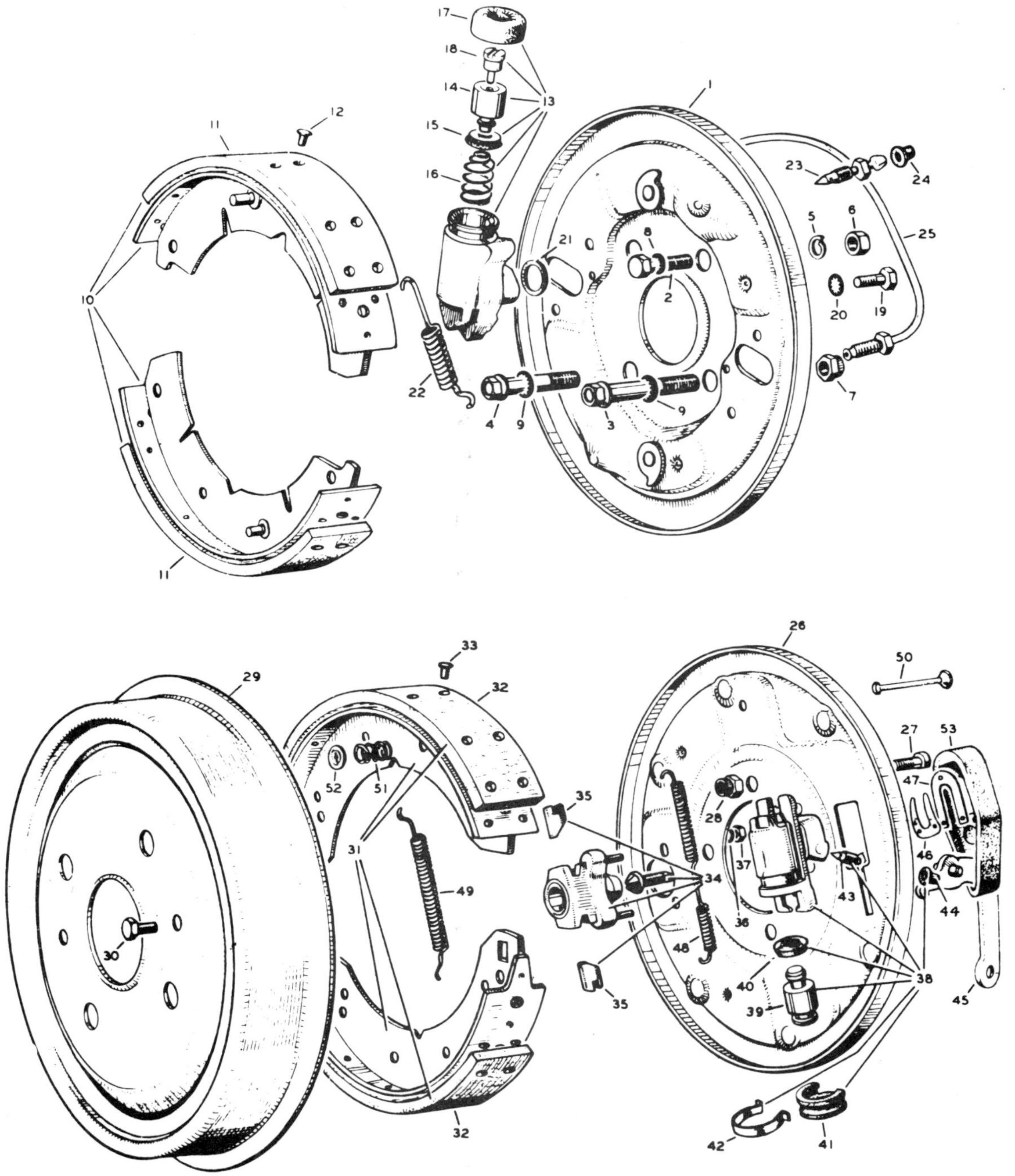

Fig 9:10 GIRLING DRUM BRAKES COMPONENTS – EXPLODED VIEW

1 Front drum backplate
2 Bolt
3 Bolt
4 Bolt
5 Lockwasher
6 Nut
7 Nut
8 Lockwasher
9 Lockwasher
10 Shoe and linings - front
11 Lining
12 Rivet
13 Cylinder assembly
14 Piston
15 Seal
16 Piston spring
17 Dust excluder
18 Piston insert
19 Cylinder fixing screw
20 Lockwasher
21 Cylinder sealing ring
22 Return spring
23 Bleed screw
24 Dust cover
25 Cylinder connecting pipe
26 Rear drum brake plate
27 Axle bolt
28 Nut
29 Brake drum (rear)
30 Bolt
31 Shoes (rear)
32 Linings
33 Rivet
34 Adjuster
35 Tappet
36 Washer
37 Nut
38 Cylinder
39 Piston
40 Seal
41 Dust cap
42 Spring clip
43 Bleed screw
44 Dust cap
45 Handbrake lever
46 Spring plate
47 Retaining plate
48 Retractor spring (double coil
49 Retractor spring (single coil)
50 Shoe securing pin
51 Pin spring
52 Cup washer
53 Dust excluder

Fig 9:11 Front brake cylinder fitting with a sealing ring (arrowed) on Girling brakes. Lockheed cylinders have a gasket

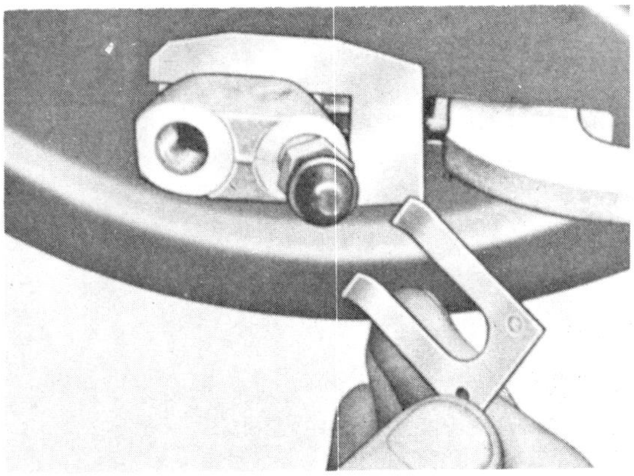

Fig 9:12 REAR WHEEL BRAKE CYLINDER – GIRLING
Spring plate in position and distance washer being inserted with closed ends near lever. Not shown is the retaining plate, fitted last, between the two, from the opposite end. The slots of the retaining plate engage the tags on the spring plate

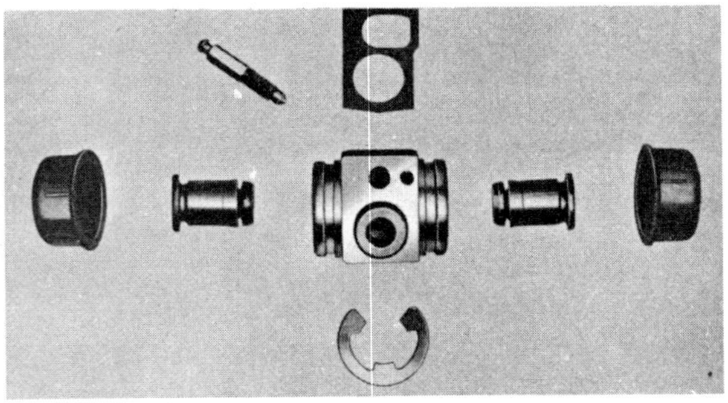

Fig 9:14 Rear wheel brake cylinder - Lockheed - exploded view

the master cylinder. Flexible hose should be checked too.
12 Replace the shoe guide clip (front) and dust cover.
13 Replace the brake shoes and drums, remove the fluid reservoir cap seal and bleed the hydraulic system (Section 14).
14 If the cylinders are to be replaced, unscrew the unions of the hydraulic pipes on the brake backplate.
15 On front brakes, undo the two cylinder fixing bolts and washers and remove the cylinder and sealing ring. Replace the cylinder with a new sealing ring. Continue reassembly as from paragraph 9, using new piston seals always (Fig 9:11).
16 On rear brake cylinders, undo the pipe union behind the backplate and also remove the bleed screw.
17 On Girling brakes, disconnect the handbrake cable from the lever by removing the clevis pin and then remove the dust cover.
18 Using a small screwdriver or spike, prise out the cylinder retaining plate followed by the spring plate and distance washer (Fig 9:12). With Lockheed rear cylinders there is an 'E' clip only to remove (Fig 9:14).
19 Lift out the cylinder assembly. Replacement is a reversal of the removal procedure. Continue reassembly as from paragraph 9, using new piston seals always.
20 Bleed the system (Section 14).

9 Disc brakes - caliper - removal, inspection and replacement

1 If the caliper pistons are suspected of malfunctioning, jack up the car and remove the relevant wheel.
2 Examine for signs of fluid leaks and, if these are apparent, it will be necessary to remove the caliper and proceed as described from paragraph 4 onwards.
3 If there are no signs of leaking, get someone to depress the brake pedal and watch how the two disc pads come up to the disc. One may move very slowly or not at all, in which case it will be necessary to remove the caliper and proceed further.
4 Remove disc pads and shims as described in Section 6.
5 Seal the reservoir cap with a piece of plastic film.
6 Undo the hydraulic pipe union from the body of the caliper and draw back the pipe.
7 Undo the two bolts holding the caliper to the steering knuckle plate. Do NOT undo the bolts which clamp the two halves of the caliper together.
8 Lift the caliper off the disc.
9 Clean the exterior of the caliper assembly and then ease each rubber piston cover out of the grooves in the piston and the caliper body, and remove them.
10 It may be possible to pull the pistons out of their bores, but if not, it will be necessary to blow them out with pressure from an air pump hose attached to the hydraulic fluid inlet port. Support one piston while the other is blown out and then block the empty cylinder with a cloth while the other comes out. If one piston moves very slowly, remove this one before the other. If one piston does not move at all, it will have seized in the cylinder. Use a hydraulic cleaning fluid, or methylated spirits, to soak it for some time in an attempt to free it. If harsher measures are needed, try to confine any damage to the piston, and not the caliper body.
11 With the pistons removed, the fluid seal rings may be eased out of the piston grooves with a small screwdriver. Make sure that the piston and groove are not damaged. Examine the bores and pistons for signs of scoring or scuffing. If severe, it is unlikely that a proper fluid seal will be possible and a new caliper assembly may be required. The part of the piston on the pad side of the seal groove may be cleaned up with steel wool if necessary. Take care to leave no traces of steel wool anywhere. Clean the cylinder bores also, using hydraulic cleaning fluid if possible, or methylated spirits otherwise.
12 Reassembly is an exact reversal of the dismantling process, taking care with the following in particular:
13 Ensure that the new fluid seal is seated properly in its groove.
14 The caliper mounting bolts have a nylon locking insert in the threads. If this is the third time of removal, then the bolts should be

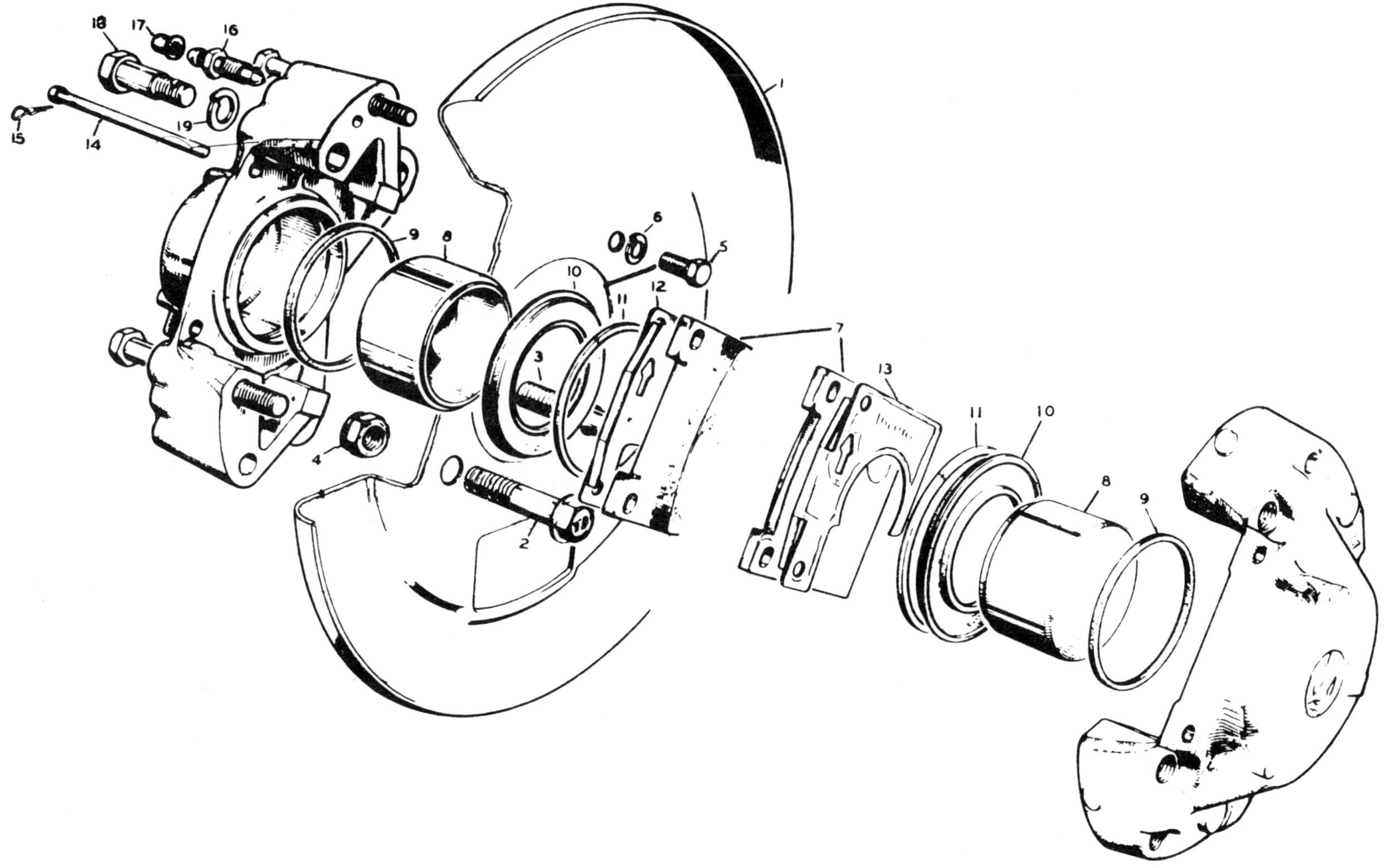

Fig 9:15 GIRLING DISC BRAKES – COMPONENTS – EXPLODED VIEW

1 Disc shield
2 Bolt, steering arm to knuckle
3 Bolt, steering arm to knuckle
4 Nut
5 Bolt, shield to knuckle
6 Lockwasher
7 Friction pad
8 Caliper piston
9 Cylinder sealing ring
10 Rubber boot
11 Retaining ring
12 Anti-squeal shim
13 Anti-squeal shim
14 Retaining pin
15 Clip
16 Bleed nipple
17 Dust cap
18 Bolt, caliper to knuckle
19 Lockwasher

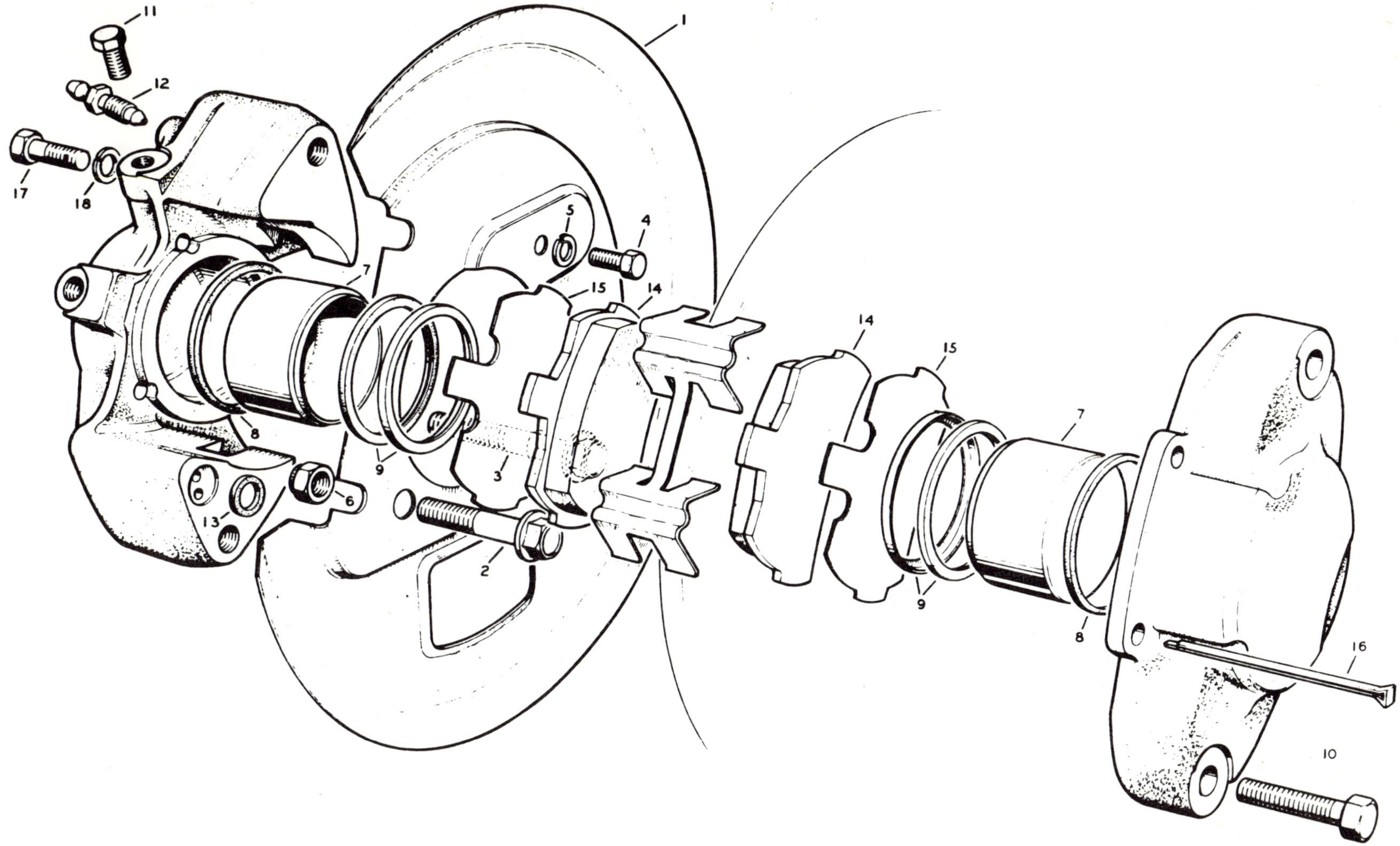

Fig 9:16 LOCKHEED DISC BRAKES – COMPONENTS – EXPLODED VIEW

1 Disc shield
2 Bolt to knuckle
3 Bolt to knuckle
4 Bolt to knuckle
5 Lockwasher
6 Locknut
7 Piston
8 Inner seal
9 Seal and retainer
10 Caliper bolt
11 Bleed drilling plug
12 Bleed screw
13 Fluid channel seal
14 Friction pad
15 Anti-squeal shim
16 Pad retaining pin
17 Caliper bolt to knuckle
18 Lockwasher

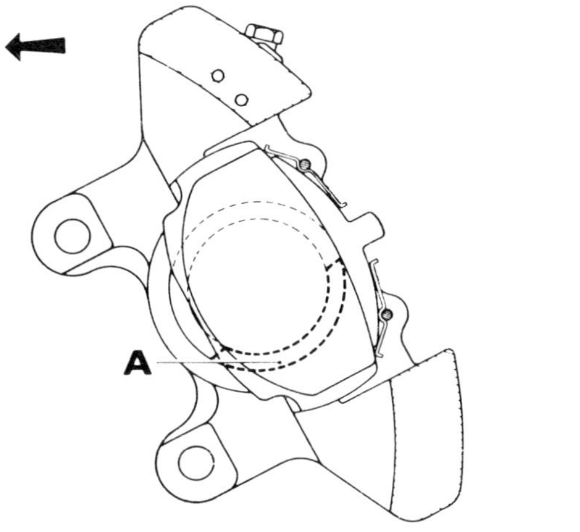

Fig 9:17 Lockheed disc brake caliper assembly showing relieved section of piston (A) in relation to front of car (arrowed)

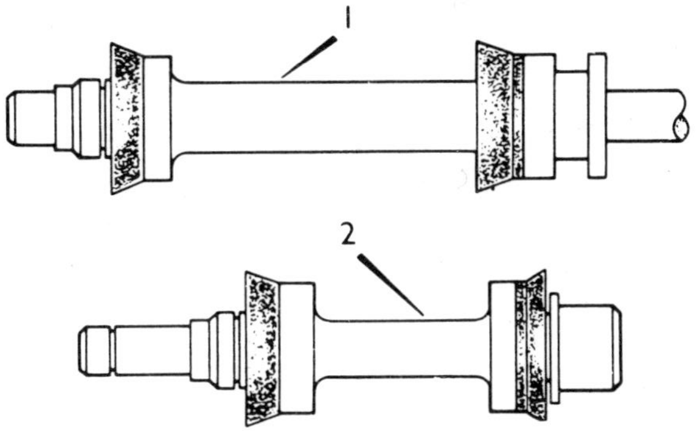

Fig 9:19 TANDEM MASTER CYLINDER PISTONS SHOWING ARRANGEMENT OF SEALS

1 Primary piston 2 Secondary piston

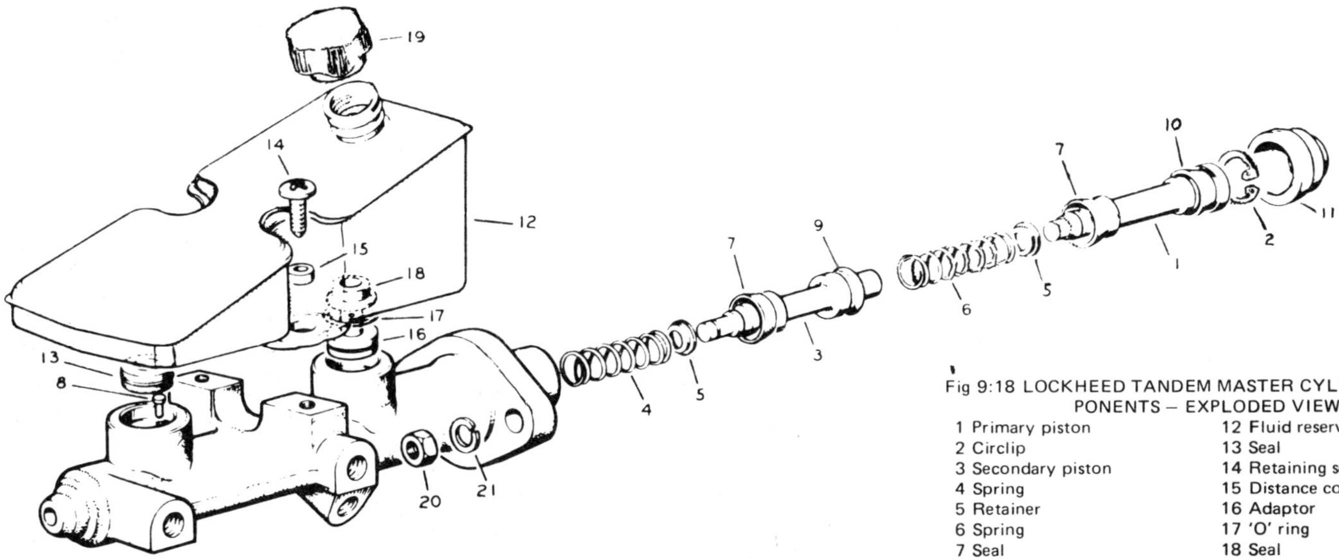

Fig 9:18 LOCKHEED TANDEM MASTER CYLINDER – COMPONENTS – EXPLODED VIEW

1 Primary piston
2 Circlip
3 Secondary piston
4 Spring
5 Retainer
6 Spring
7 Seal
8 Stop pin
9 Seal
10 Seal
11 Pushrod seal
12 Fluid reservoir
13 Seal
14 Retaining screw
15 Distance collar
16 Adaptor
17 'O' ring
18 Seal
19 Filler cap
20 Nut
21 Spring washer

renewed. Tighten the bolts to the specified torque of 33 lb/ft. Replace the pads as described in Section 6, remove the reservoir cap seal and bleed the system.

15 Cars fitted with disc brakes made by Lockheed are basically the same in operating principle. Two particular items to be noted on reassembly are the piston dust seal and the piston itself. The grooved ubber dust seal is located in a recess in the mouth of the cylinder ore and is kept there by a retaining ring. The piston has a cut-away ortion on its outer edge, and this must face down and rearwards as hown in Fig 9:17, when the piston is pressed back into the cylinder. Care must be taken, when refitting the dust seal and retainer, to ensure hat they go in square and undistorted.

0 Disc brakes, disc run-out check

If the disc does not run true, then it will tend to push the disc ads aside and force the pistons further into the caliper. This will ncrease the brake pedal travel necessary to apply the brakes, apart rom impairing brake efficiency and the life of the pads.

To check the disc run-out (trueness), jack up the car and remove he wheel. Ensure the hub bearing has no free float in it.

Set a dial gauge micrometer on a firm stand up to a friction ace of the disc, near the outside edge, so that a reading above 004 in is registered on the gauge.

Spin the hub and if the gauge registers more than ± .004 in., the isc is warped and needs replacement.

1 Drum and disc brakes - hydraulic tandem master cylinder

All models are fitted with dual braking systems and this is provided y a tandem master cylinder. Should the hydraulic system fail on ther the front or rear brakes, it means that the car may still be alted as the pressure in each system is independent of the other.

Unless there are obvious signs of leakage, any defects in the master ylinder are usually the last to be detected in a hydraulic system.

Before assuming that a fault in the system is in the master cylinder e pipes and wheel cylinders should all be checked and examined as escribed in Sections 7, 8 and 9.

To remove the master cylinder, first disconnect the pipes which ome from it. It should be noted that all pipe unions to the master ylinder have metric threads (10 mm) and are black in colour for entification.

Remove the clevis pin from the brake pedal which attaches the ushrod from the master cylinder.

Remove the bolts and nuts which hold the master cylinder to the edal support bracket and lift it out. (On models fitted with a servo it the master cylinder is bolted to the servo body. Take care not to ill fluid when draining the unit and drain it all into a container soon as possible afterwards).

Pull off the fluid reservoir from the cylinder body by lifting it aight up.

To dismantle the cylinder, remove the circlip from the end of the re. The primary piston may then be drawn out. On cars with a ser- unit, a spring follows and behind this is the secondary piston.

Before the secondary piston can come out, the rubber seal and op pin must be lifted out of the secondary inlet port. The pin is a ose fit, but the piston will need pushing forward a fraction against e spring to release it. The secondary piston and spring may then be oped out.

Examine the cylinder bore for signs of scoring or other deteriora- n and, if in doubt, renew it. It is assumed that the seals on the pis- ns will be renewed anyway - it would be pointless not to renew em.

When fitting new seals make sure that you do not get them uddled up. The front seals on the primary and secondary pistons not the same. See that the lips of the seals on the primary piston th face into the cylinder, whereas those on the secondary piston e outwards from the centre of the piston (Fig 9:19).

12 The piston spring(s) have washers fitted at one end and these should go over the spigots on the ends of the pistons.

13 Some master cylinders are fitted with a pressure warning lamp device. This is a piston maintained in equilibrium by the two halves of the pressurised system. If the pressure of one half should drop, the piston moves and operates a switch.

To remove the piston assembly, undo the plug in the end of the master cylinder and also the actuator switch. The piston can then be tapped out (Fig 9:20).

14 Reassembly of the cylinder is a reversal of the removal procedure. Take care to fit the pistons correctly, and do not let the lips of the seals turn back when guiding them into the cylinder. The cylinder bore should be lubricated with fluid.

15 The fluid reservoir should be fitted after the stop pin and seal have first been put into the secondary piston inlet port, and the adaptor with 'O' ring into the primary inlet port. The seal which fits over the adaptor is best put into the reservoir aperture first. The seals and pipe on the reservoir should be well lubricated with fluid before pressing the reservoir into place (Fig 9:21).

16 Replace the unit to the mounting as appropriate and, when con- necting up the outlet pipe unions, make sure that the rear brake connection goes to the single port at the end of the cylinder, the pipe from the left hand front brake to the upper of the two remaining ports, and pipe from right front brake to lower port.

17 Bleed the brake system and check for leaks at any of the unions.

12 Vacuum servo unit - general description

1 The vacuum servo unit is incorporated with a master cylinder and is operated directly from the brake pedal. It acts on all four wheels. When the pedal is depressed, suction from the induction manifold of the engine is applied to the piston inside the large cylindrical bowl of the servo unit. The piston then moves a plunger, attached to its centre, applying additional pressure to the hydraulic master cylinder piston. If the servo unit should fail to operate, pressure from the foot pedal is still applied to the system but, of course, the pedal pressure required will be more than it would be if the servo was functioning. The unit is mounted on brackets fitted to the pedal supports.

2 Servicing can be carried out but certain special tools are needed, although these can, of course, be made up by an enthusiast with reasonable facilities. The tools needed are for correct removal of the cover which has a bayonet type fixing into the shell, and a flange on which the master cylinder can be mounted to hold the whole unit firmly in a vice (Figs 9:23, 9:24, 9:25).

13 Vacuum servo unit - checking, dismantling, servicing and reassembly

1 The filter can be changed by drawing back the rubber boot from the end cover and withdrawing the retainer ring. The filter pad must be cut to remove it from the pushrod, as must the new one prior to fitting. Make sure the retainer is pressed back fully and the rubber boot properly engaged with five lugs on the end cover (Fig 9:26).

To remove the unit, disconnect the vacuum pipe and the hy- draulic pipes from the master cylinder and undo the mounting stud units.

2 Before beginning dismantling, obtain a complete repair kit which will contain the necessary seals and diaphragm needed to recondition the unit. If in any doubt, it may be preferable to obtain an exchange unit.

3 Having removed the unit from the car together with the master cylinder, attach the cover removal tool to the four mounting studs and clamp one leg of the tool in the vice. Undo the nuts securing the master cylinder and take it off. It is possible at this stage to check the setting of the servo pushrod. It is critical and needs careful measurement with a suitable straight edge or gauge. The correct projection is .408 inch and must be accurate to within .005 inch.

If incorrect, proceed with dismantling in order to adjust it. Remove

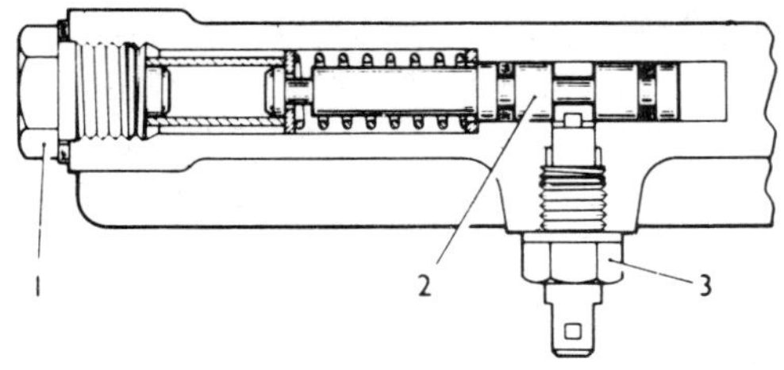

Fig 9:20 HYDRAULIC PRESSURE WARNING INDICATOR AND SWITCH
1 Plug 3 Switch
2 Actuator piston

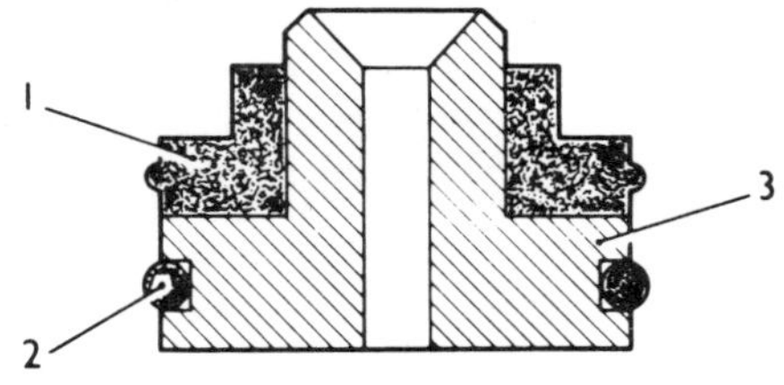

Fig 9:21 PRIMARY PISTON INLET PORT ADAPTOR AND SEALS – CROSS SECTION
1 Reservoir sealing ring 3 Adaptor
2 'O' ring

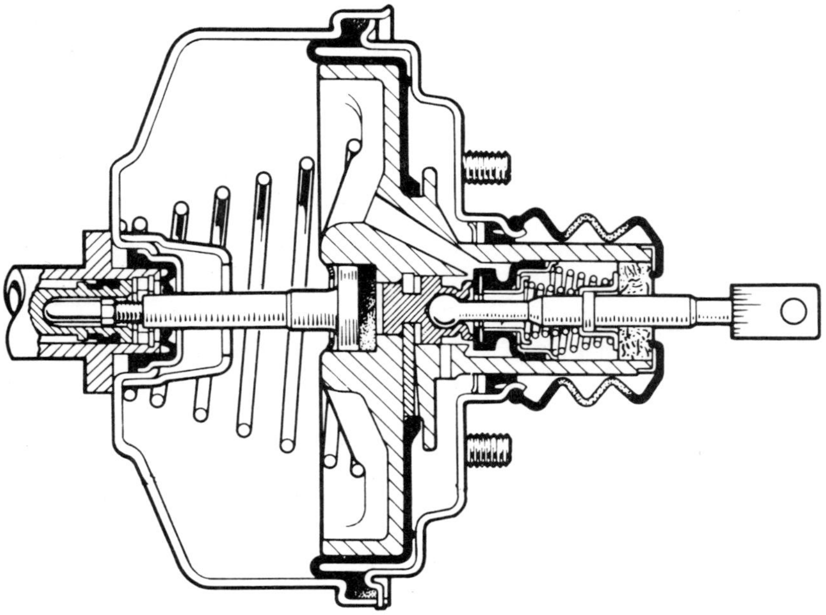

Fig 9:22 Servo unit - cross section

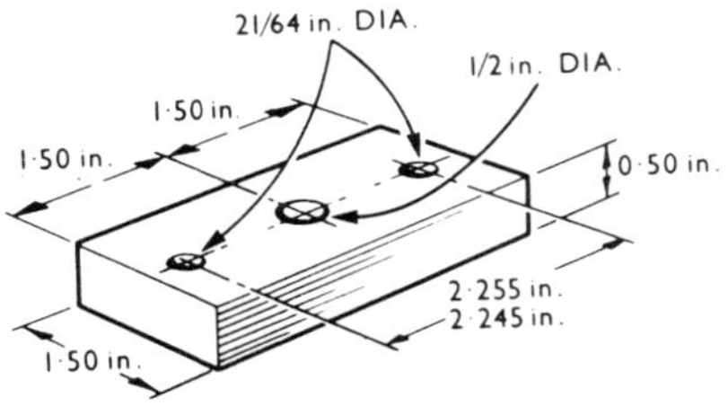

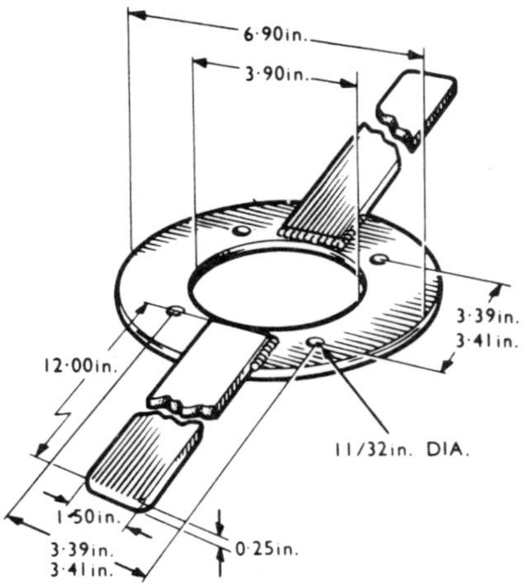

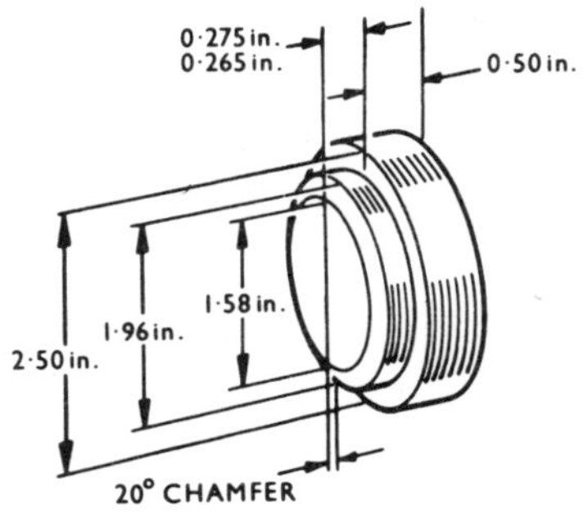

Figs 9:23, 9:24, 9:25 TOOL DIMENSIONS TO ASSIST SERVICE OPERATIONS ON THE SERVO UNIT
Left: Slave flange Centre: Cover removal tool Right: Cover seal replacement mandrel

the rubber mounting and then fit the slave flange on the studs in its place and tighten the nuts to 11 lb/ft.
4 Release the cover removal tool from the vice and clamp the whole assembly instead, by gripping the slave flange in the vice.
5 Mark the relative positions of the shell and end cover and turn the cover removal tool anticlookwise whilst maintaining pressure, until the cover can be released and the spring pressure relieved.
6 Remove the return spring stop from the shell. Leave the pushrod where it is.
7 Remove the slave flange from the master cylinder mounting studs and carefully push out the master cylinder seal and seal retainer. Take off the cover removal tool, and the pushrod retainer can be prised out of the valve body to release the rod. Pull off the rubber boot and take the filter retainer out of the valve body. Then separate the filter pad and valve from the end cover.
8 The valve bearing, seal and retainer may be prised out of the end cover.
9 Remove the diaphragm from the valve body and by holding it with the key downwards, the valve rod and plunger assembly can be released when pressing the valve rod in. If this assembly is suspect it must be renewed.
10 Push out the reaction disc with a pencil from the valve body.
11 The pushrod adjuster can be turned if the stem is gripped in a vice near the head. One complete revolution represents approximately .035 inch. If the adjuster requires less than 5 lb/in to turn it, it is too loose and a new one is required.
12 All parts should be cleaned with brake fluid and kept scrupulously clean and free from any bits of fluff that could come from certain types of cleaning cloth.
13 Where lubricant is necessary, use only Lockheed Disc Brake lubricant. Lubricate valve rod and plunger, check that it is seating centrally and insert assembly into the valve body.
14 Refit the key, using light pressure on the valve rod to secure it in the body.
15 Fit the new diaphragm to the valve body making sure the centre sits properly in its groove.
16 Lubricate the reaction disc and put it back in the valve bore using the pushrod to seat it.
17 To fit the new end cover seal, bearing and retainer, it is best to use a mandrel of the type illustrated (Fig 9:25). The lips of the seal should face away from the bearing ring and the flat side of the retainer goes in first.
18 Put the valve body into the end cover and put the filter over the forked end of the valve rod. Refit retainer and boot.
19 Fit the tool to the cover, once more, and the slave flange. Engage the spring stop in swollen end of the spring, and put it in the shell so that the recesses lock in position on the heads of the master cylinder mounting studs.
20 Lubricate the edge of the diaphragm where it will contact the lip of the shell and cover, and then line up the cover to the shell on the marks made previously. Dress the cover down carefully so as not to trap the diaphragm and turn it clockwise until it is up against the stops.
21 Remove the slave flange and insert the pushrod head into the valve body applying only enough pressure to seat it against the reaction disc.
22 Check the pushrod projection, as mentioned earlier, and if correct, press in the retainer using a piece of tube. Do not use excessive pressure. Then put the metal retainer in the shell over the pushrod, guarding against scratching the rod. Smear the rod with silicone grease and insert the rubber seal so that the pushrod adjuster and threads protrude. If this is not done properly, the shell could leak. It is most important also that the vent holes in the rubber seal are not blocked with any traces of dirt or lubricant, otherwise air might find its way into the master cylinder.
23 Before fitting the master cylinder in position, note the position of the vacuum hose connection which should be at 5 o'clock with the reservoir in the upright position. Press the master cylinder well home into its rubber seal before fitting the washers and nuts (Fig 9:27).

14 Hydraulic system - bleeding

1 The system should need bleeding only when some part of it has been dismantled which would allow air into the fluid circuit, or if the reservoir level has been allowed to drop so far that air has entered the master cylinder. If the vehicle has been standing unused for any length of time it is possible that air bubbles may have developed in the system, due to the air absorbing nature of hydraulic fluid. Bleed nipples are found on each of the front wheels and on the rear right wheel only. The line goes through the left hand rear cylinder on route to the right rear cylinder.
2 Ensure that a supply of clean non-aerated fluid of the correct specification is to hand in order to replenish the reservoir during the bleeding process. It is advisable, if not essential, to have someone available to help, as one person has to pump the brake pedal while the other attends to each wheel. The reservoir level has also to be continuously watched and replenished. Fluid bled out should not be re-used. A clean glass jar is required, together with a 9–12 inch length of $1/8$ inch internal diameter rubber tube that will fit tightly over the bleed nipples.
3 The order of bleeding the wheels is to start with the longest line which in the Viva is the right rear, followed by right front and left front, for drum brakes.
4 For disc brakes, the engine must not be running and the pedal should be operated three or four times to make sure that there is no residual vacuum in the servo unit. The order of bleeding should be right front, left front, right rear. The lengths of the runs to the front wheels are no greater, but the fluid capacity of the disc caliper cylinders is much larger.
5 Make sure the bleed nipple is clean and put a small quantity of fluid in the bottom of the jar. Fit the tube onto the nipple and place the other end in the jar under the surface of the fluid. Keep it under the surface throughout the bleeding operation.
6 Unscrew the bleed screw ½ turn and get your assistant to depress and release the brake pedal in short sharp bursts when you direct him. Short sharp jabs are better, because they will force any air bubbles along the line with the fluid rather than pump the fluid past them. It is not essential to remove all the air first time. If the whole system has to be bled, attend to each wheel for three or four complete pedal strokes and then repeat the process. On the second time round, operate the pedal sharply in the same way until no bubbles come out of the pipe into the jar. With the brake pedal in the fully depressed position, the bleed screw should be tightened. Do not forget to keep the reservoir topped up throughout.
7 When all wheels have been bled satisfactorily, re-adjust the shoes (Section 2).
8 If the reason for bleeding has been a repair to a pipe or cylinder near a wheel, then it should be normally necessary to bleed only the wheel of the line in question - PROVIDED that no fluid has been allowed to drain out of the disconnected line. If in any doubt, bleed the whole system.
9 Depress the brake pedal which should offer a firm resistance with no trace of 'sponginess'. The pedal should not continue to go down under sustained pressure. If it does, there is a leak, or the master cylinder seals are worn out.

15 Brake pedal - removal and replacement

1 If for any reason (such as worn out bushes), the brake pedal needs to be removed, follow the procedures as described for the clutch pedal in Chapter 5.

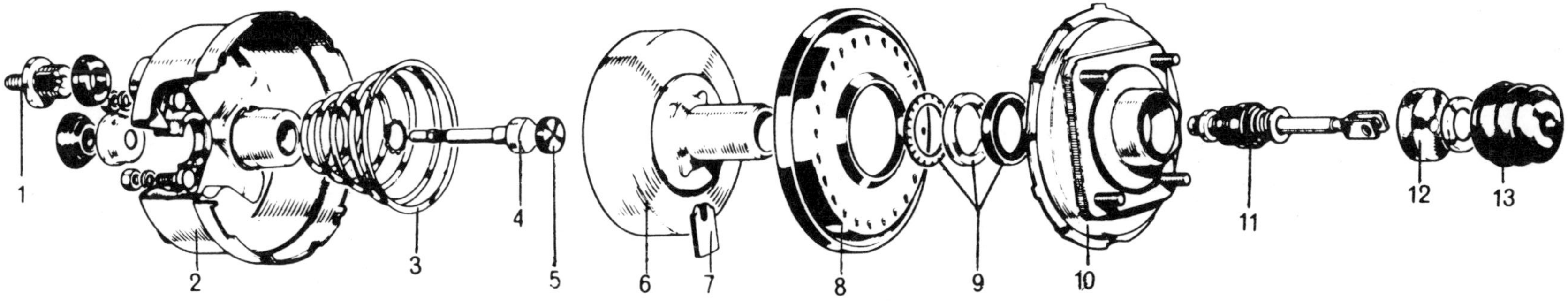

Fig 9:26 SERVO UNIT – EXPLODED VIEW

1 Suction pipe union and non-return valve
2 Body shell
3 Return spring
4 Piston rod
5 Seal
6 Piston
7 Piston rod locking plate
8 Diaphragm
9 Seal assembly
10 End cover
11 Push rod assembly
12 Filter
13 Push rod and filter cover

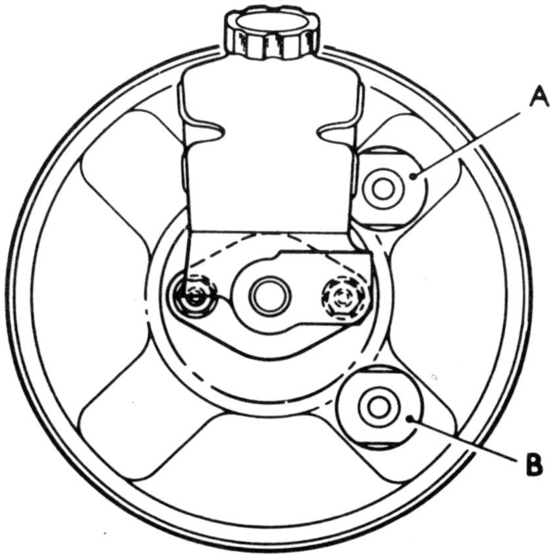

Fig. 9:27 Master cylinder/servo assembly showing correct position of vacuum hose connection at point 'B'. (Point 'A' is for 1600 cc engine models)

Before diagnosing faults from the following chart, check that any braking irregularities are not caused by:

1 Uneven and incorrect tyre pressures.
2 Incorrect 'mix' of radial and cross-ply tyres.
3 Wear in the steering mechanism.
4 Defects in the suspension and dampers.
5 Misalignment of the body frame.

NOTE: For vehicles fitted with disc brakes at the front the references in the chart to front wheel shoe adjustments do not apply. The 'Reason/s' referring to hydraulic system faults or wear to the friction material of the linings still apply, however. Disc pads also come in different material and references to variations are also relevant.

Fault Finding Chart – Braking System

Symptoms	Reason/s	Remedy
Pedal travels a long way before the brakes operate	Brake shoes set too far from the drums	Adjust the brake shoes to the drums, or check operation of automatic adjuster.
Stopping ability poor, even though pedal pressure is firm	Linings and/or drums badly worn or scored	Dismantle, inspect and renew as required.
	Failure of one circuit in the dual hydraulic system	Check both circuits for hydraulic leaks and repair
	One or more wheel hydraulic cylinders seized, resulting in some brake shoes not pressing against the drums (or pads against discs)	Dismantle and inspect wheel cylinders. Renew as necessary.
	Brake linings contaminated with oil	Renew linings and repair source of oil contamination.
	Wrong type of linings fitted (too hard)	Verify type of material which is correct for the car, and fit it.
	Brake shoes wrongly assembled	Check for correct assembly.
	Servo unit not functioning (disc brakes)	Check and repair as necessary.
Car veers to one side when the brakes are applied	Brake pads or linings on one side are contaminated with oil	Renew pads or linings and stop oil leak.
	Hydraulic wheel cylinder(s) on one side partially or fully seized	Inspect wheel cylinders for correct operation and renew as necessary.
	A mixture of lining materials fitted between sides	Standardise on types of linings fitted.
	Unequal wear between sides caused by partially seized wheel cylinders	Check wheel cylinders and renew linings and drums as required
Pedal feels spongy when the brakes are applied	Air is present in the hydraulic system	Bleed the hydraulic system and check for any signs of leakage.
Pedal feels springy when the brakes are applied	Brake linings not bedded into the drums (after fitting new ones)	Allow time for new linings to bed in after which it will certainly be necessary to adjust the shoes to the drums as pedal travel will have increased.
	Master cylinder or brake backplate mounting bolts loose	Retighten mounting bolts.
	Severe wear in brake drums causing distortion when brakes are applied	Renew drums and linings.
Pedal travels right down with little or no resistance and brakes are virtually non-operative (With dual braking systems this would be extraordinary as both systems would have to fail at the same time).	Leak in hydraulic systems resulting in lack of pressure for operating wheel cylinders	Examine the whole of the hydraulic system and locate and repair source of leaks. Test after repairing each and every leak source.
	If no signs of leakage are apparent all the master cylinder internal seals are failing to sustain pressure	Overhaul master cylinder. If indications are that seals have failed for reasons other than wear all the wheel cylinder seals should be checked also and the system completely replenished with the correct fluid.
Binding, juddering, overheating	One or a combination of causes given in the foregoing sections..	Complete and systematic inspection of the whole braking system.

Chapter 10 Electrical system

Contents

Specifications

Battery

Standard	Exide 6 VTA7 BR or Lucas BH7/9 32 amp hours at 20 hr rate
Heavy duty	Exide 6 VTA 11BR or No 161 55 amp hours at 20 hr rate
Earth	Negative

Starter Motor

Type	Lucas M35 G/I
Brush length minimum	.30 in
Brush spring tension	34–46 oz new brushes 25 oz minimum length
Minimum commutator diameter	1.28 in
Free running current	45 amps at 9500–11000 rpm
Lock torque	10 lb/ft at 420–440 amps 7.8–7.4 volts

Type	Lucas M35 J/I
Brush length	.38 in (9.5 mm) minimum
Spring pressure on brushes	At .06 in (1.5 mm) protrusion - 28 oz (760 gms)
Minimum commutator thickness	.080 in (2 mm)
Armature shaft end float	.010 in (.25 mm) maximum
Free running current	68 amps at 8000–10000 rpm
Lock torque	7 lb/ft at 350–375 amps

Type	Lucas M35 K/PE
Brush length	.38 in (9.5 mm) minimum
Spring pressure on brushes	At .06 in (1.5 mm) protrusion - 28 ozs (790 gms)
Minimum commutator thickness	.080 in (2 mm)
Armature shaft end float	.010 in maximum
Free running current	70 amps at 8000–11500 rpm
Lock torque	8 lb/ft at 350–400 amps

Solenoid switch test information		
Series winding resistance	.21–.25 ohms	
Shunt winding resistance	.9–1.1 ohms	
Alternators		
Type	Lucas 10 AC	11 AC
Voltage	12	12
Output	35 amps	45 amps
Field resistance (±5%)	3.5 ohms	3.7 ohms
Brushes - minimum length	.16 in	.16 in
Regulator type	4 TR or Prestolite SR 12	
Voltage setting	13.9–14.3 volts	
Relay - Type	6 RA	
Warning lamp contact type	3 AW	
Type	Lucas 15 ACR	17 ACR
Voltage	12	12
Output	28 amps	36 amps
Field resistance (± 5%)	4.3 ohms	4.16 ohms
Brushes - minimum length	.20 in	.20 in
Brush spring pressure	7–10 oz	7–10 oz
Regulator (incorporated)	8 TRD	8 TRd
Type	AC Delco 'Delcotron' DN 460	
Voltage	12	
Output	35 amps	
Field resistance (± 5%)	2.8 ohms	
Brushes - minimum length	0.2 in (5.0 mm)	
Brush spring pressure	8 - 13 oz (2.2 - 3.6 N)	
Windscreen wiper motors		
Total light running current consumption - warm		
Single speed	1 amp	
Two speed - high	2 amps	
- low	1.2 amps	
Cross shaft end float	0–.0005 in	
Worm wheel end float	.002 in max	
Wiper arm spring tension	16–18 oz	

Fuses

A fuse block is mounted in the engine compartment on the bulkhead behind the dash panel and contains four 35 amp fuses and provision for two spares

No 4 fuse is fed via the thermal interruption and lighting switch. A fusible link is fitted between the positive battery terminal and the main feed protecting the whole of the wiring system except the starter.

Fuse No 1 protects	Horn, interior lamp, headlamp flasher
Fuse No 2 protects	Stop lamp, turn signal lamp and warning lamps, oil and alternator warning lamps, fuel and temperature gauges, heater motor and reverse lights
Fuse No 3 protects	Windscreen wipers, radio, cigarette lighter
Fuse No 4 protects	Instrument lamps, rear lamps, number plate lamps, boot interior lamp and cigarette lighter bulb

Bulbs

Headlamps	45/40 watt Unified European cap
Side lamps, side repeater lamps, number plate and boot interior lamp	6 watt miniature centre contact
Tail/stop lamps	5/21 watt, small bayonet cap
Reverse lamps and direction indicator flasher lamps	21 watt, single centre contact
Interior courtesy light	6 watt festoon
Ignition and illumination lamps	3 watt wedge-base capless
Indication and warning lamps	1.5 watt wedge-base capless
Automatic transmission selector lamp	1.2 watt, peanut
Cigarette lighter lamp	2.2 watt miniature
Fog lamps	55 watt quartz halogen
Hazard warning switch lamp	2 watt peanut

1 General description

1 The electrical system is of the 12-volt type and the major components comprise: A 12-volt battery with the negative terminal earthed; a starter motor which is fitted to the clutch bellhousing on the right hand side of the engine; an alternator fitted to the front right hand side of the engine, driven by the fan belt from the crankshaft pulley; and a transistorised voltage control unit.
2 The 12-volt battery supplies a steady amount of current for the ignition, lighting, and other electrical circuits, and provides a reserve of electricity when the current consumed by the electrical equipment exceeds that being produced by the alternator. The output from the alternator is controlled by the voltage regulator, which ensures a high output if the battery is in a low state of charge or the demands from the electrical equipment high, and a low output if the battery is fully charged and there is little demand from the electrical equipment.
3 The voltage regulator is integral in later type alternators.

2 Battery - removal and replacement

1 Disconnect the negative (earth) lead from the battery terminal post and then the positive lead similarly. The leads are held by either a clamp, which necessitates slackening the clamp bolt and nut, or by a screw driven through an all enclosing shroud.
2 Remove the battery clamp and carefully lift the battery out of its compartment. Hold the battery vertical to ensure that none of the electrolyte is spilled (photo).
3 Replacement is a direct reversal of this procedure. NOTE: Replace the positive lead before the earth (negative) lead and smear the terminals with petroleum jelly (vaseline) to prevent corrosion. NEVER use an ordinary grease as applied to other parts of the car.

3 Battery - maintenance and inspection

1 Normal weekly battery maintenance consists of checking the electrolyte level of each cell to ensure that the separators are covered by ¼ inch of electrolyte. If the level has fallen, top up the battery using distilled water only. Do not overfill. If a battery is overfilled or any electrolyte spilled, immediately wipe away the excess, as electrolyte attacks and corrodes any metal it comes into contact with very rapidly.
2 As well as keeping the terminals clean and covered with petroleum jelly, the top of the battery, and especially the top of the cells, should be kept clean and dry. This helps prevent corrosion and ensures that the battery does not become partially discharged by leakage through dampness and dirt.
3 Once every three months, remove the battery and inspect the battery securing bolts, the battery clamp plate, tray and battery leads for corrosion (white fluffy deposits on the metal which are brittle to touch). If any corrosion is found, clean off the deposits with ammonia and paint over the clean metal with an anti-rust/anti-acid paint.
4 At the same time inspect the battery case for cracks. If a crack is found, clean and plug it with one of the proprietary compounds marketed by firms, such as Holts, for this purpose. If leakage through the crack has been excessive then it will be necessary to refill the appropriate cell with fresh electrolyte as detailed later. Cracks are frequently caused to the top of the battery cases by pouring in distilled water in the middle of winter AFTER instead of BEFORE a run. This gives the water no chance to mix with the electrolyte and so the former freezes and splits the battery case.
5 If topping up the battery becomes excessive and the case has been inspected for cracks that could cause leakage, but none are found, the battery is being overcharged and the voltage regulator will have to be checked and reset.
6 With the battery on the bench at the three monthly interval check, measure its specific gravity with a hydrometer to determine the state of charge and condition of the electrolyte. There should be very little variation between the different cells and if a variation in excess of .025 is present it will be due to either:
a) Loss of electrolyte from the battery at some time caused by spillage or a leak, resulting in a drop in the specific gravity of the electrolyte when the deficiency was replaced with distilled water instead of fresh electrolyte.
b) An internal short circuit caused by buckling of the plates or a similar malady pointing to the likelihood of total battery failure in the near future.
7 The specific gravity of the electrolyte for fully charged conditions at the electrolyte temperature indicated, is listed in Table A. The specific gravity of a fully discharged battery at different temperatures of the electrolyte is given in Table B.

Table A
Specific Gravity - Battery fully charged

1.268 at 100°F or 38°C	electrolyte	temperature
1.272 at 90°F or 32°C	"	"
1.276 at 80°F or 27°C	"	"
1.280 at 70°F or 21°C	"	"
1.284 at 60°F or 16°C	"	"
1.288 at 50°F or 10°C	"	"
1.292 at 40°F or 4°C	"	"
1.296 at 30°F or -1.5°C	"	"

Table B
Specific Gravity - Battery fully discharged

1.098 at 100°F or 38°C	electrolyte	temperature
1.102 at 90°F or 32°C	"	"
1.106 at 80°F or 27°C	"	"
1.110 at 70°F or 21°C	"	"
1.114 at 60°F or 16°C	"	"
1.118 at 50°F or 10°C	"	"
1.122 at 40°F or 4°C	"	"
1.126 at 30°F or -1.5°C	"	"

4 Battery - electrolyte replenishment

1 If the battery is in a fully charged state and one of the cells maintains a specific gravity reading which is .025 or more lower than the others, and a check of each cell has been made with a voltage meter to check for short circuits (a four to seven second test should give a steady reading of between 1.2 to 1.8 volts), then it is likely that electrolyte has been lost from the cell with the low reading at some time.
2 Top the cell up with a solution of 1 part sulphuric acid to 2.5 parts of water. If the cell is already fully topped up draw some electrolyte out of it with a pipette.
3 When mixing the sulphuric acid and water NEVER ADD WATER TO SULPHURIC ACID - always pour the acid slowly onto the water in a glass container. IF WATER IS ADDED TO SULPHURIC ACID IT WILL EXPLODE.
4 Continue to top up the cell with the freshly made electrolyte and then recharge the battery and check the hydrometer readings.

5 Battery - charging

1 In winter time when heavy demand is placed upon the battery, such as when starting from cold, and much electrical equipment is continually in use, it is a good idea occasionally to have the battery fully charged from an external source at the rate of 3.5 to 4 amps.
2 Continue to charge the battery at this rate until no further rise in specific gravity is noted over a four hour period.
3 Alternatively, a trickle charger, charging at the rate of 1.5 amps can be safely used overnight.
4 Specially rapid 'boost' charges which are claimed to restore the

power of the battery in 1 to 2 hours are most dangerous as they can cause serious damage to the battery plates through overheating.
5 While charging the battery note that the temperature of the electrolyte should never exceed 100°F.

6 Alternators - general description

1 The Viva HC is one of the first medium priced cars to fit alternators as standard equipment on all models in the range.

Basically the alternator, as its name implies, generates alternating current rather than direct current. This current is rectified (by diodes) into direct current so that it can be stored by the battery. The transistorised regulators are self-limiting in current output so they control only the voltage.

Apart from the renewal of the rotor slip ring brushes and rotor shaft bearings, there are no other parts which need periodic inspection. All other items are sealed assemblies and must be replaced if indications are that they are faulty.

7 Alternators - safety precautions

If there are indications that the charging system is malfunctioning in any way, care must be taken to diagnose faults properly, otherwise damage of a serious and expensive nature may occur to parts which are in fact quite serviceable.

The following basic requirements must be observed at all times, therefore, if damage is to be prevented:
1 ALL alternator systems use a NEGATIVE earth. Even the simple mistake of connecting a battery the wrong way round could burn out the alternator diodes in a few seconds.
2 Before disconnecting any wires in the system the engine and ignition circuits should be switched off. This will minimise accidental short circuits.
3 The alternator must NEVER be run with the output wire disconnected.
4 Always disconnect the battery from the car's electrical system if an outside charging source is being used.
5 Do not use test wire connections that could move accidentally and short circuit against nearby terminals. Short circuits will not blow fuses - they will blow diodes or transistors.
6 Always disconnect the battery cables and alternator output wires before any electric welding work is done on the car body.

8 Lucas 10 AC and 11 AC alternator systems - fault diagnosis

1 It is essential that when a fault occurs the correct procedure is followed to diagnose it. If it is not, the likelihood of damage is high. The safety precautions as described in Section 7 should always be observed.
2 No proper diagnosis is possible without an ammeter (0-100 amps range) a voltmeter (0-50 volts range) and a test lamp (12v 6watt) being available. If you are unable to acquire these then leave the circuit checking to a competent electrician.
3 Check the obvious first, ie battery, battery terminals, fan belt tension and disconnected wires.
4 Follow the line of diagnosis as shown in the accompanying tables (Fig 10:4).

9 Lucas 15 ACR or 17 ACR alternator systems - fault diagnosis

1 Paragraphs 1-3 in the previous section apply. Thereafter follow the line of diagnosis as shown in the relevant table (Fig 10:5).

10 AC Delco alternator systems - fault diagnosis

1 Fault diagnosis procedures in the manufacturer's service schedule are not readily interpreted into 'Do-it-yourself' terms. If a fault develops it is recommended that proper checks are made by a service station which you know is equipped to deal with this make of alternator.

11 Alternators - removal, replacement and belt adjustment

1 Details of the alternator mountings and the procedures to be adopted when removing or adjusting them are given in Chapter 2, 'Fan belt - adjustment'.
2 Note that one of the mounting bolt holes in the alternator is fitted with a detachable split sliding bush. This enables the alternator to be put in position on the mounting brackets, without imposing any strain on the lugs when the bolts are tightened. It is essential that the lug **without** the bush (the front one) is tightened up first.
3 Electrical connections to the alternator are made through two multi-socket connectors (photos).

12 Alternators - dismantling and inspection

1 If tests indicate that the alternator is faulty, it is possible that the slip ring brushes and slip rings may be the cause. In this case the remedy on the 10AC and 11AC units does not require any operations likely to cause damage to the rest of the unit.
2 After removing the fan and pulley nut, the through bolts may be removed. Mark the position of the end covers relative to the stator and withdraw the drive end cover and rotor together. The brushes may be checked for length and the slip rings cleaned up with fine glass paper.
3 15ACR and 17ACR alternators require the unsoldering of the stator connections to get at the brushes and slip ring, and this is not recommended. If the diodes to which they are attached are overheated they could be damaged.
4 With AC Delco models the brushes are not soldered.
5 Although, therefore, there is the possibility of an owner successfully rectifying a fault, we do not recommend dismantling as a general principle, as more damage could be caused to the system - not just the alternator, if a mistake is made.
6 When an alternator is diagnosed as unserviceable, it should only be as a result of a thorough check of the complete system. If this is not done, a new unit could be completely ruined immediately following installation if something is also at fault elsewhere.

13 Starter motors - testing in the car

1 If the starter motor fails to operate, then check the condition of the battery by turning on the headlamps. If they glow brightly for several seconds and then gradually dim, the battery is in an uncharged condition.
2 If the headlamps glow brightly and continue to glow, and it is obvious that the battery is in good condition, then check the tightness of the battery wiring connections (and in particular the earth lead from the battery terminal to its connection on the bodyframe). Check the tightness of the connections at the relay switch and at the starter motor.
3 If the starter motor still fails to turn, the fault lies in the wiring, the solenoid switch or the motor itself. The following procedure will determine where the fault lies. (To prevent inadvertent starting of the engine remove the HT lead from the coil):
4 Connect a voltmeter (or 12-volt bulb) to the terminal to which the white/red wire is connected [Fig 10:9(A)]. If there is no reading (12 volts) when the starter switch is operated, there is a fault in the starter switch or white/red wire.

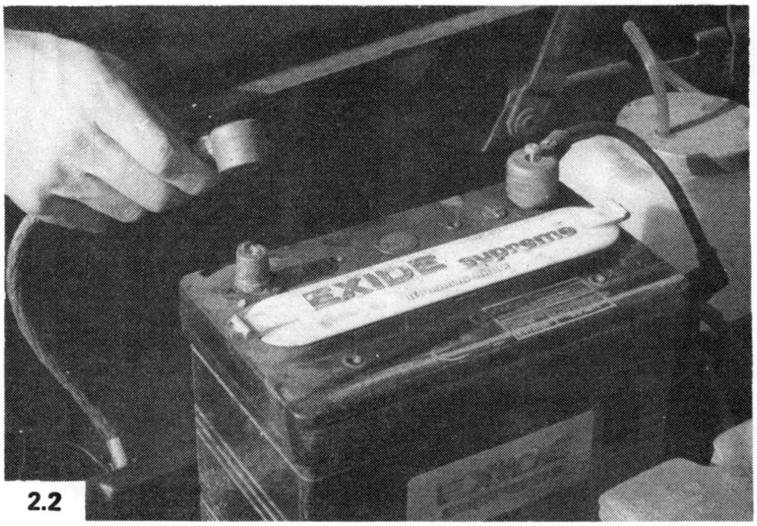

Battery removal

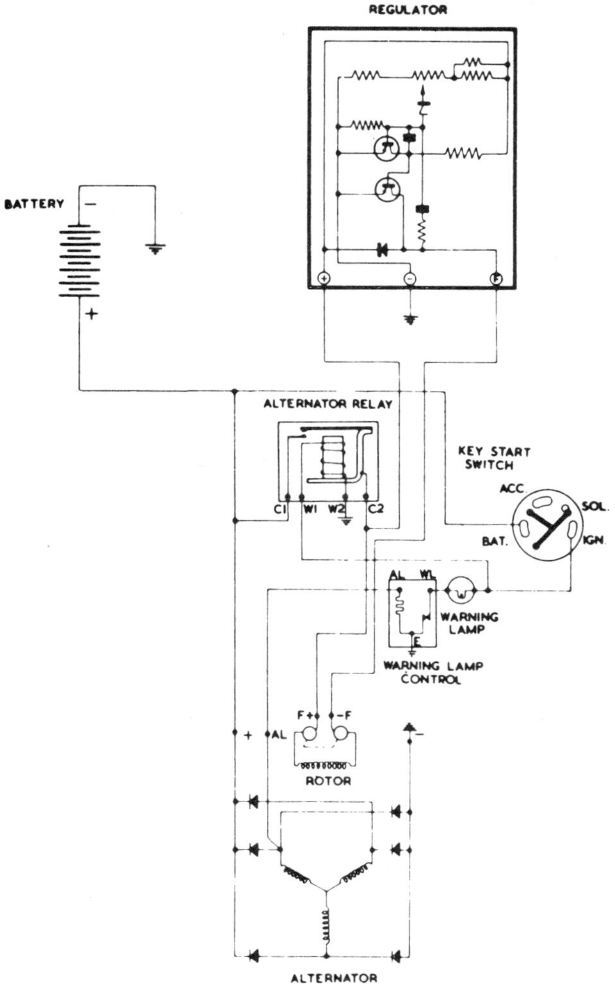

Fig 10:1 Lucas alternator circuit (10AC and 11AC)

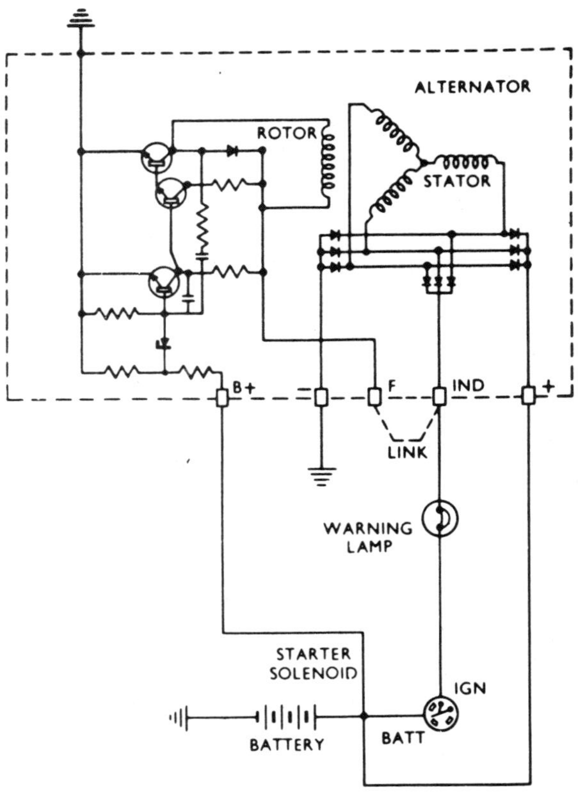

Fig 10:2 Lucas alternator circuit (15ACR and 17ACR)

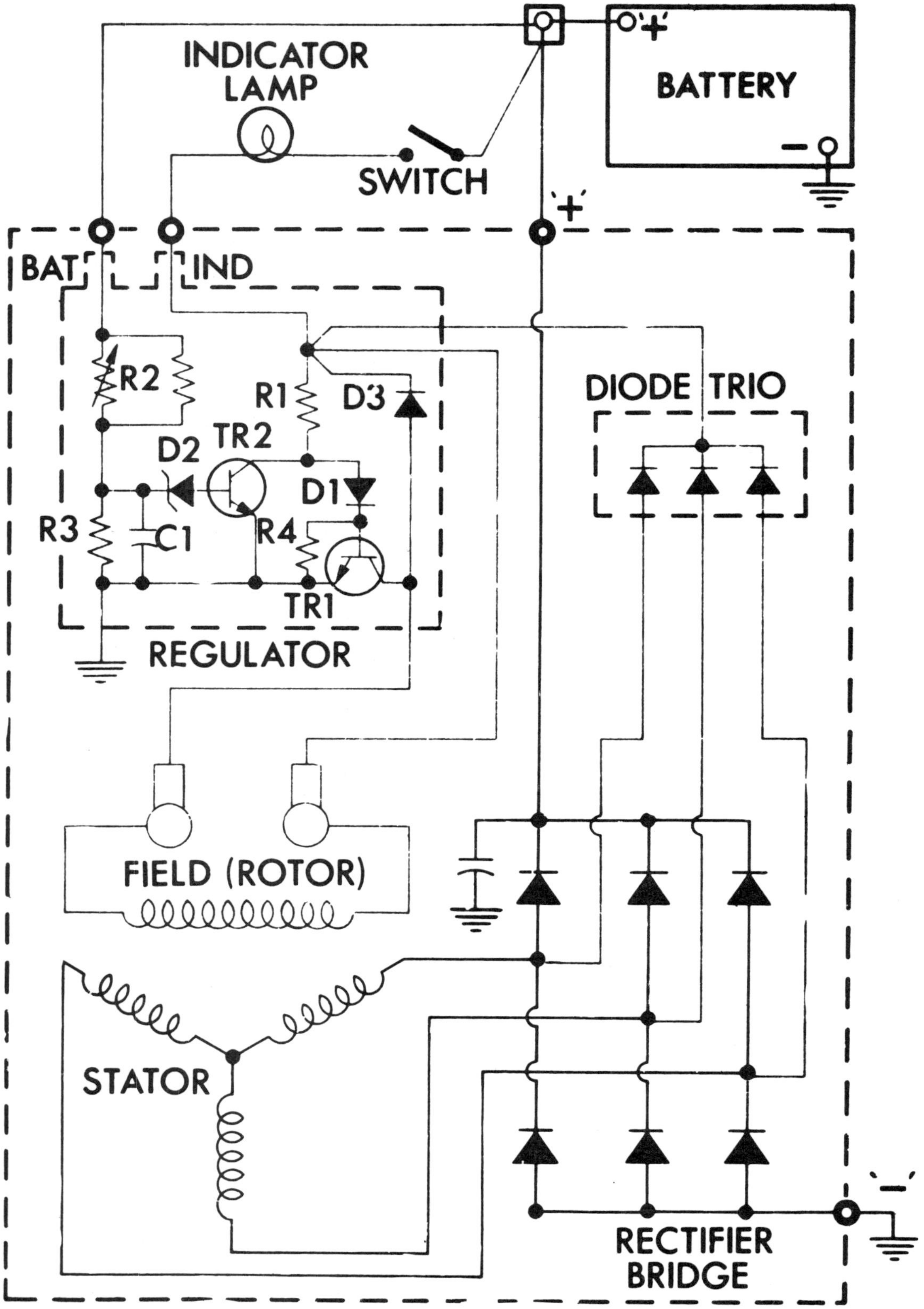

Fig 10:3 AC-Delco alternator - internal wiring circuit

1 Check fan belt(s) for tension and condition.
2 Disconnect battery ground terminal. Disconnect wire (brown) from alternator positive terminal and connect ammeter between wire and terminal. Re-install battery terminal.
3 Remove warning lamp control three-way connector and connect voltmeter between 'AL' wire (brown/purple) and ground.
4 Install new warning lamp control.
5 Install new or repair alternator.
6 Connect voltmeter between alternator 'AL' and ground.
7 Check wire (brown/purple) between alternator 'AL' and warning lamp control 'AL' for continuity.
8 Disconnect 'C1' wire (brown) and 'C2' wire (brown/white) from relay and join together.
9 Reconnect 'C1' wire (brown) and 'C2' wire (brown/white) to relay and check 'W2' terminal to ground.
10 Check all relay connections for cleanliness and security.
11 Install new relay.
12 With 'C1' and 'C2' relay wires joined together, remove three-way connector from regulator and bridge ground wire (black) and field wire (brown/green) in connector.
13 Install new regulator.
14 Reconnect wires to 'C1' and 'C2' terminals on relay.
15 Remove field double connector from alternator and connect test lamp between terminals of connector.
16 Check field wires and connections between: starter solenoid and relay 'C1' (brown); relay 'C2' and regulator positive and alternator 'F' positive (brown/white); regulator 'F' terminal and alternator 'F' negative (brown/green); regulator negative terminal to ground (black).
17 Reconnect three-way connector to regulator.
18 Connect voltmeter between battery positive and negative terminals, switch on lights and increase engine speed to approximately 1500 rpm. Voltmeter should read 13.9 to 14.3 volts.
19 Disconnect voltmeter from between battery positive and negative terminals: connect voltmeter between regulator positive terminal and ground.
20 Remove three-way connector from regulator and bridge field wire (brown/green) and negative wire (black) in regulator connector; increase engine speed.
21 If fan belt(s) tension and condition are satisfactory, a faulty battery or an overloaded system is indicated. A comparison should be made between electrical loading and alternator output (10AC: 35 amp; 11AC: 45 amp).
22 Clean connections 'C1' and 'C2' on relay: clean positive and negative connections on regulator and regulator ground connections: clean battery posts and terminals.
23 Remove warning lamp control three-way connector and bridge ground wire (black) and 'WL' wire (brown/yellow) in connector.
24 Bridge 'C1' and 'C2' terminals on relay.
25 Remove bridge from between 'C1' and 'C2' terminals on relay and link regulator 'F' and negative wires.
26 Check warning lamp bulb. Check bulb-holder for loose connection.
27 Check continuity of warning lamp circuit, ie battery to warning lamp (green): warning lamp to warning lamp control 'WL' (brown/yellow): warning lamp control 'E' terminal to ground (black).
28 Check that charging system operates satisfactorily by connecting voltmeter across battery terminals and ammeter in series with alternator output circuit. Impose approximate 35 amp (10AC) or 45 amp (11AC) load on battery, start engine and increase engine speed until ammeter reads maximum charge (35 amp and 45 amp respectively). Remove load from battery. Ammeter needle should then drop slowly back to show trickle charge. Voltmeter should show 13.9 to 14.3 volts.

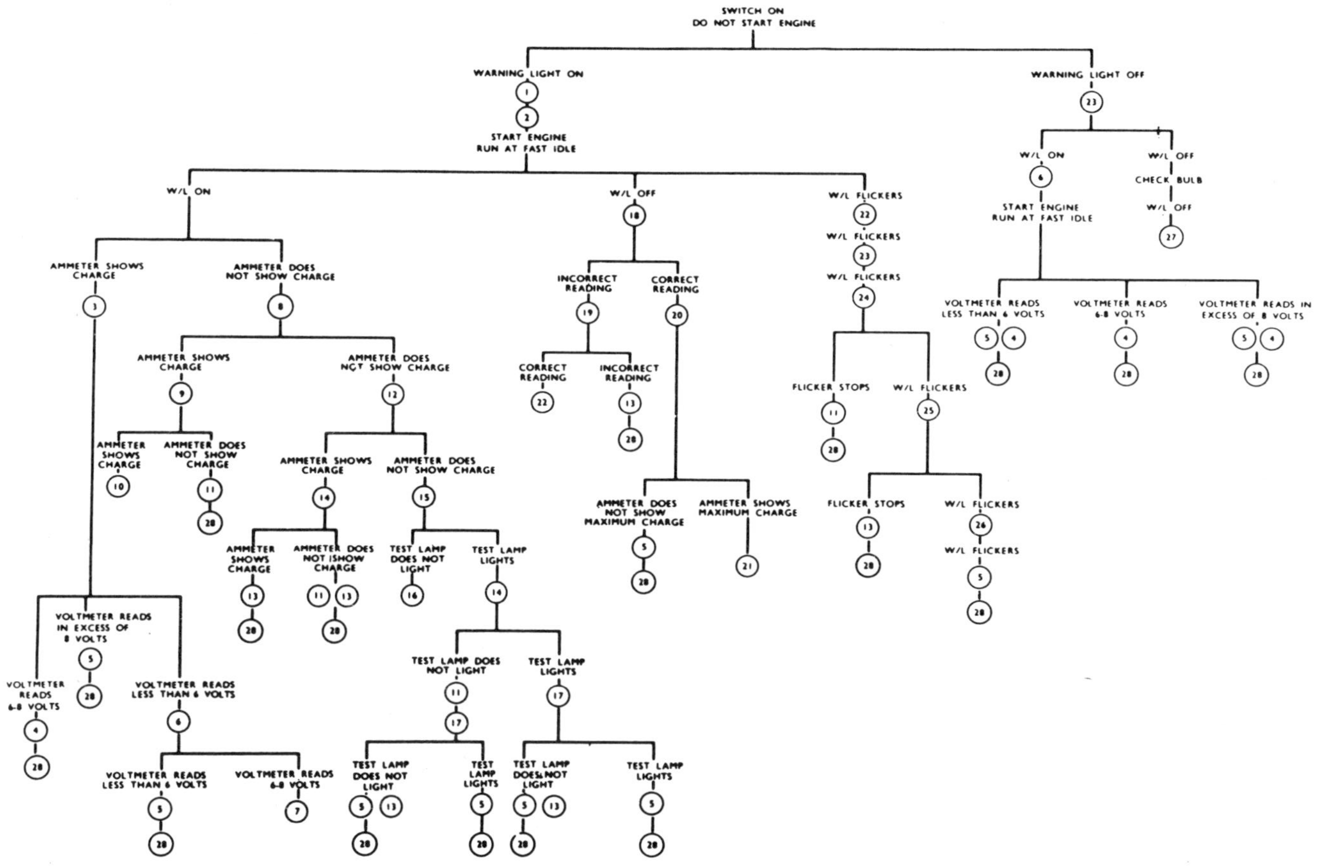

Fig 10:4 Lucas 10/11 AC alternators - fault diagnosis procedure

1 Check fan belt(s) for tension and condition.
2 Disconnect main output connector and auxiliary connector. Install slave wire with a male Lucar terminal and a female Lucar terminal between alternator negative terminal and socket. Connect ammeter between alternator positive terminal and socket removed from this terminal. Reconnect auxiliary connector.
3 Remove rear cover from alternator. Re-install auxiliary connector, slave and ammeter wires. Bridge outer brush contact strip to ground. Adjust engine speed to give maximum output.
4 Install new or repair alternator.
5 Remove connector from field and sensing terminals (IND and B+). 'Switch On' but do not start engine.
6 Check for short circuit in wire between alternator indicator terminal and warning light bulb.
7 Connect voltmeter between battery positive and negative terminals; increase speed to approximately 1500 rpm; Voltmeter should read 14.1 to 14.5 volts, ammeter reading 7.5 amp maximum. Higher amperage which would probably give lower voltage readings could indicate need to recharge battery before continuing with test.
8 If fan belt tension and condition are satisfactory, faulty battery or overloaded system is indicated. Comparison should be made between electrical loading and alternator output. 15ACR: 28 amp. 17ACR: 36 amp.
9 Remove voltmeter from battery and connect it between battery sensing terminal on alternator (B+) and ground (this should be done without moving socket connector by inserting a probe in rear of connector to contact Lucas female blade terminal of brown 14/.012 wire).
10 Check battery terminals, ground strap connections, and wiring between battery and alternator for poor connection and resistive circuits.
11 Install new 8TR regulator.
12 Check wire from alternator B+ to starter solenoid for continuity.
13 Remove connector from alternator indicator socket and bridge double wires (brown/yellow 9/.012) in connector to ground.
14 Remove rear cover from alternator, re-install socket connectors. Disconnect yellow wire from field diode heat sink.
15 Reconnect yellow wire to field diode heat sink. Connect slave wire between outer brush contact strip and ground.
16 Disconnect slave wire from outer brush contact strip and connect between inner brush contact strip and ground
17 Check connecting wire between indicator and field terminals in socket connector for continuity.
18 Check warning lamp bulb. Check bulb-holder for loose connection. Check wire (brown/yellow) between alternator indicator terminal, warning lamp bulb and key-start switch for continuity. Note: warning lamp bulb must be 12 volt 2.2 watt.
19 Check that charging system operates satisfactorily by connecting voltmeter across battery terminals and ammeter in series with alternator output circuit. Impose approximate 28 amp (15ACR) or 36 amp (17ACR) load on battery, start engine and increase engine speed until ammeter reads maximum charge, 28 amp and 36 amp respectively. Remove load from battery. Ammeter should then drop slowly back to show trickle charge. Voltmeter should show 14.1 to 14.5 volts.

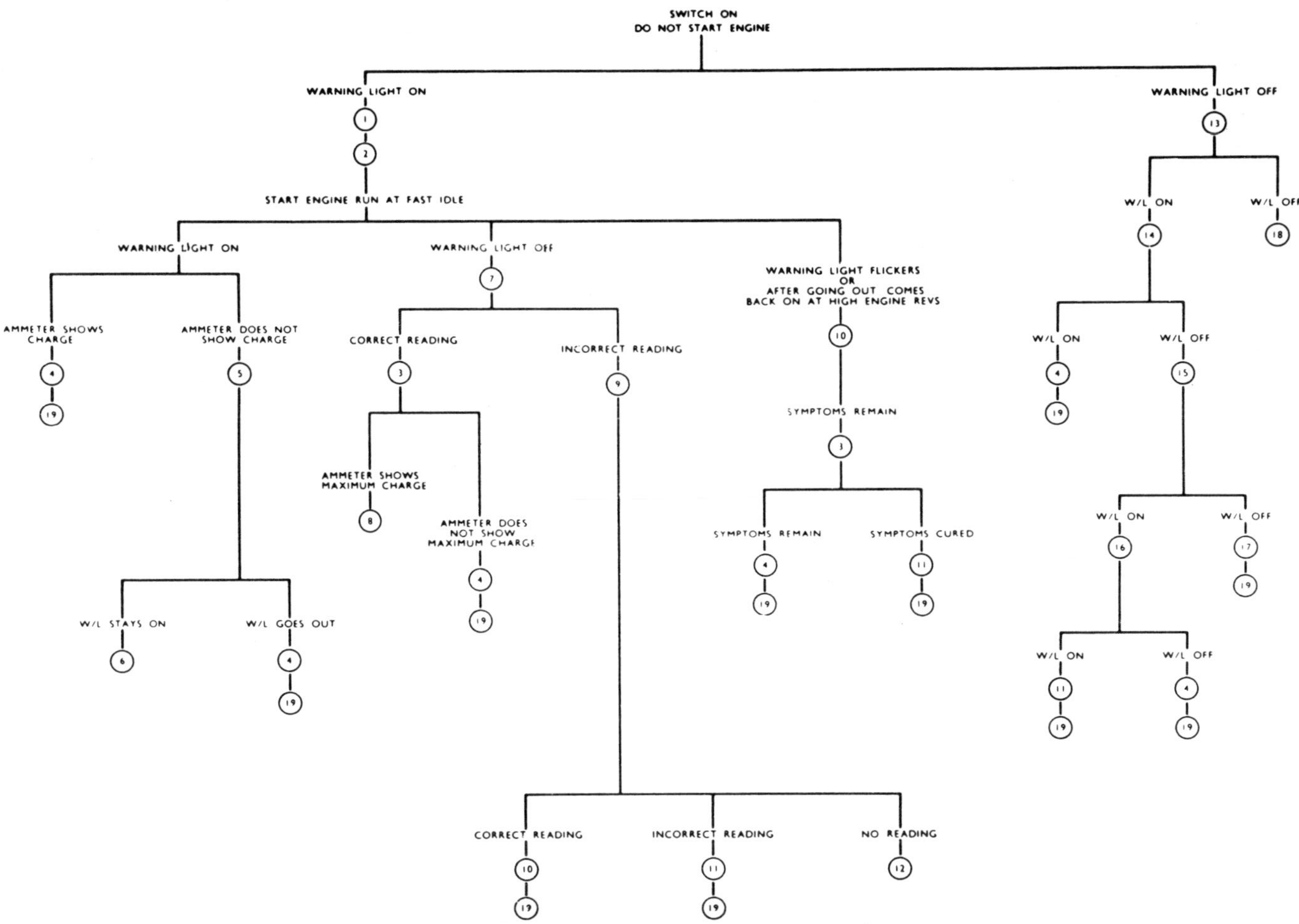

Fig 10:5 Lucas 15/17ACR alternators - fault diagnosis procedure

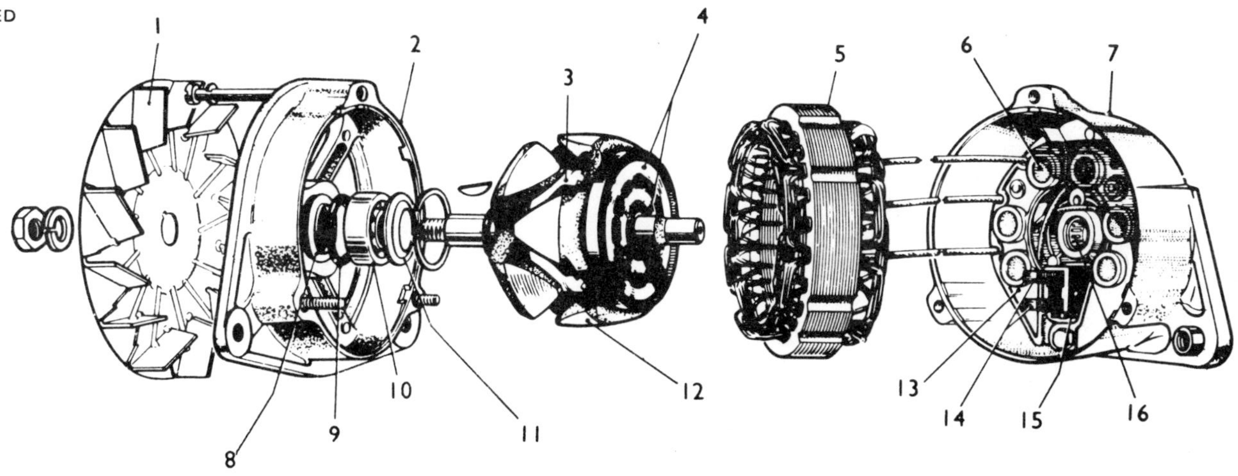

Fig 10:6 LUCAS 10AC/11AC ALTERNATORS – EXPLODED ILLUSTRATION

1 Fan
2 Drive end cover
3 Rotor field windings
4 Slip rings
5 Stator
6 Diodes
7 Slip ring end cover
8 'O' ring retainer washer
9 'O' ring oil seal
10 Bearing
11 Bearing retaining plate
12 Rotor
13 Diode heat sink
14 Brushes
15 Brush holder
16 Needle roller bearing

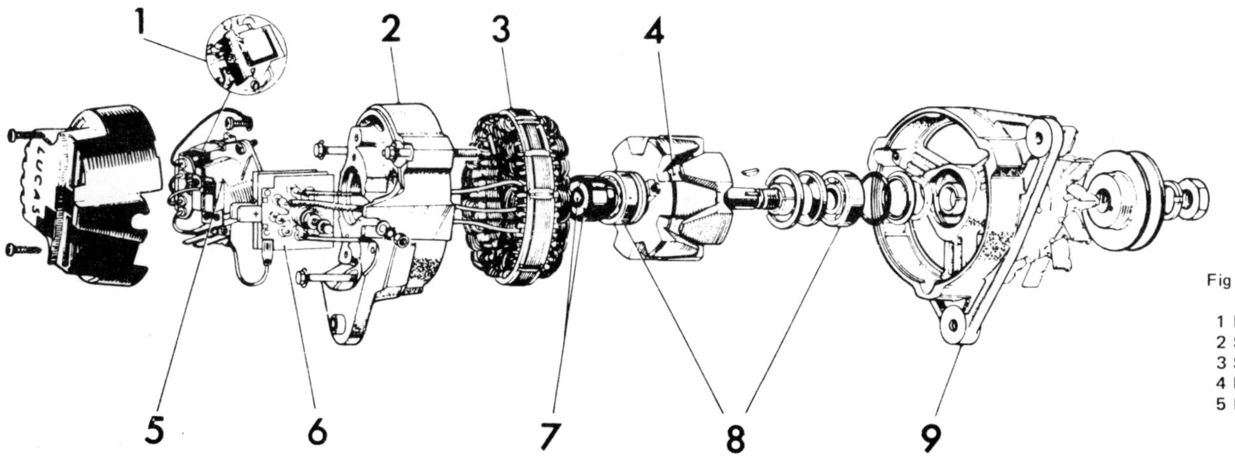

Fig 10:7 LUCAS 15ACR/17ACR ALTERNATORS – EXPLODED ILLUSTRATION

1 Regulator
2 Slip ring end cover
3 Stator
4 Rotor
5 Brush housing
6 Rectifier (diodes and heat sink)
7 Slip rings
8 Bearings
9 Drive end cover

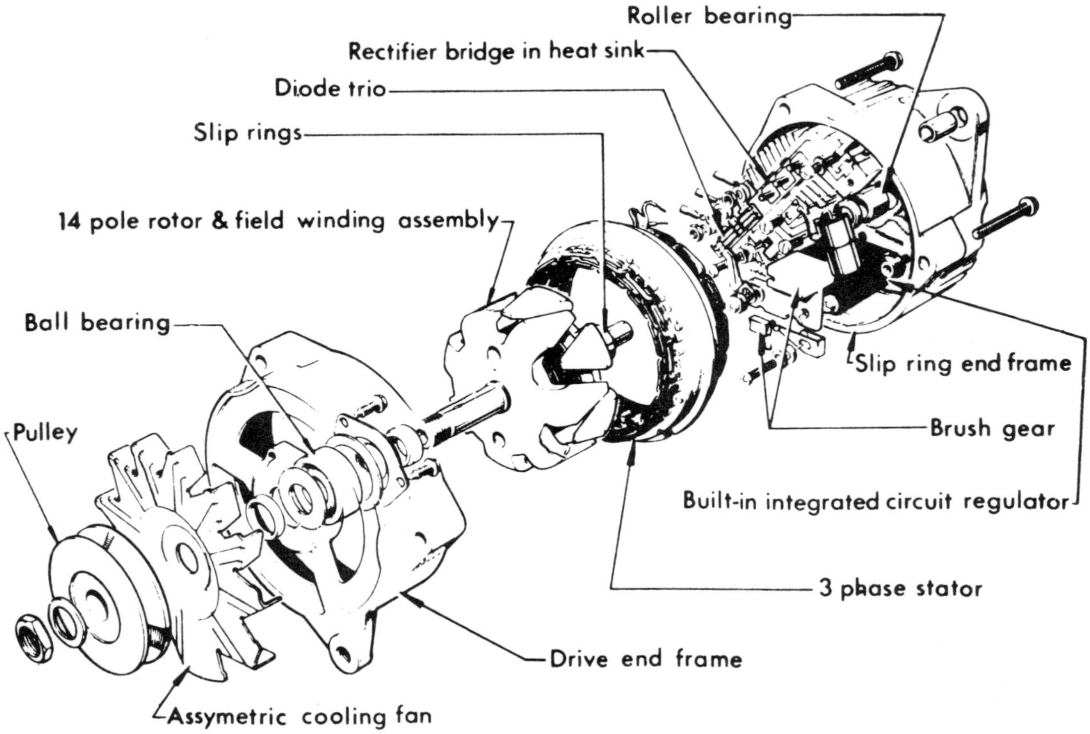

Fig 10:8 AC-Delco 'Delcotron' DN 460 Alternator - exploded illustration

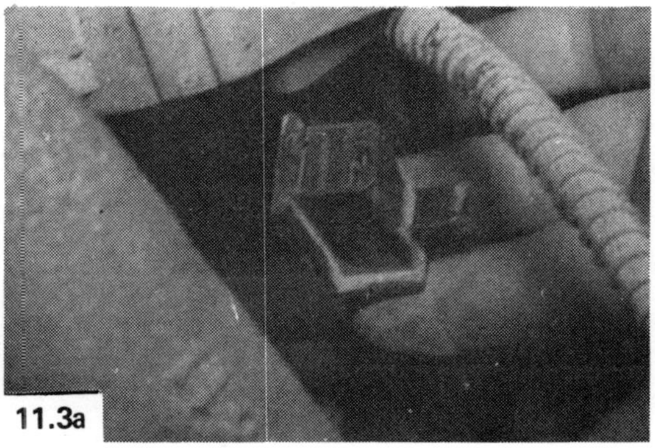

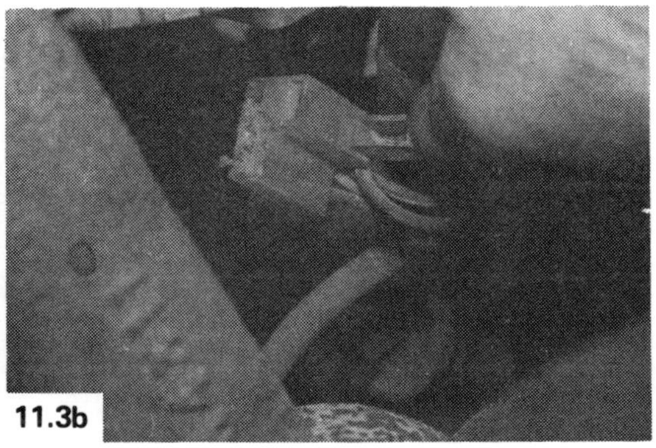

Alternator connector sockets

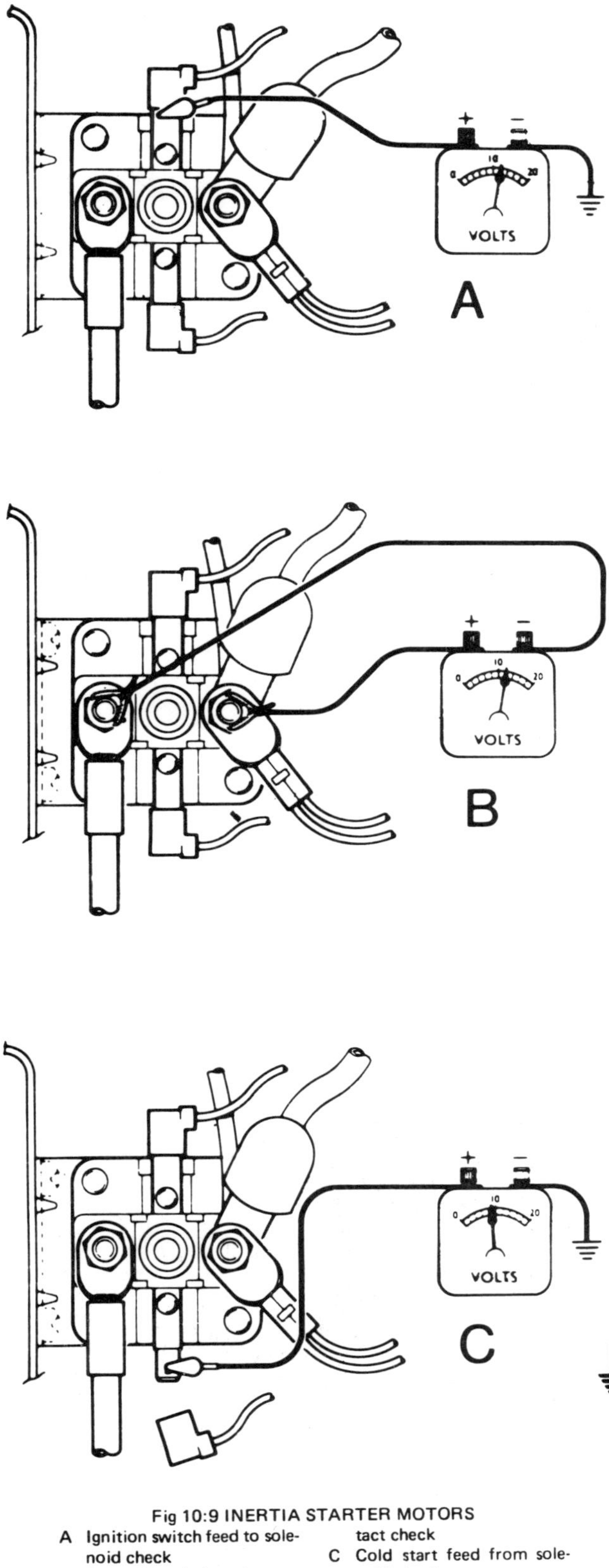

Fig 10:9 INERTIA STARTER MOTORS

A Ignition switch feed to solenoid check
B Solenoid switch main contact check
C Cold start feed from solenoid check

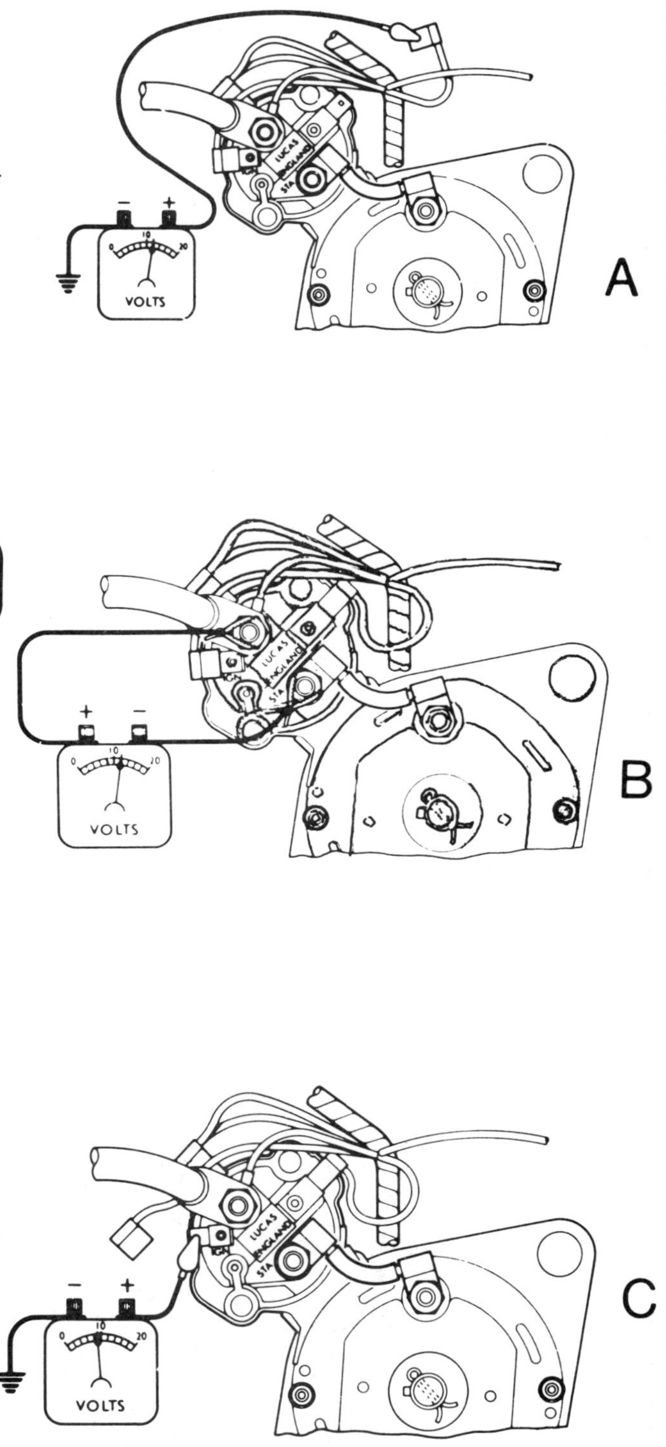

Fig 10:10 PRE-ENGAGED STARTER MOTORS

A Ignition switch feed to solenoid check
B Solenoid switch main contact check
C Cold start feed from solenoid check

5 If the previous test is favourable connect the voltmeter (or 12 volt bulb) to the two main terminals as shown in Fig 10:9(B). A 12 volt reading should be obtained (without touching the starter switch). If the starter switch is operated the reading should fall to zero (bulb goes out). If it does not, the solenoid switch needs renewal.
6 The terminal on the solenoid switch, to which the white lead is attached, is the cold start feed to the coil which directs current to the coil only when the solenoid is operating. It can be tested by connecting a voltmeter (or 12 volt bulb) to the terminal, after disconnecting the lead, as shown in Fig 10:9(C). When the starter is operated there should be a 9v reading at least (bulb glows). If not there must be a fault in the internal solenoid connection.
7 On pre-engaged starters the solenoid switch is mounted directly on top of the starter motor. For these units the procedure is as follows:
8 Disconnect the white/red wire from the solenoid and connect the voltmeter (or bulb) from the wire to earth. When the key start switch is operated the voltage should read 12v. Otherwise the wiring or start switch is faulty [Fig 10:10(A)].
9 Connect the voltmeter across the two main terminals of the solenoid and a 12v reading should be given. When the key start is operated the voltage should drop to zero. If otherwise, the solenoid needs renewal [Fig 10:10(B)].
10 To check the cold start feed to the coil, disconnect the white wire from the 'IGN' terminal and connect the voltmeter from the terminal to earth. When the start switch is operated the reading should be at least 9 volts [Fig 10:10(C)].
11 On either type of starter a solenoid fault confined to the non-functioning of the coil feed wire does not necessarily mean that the car will not start, although difficulty will certainly occur in cold weather of if the battery charge is low.
12 NOTE: When a new or reconditioned engine has been fitted it will be initially very stiff to turn. This could result in a very rapid discharge of the battery and slow turning of the engine before the engine has been successfully started. It is always advisable to have an additional battery available (in someone elses car perhaps) with a pair of jumper leads so that the extra power is available when it is really needed.
13 If the starter motor is the faulty item, it must be removed from the car for inspection. Make sure that it is not merely jammed. This can be ascertained by putting a spanner on the square end of the shaft which protrudes. If it turns easily, the starter is free. Otherwise use the spanner to turn the shaft in either direction until it is completely free.
This latter facility only applies to inertia type starters.

14 Starter motors - removal and replacement

1 Disconnect the battery earth lead from the negative terminal.
2 Take the alternator off (except with a pre-engaged type starter). Otherwise you will not be able to draw the starter motor far enough forward to disengage it from the bellhousing. With the pre-engaged starter, it will be necessary to remove the alternator, and the engine mounting and bracket from the side of the crankcase with all that this entails, as mentioned in Chapter 1.
3 Next disconnect the lead at the solenoid switch which goes to the starter. Where the solenoid is fitted directly on the starter, disconnect the other small leads also.
4 Remove the two bolts and lockwashers securing the starter motor to the clutch housing and lift out the starter.
5 Replacement is the reverse procedure of removal.

15 Starter motor M35 G/I - dismantling and reassembly

1 With the starter motor on the bench, loosen the screw on the cover band and slip the cover band off. With a piece of wire bent into the shape of a hook, lift back each of the brush springs in turn and check the movement of the brushes in their holders by pulling on the flexible connectors. If the brushes are so worn that their faces do not rest against the commutator, or if the ends of the brush leads are exposed on their working face, they must be renewed.
2 If any of the brushes tend to stick in their holders, then wash them with a petrol moistened cloth and, if necessary, lightly polish the sides of the brush with a very fine file, until the brushes move quite freely in their holders.
3 If the surface of the commutator is dirty or blackened, clean it with a petrol dampened rag. Secure the starter motor in a vice and check it by connecting a heavy gauge cable between the starter motor terminal and a 12-volt battery.
4 Connect the cable from the other battery terminal to earth in the starter motor body. If the motor turns at high speed it is in good order.
5 If the starter motor still fails to function, or if it is wished to renew the brushes, then it is necessary to further dismantle the motor.
6 Lift the brush springs with the wire hook and lift all four brushes out of their holders one at a time.
7 Remove the terminal nuts and washers from the terminal post on the commutator end bracket.
8 Unscrew the two through bolts which hold the end plates together and pull off the commutator end bracket. Also remove the driving end bracket which will come away complete with the armature.
9 At this stage if the brushes are to be renewed, their flexible connectors must be unsoldered and the connectors of new brushes soldered in their place. Check that the new brushes move freely in their holders as detailed above. If cleaning the commutator with petrol fails to remove all the burnt areas and spots, then wrap a piece of glass paper round the commutator and rotate the armature.
10 If the commutator is very badly worn, remove the drive gear as detailed in the following section. Then mount the armature in a lathe and, with the lathe turning at high speed, take a very fine cut out of the commutator and finish the surface by polishing with glass paper. DO NOT UNDERCUT THE MICA INSULATORS BETWEEN THE COMMUTATOR SEGMENTS.
11 With the starter motor dismantled, test the four field coils for an open circuit. Connect a 12-volt battery with a 12-volt bulb in one of the leads between the field terminal post and the tapping point of the field coils to which the brushes are connected. An open circuit is proved by the bulb not lighting.
12 If the bulb lights, it does not necessarily mean that the field coils will be earthing to the starter yoke or pole shoes. To check this, remove the lead from the brush connector and place it against a clean portion of the starter yoke. If the bulb lights, the field coils are earthing. Replacement of the field coils calls for the use of a wheel operated screwdriver, a soldering iron, caulking and riveting operations and is beyond the scope of the majority of owners. The starter yoke should be taken to a reputable electrical engineering works for new field coils to be fitted. Alternatively, purchase an exchange Lucas starter motor.
13 If the armature is damaged, this will be evident after visual inspection. Look for signs of burning, discolouration, and for conductors that have lifted away from the commutator. Reassembly is a straightforward reversal of the dismantling procedure.

16 Inertia starter pinion (M35 G/I and J/I) - dismantling and reassembly

1 The starter motor drive is of the outboard type. When the starter motor is operated the pinion moves into contact with the flywheel gear ring by moving in towards the starter motor.
2 If the engine kicks back, or the pinion fails to engage with the flywheel gear ring when the starter motor is actuated, no undue strain is placed on the armature shaft, as the pinion sleeve disengages from the pinion and turns independently.
3 It is essential to obtain a press or clamp which can safely compress the heavy spring at the end of the shaft. Fig 10:13 shows the official tool (JWP 376) used by Vauxhall agents for this, but if one possesses a vice and a little ingenuity it can be done in other ways. As soon as

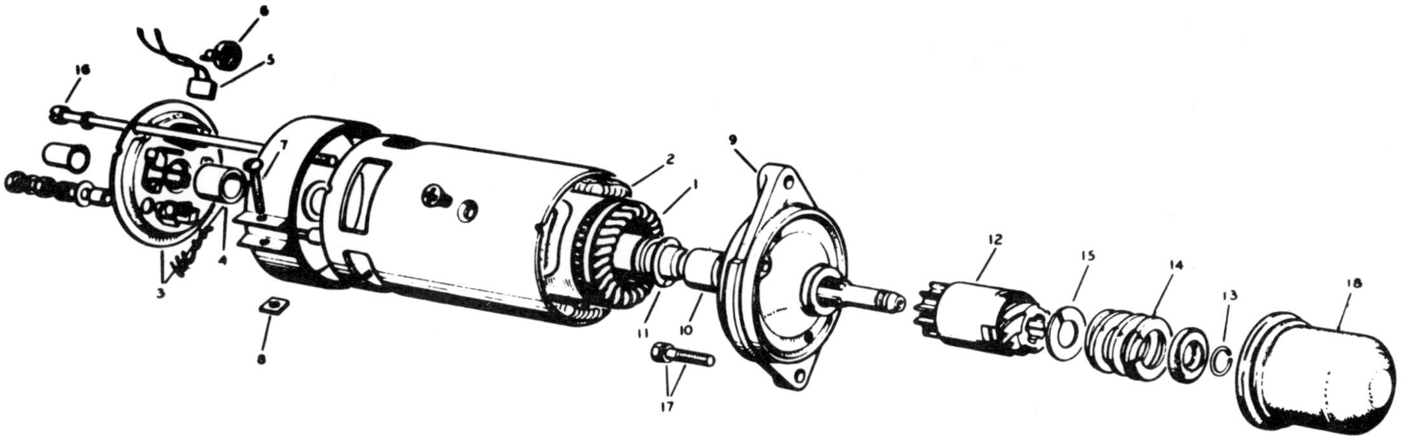

Fig 10:11 STARTER MOTOR LUCAS M35G/I – EXPLODED VIEW

1 Armature
2 Field coils
3 Commutator end bracket
4 End bracket bush
5 Brush
6 Brush spring
7 Cover screw
8 Nut
9 Drive end bracket
10 End bracket bush
11 Thrust washer
12 Drive pinion and slave
13 Jump ring
14 Main spring
15 Thrust washer
16 Through bolt
17 Mounting bolt
18 Rubber boot

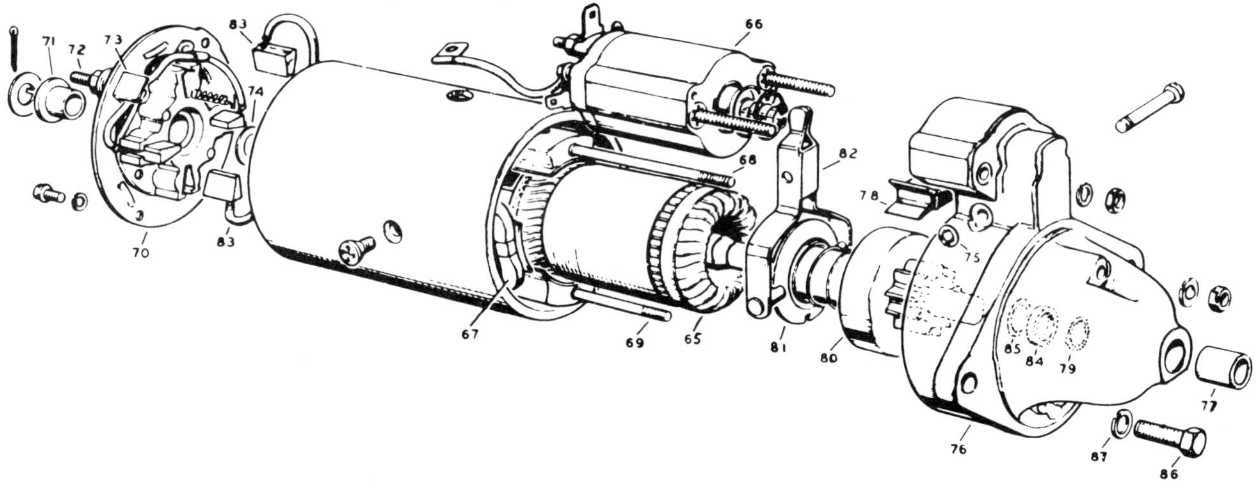

Fig 10:12 STARTER MOTOR M35K/PE – EXPLODED VIEW

65 Armature
66 Solenoid
67 Field coil
68 Pole piece and long stud
69 Pole piece and short stud
70 Commutator end bracket
71 Commutator end bracket bush
72 Field terminal
73 Terminal insulating bush
74 Thrust plate
75 Pivot pin retaining clip
76 Drive end bracket
77 End bracket bush
78 Grommet
79 Jump ring
80 Roller clutch drive
81 Bearing bush
82 Lever and pivot assembly
83 Brush
84 Thrust collar
85 Shim
86 Fixing bolt
87 Lockwasher

the pressure is taken off the circlip it can be removed with a screwdriver. Then release the pressure on the spring.

4 Remove the spring collar and spring.

5 Slide the remaining parts with a rotary action off the armature shaft.

6 Reassembly is a straightforward reversal of the above procedure. NOTE: It is most important that the drive gear is completely free from oil, grease and dirt. With the drive gear removed, clean all the parts thoroughly in paraffin. UNDER NO CIRCUMSTANCES OIL THE DRIVE COMPONENTS Lubrication of the drive components causes dust to adhere and results in sticking.

17 Starter motor bushes (M35 G/1) - inspection, removal and replacement

1 With the starter motor stripped down check the condition of the bushes. They should be renewed when they are sufficiently worn to allow visible side movement of the armature shaft.

2 The old bushes are simply driven out with a suitable drift and the new bushes inserted by the same method. As the bearings are of the phospher bronze type it is essential that they are properly saturated with engine oil before fitting.

18 Starter motor (M35 J/1)

The M35 J/1 starter is different from the G/1 only in so far as a face end commutator is used. In this respect it is almost identical to the M35 K/PE starter dealt with in the next sections.

19 Pre-engaged starter motor (M35K/PE) - general description

Some Vivas are fitted with pre-engaged type starters. The M35 K/PE model is a heavy duty version. End face commutators and four wedge section brushes are used. The drive pinion engagement is by means of a solenoid actuator and starter switch mounted on the starter motor. The starter motor does not turn until the pinion is engaged with the flywheel ring gear. To provide for the possibility of over-run the pinion is driven through a one way roller clutch.

20 Pre-engaged starter motor (M35K/PE) - dismantling and reassembly

1 It will be necessary to dismantle the starter motor if checks indicate that failure to turn the engine is due to faults with it. After some time the brushes will also wear sufficiently to warrant renewal.

2 The solenoid may be removed after detaching the short connecting cable from the other main terminal and removing the two securing nuts. If this is all that needs replacing, a new one can be fitted over the existing plunger now.

3 To dismantle the motor further, remove the short bolts, or through bolts, which hold the commutator end bracket with the brush gear to the main yoke. Also remove the split pin, washers and shims from the end of the shaft. The end bracket may then be carefully removed. Do not lose the thrust washer which is located over the end of the shaft inside.

4 Next remove the bolts holding the drive end cover in position (if necessary) and the end cover complete with armature and shaft may be drawn out of the yoke.

5 To separate the end cover from the armature it will be necessary to drive out the pin on which the engagement lever pivots (see Fig 10:14).

6 To take the drive pinion and clutch assembly off the armature shaft it will be necessary to drive the thrust collar down the shaft with a piece of suitable tube and then remove the circlip which is exposed (Fig 10:15). If the driving gear assembly is worn, the whole unit should be renewed. To check that the roller clutch is in good condition, it should lock and take up the drive in one direction immediately it is turned. When turned in the opposite direction it should rotate smoothly and evenly. The whole clutch unit should also slide easily and without excessive play along the splines of the armature shaft.

7 Examine the brushes to ensure that they are not less than the permitted minimum length of .38 inch (9.5 mm). If they need renewal, first obtain the new ones. Two will be supplied complete with their terminal post and the other two separately for soldering to the field coil end tags. The field coils are made from copper strip and the new brushes may be soldered directly on to the tag ends (Fig 10:16).

8 Clean up the face of the commutator with a petrol moistened rag. Light scoring may be cleaned up with fine glass paper. If there is deep scoring the commutator face may be skimmed in a lathe, provided it does not diminish in thickness below .080 inch (2 mm) (Fig 10:17). Do not undercut the segment insulation.

9 If it is suspected that the field coil insulation is gone, it is recommended that it is removed, checked and repaired by specialists with the proper equipment.

10 Check that the shaft is a good fit into each of the end plate bushes and renew the bushes if necessary.

11 Assembly is a procedure that must be carried out in sequence with attention to several points to ensure that it is correct.

12 First assemble the engagement lever to the solenoid plunger so that the chamfered corner faces the solenoid. Then make sure that the retaining plate is correct, relative to the lever (Fig 10:18).

13 Next fit the drive pinion and clutch assembly on the armature shaft, fit the engagement lever fork to the clutch and assemble the whole lot together to the drive end cover. Then fit a new lever pivot pin and peen over the end to prevent it coming out.

14 Replace the yoke, and then lightly screw up the end cover bolts (if fitted).

15 Next place the thrust washer over the commutator end of the shaft, fit all the brushes into their appropriate holders in the end cover and replace the end cover on to the shaft (Fig 10:19). Both end covers have locating pips to ensure they are fitted correctly to the yoke.

16 Replace the through bolts and tighten them up at both ends.

17 Replace the thrust washer and shims to the end of the shaft, install a split pin, and then measure the end float gap between the thrust washer and the end cover with a feeler gauge. It should be no more than .010 inch (.25 mm) (Fig 10:20). Additional shims should be added to reduce the end float as required.

18 Next replace the rubber pad between the drive end bracket (under the solenoid plunger housing) and the yoke, and refit the solenoid. Reconnect the short cable to the solenoid terminal.

19 Before replacing the starter in the car after reassembly, it is a good idea to check that it is functioning properly by connecting it temporarily to the battery.

21 Fusible link, fuses and thermal circuit breaker

1 So that complete protection of the wiring circuit is assured the main feed cable from the battery has a section consisting of a fusible link which will burn out in the event of the fuse circuits failing or being bridged. It runs from a connector on the battery tray to the feed terminal on the solenoid switch, or a special connector into the battery cable on those cars fitted with pre-engaged starter motors.

2 Four fuses are fitted in a separate fuse holder positioned on the engine side of the dash panel. The area adjacent to each fuse is marked from 1 to 4 for easy identification.

3 35 amp fuses are used. They protect the circuits as listed in the 'Specifications' at the beginning of the Chapter.

4 If any of the fuses blow, check the circuits on that fuse to trace the fault, before renewing the fuse.

5 Headlamp and sidelamp circuits are protected by a thermal circuit breaker. It can be tested by putting an ammeter and variable resistance

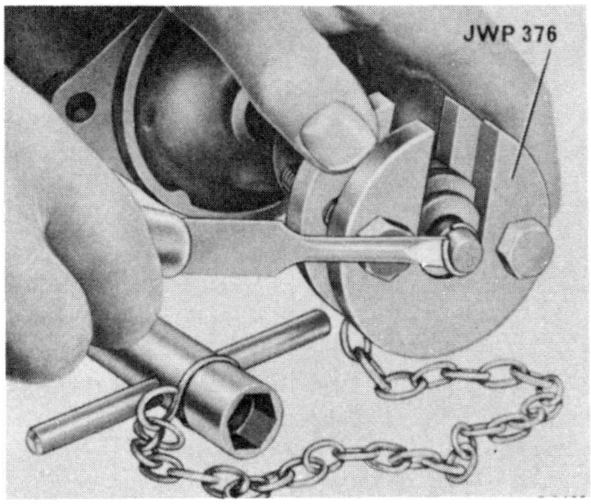

Fig 10:13 Inertia starter pinion - spring clamp for removal of jump ring

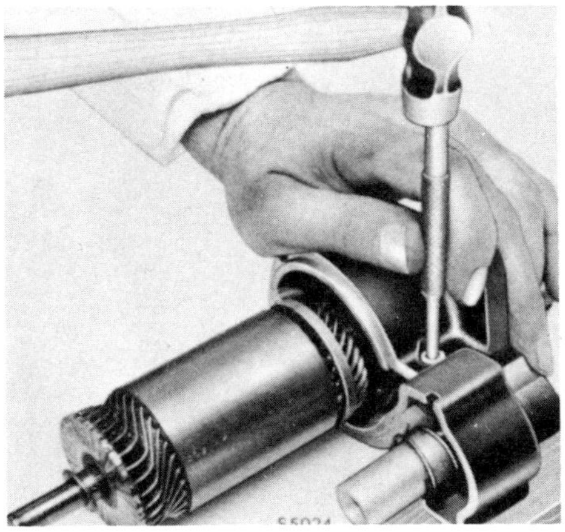

Fig 10:14 Pre-engaged starter pinion - driving out the pivot pin

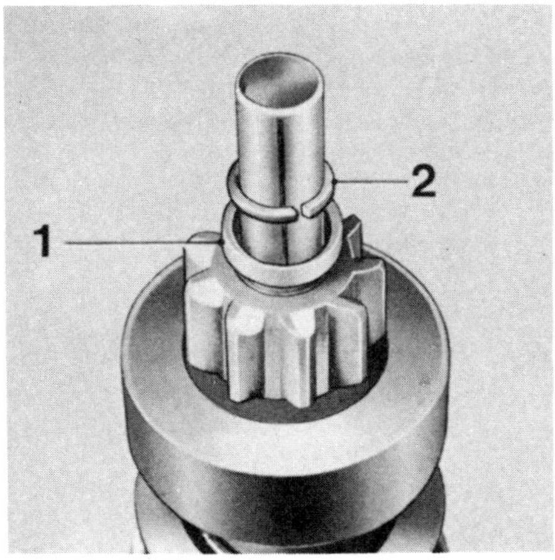

Fig 10:15 PRE-ENGAGED STARTER PINION – PINION REMOVAL
1 Thrust collar 2 Jump ring

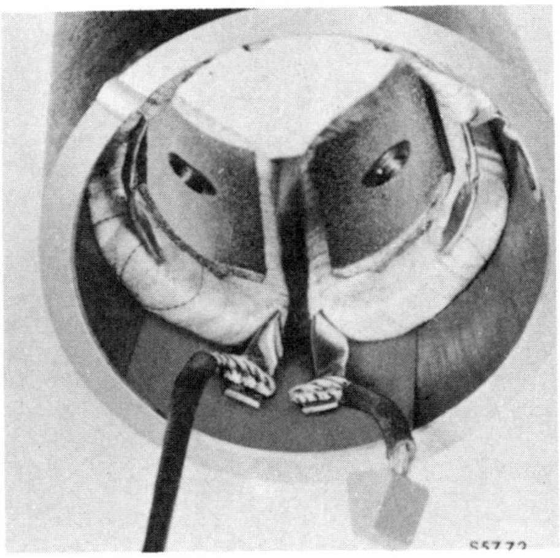

Fig 10:16 M35K/PE starter motor - field coil brushes

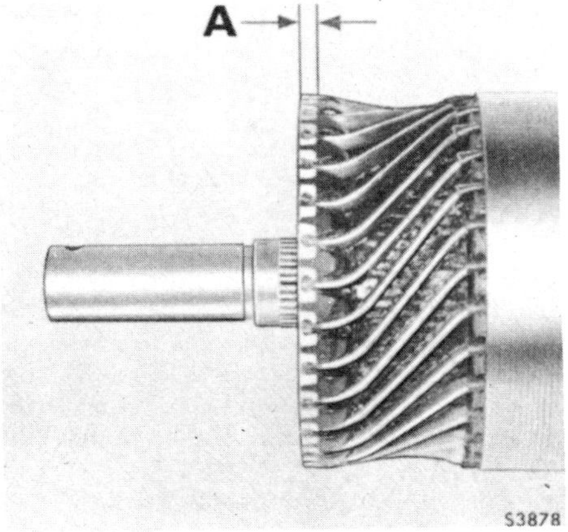

Fig 10:17 M35K/PE STARTER – FACE END COMMUTATOR THICKNESS
A = .080 inch (2 mm) minimum

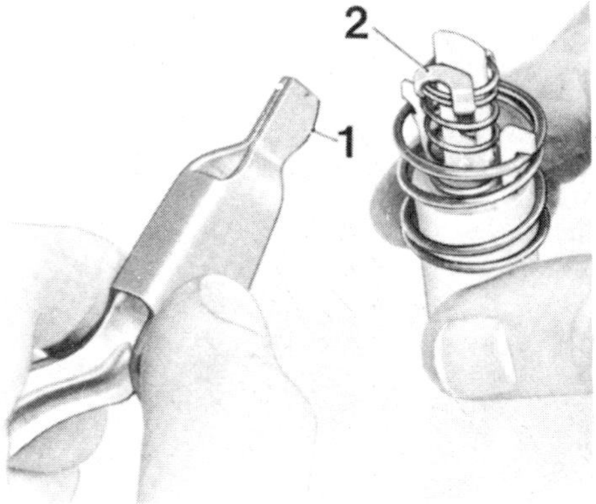

Fig 10:18 PRE-ENGAGED STARTER PINION ASSEMBLY – FITTING LEVER TO SOLENOID PLUNGER
1 Radiused edge of lever 2 Retaining plate

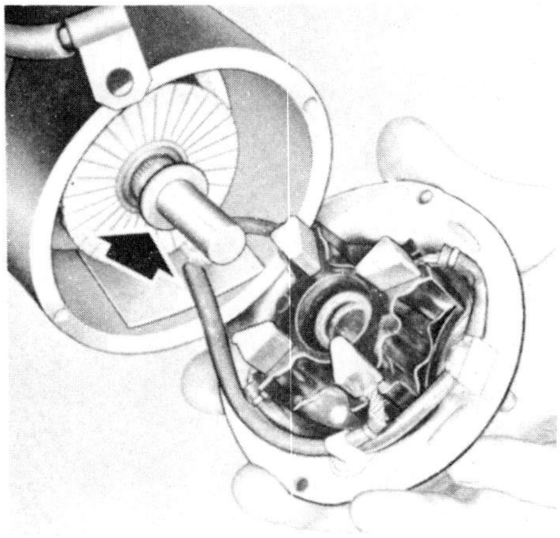

Fig 10:19 M35K/PE starter motor - assembling commutator end cover. Note thrust washer (arrowed)

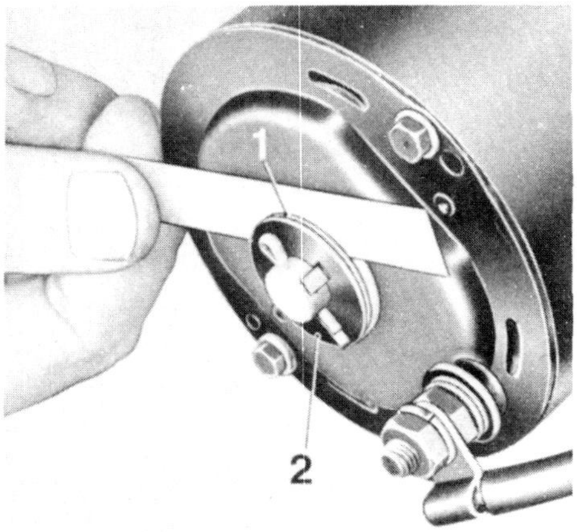

Fig 10:20 M35K/PE STARTER MOTOR – CHECKING ARMATURE END FLOAT
1 Thrust plate 2 Shims

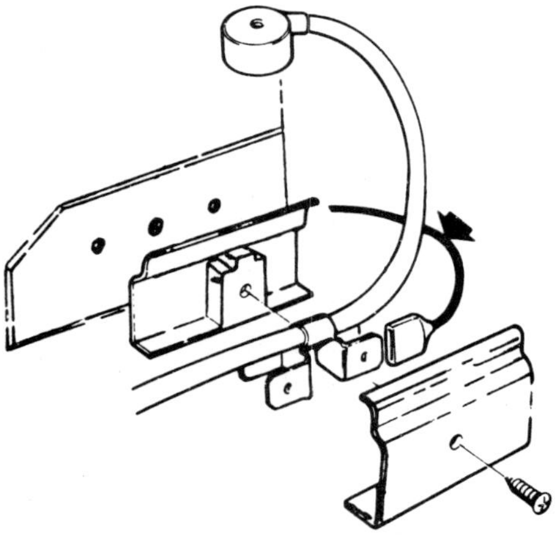

Fig 10:21 Fusible link (arrowed) - models with pre-engaged starters connect to the battery cable (instead of the starter solenoid) by a special connector on the battery box

in the circuit in series. The breaker should open at a current of 33 amps after ½–3 minutes at 20°C.

6 The fuse block is mounted on the right hand side of the front wheel arch. The connectors to the fuse terminals, however, are inside underneath. The thermal circuit breaker is alongside the terminals.

22 Electrical wiring - general

1 The modern developments in automobile wiring systems are perhaps more apparent in the Viva than in many other cars. The total absence of a maze of wires behind the instrument panel is due to the use of a printed circuit for all lights and instruments on the panel.

All individual circuit wires are gathered into neat harnesses secured to the body panels with clips. Multi-socket connectors joint the three main harnesses together. The three being: engine harness, engine compartment harness, and a lighting and instrument harness.

2 The complete wiring protection offered by the fusible link, thermal link and fuses, means that any faults developing in any electrical component will not overload the wiring sufficiently to cause it to need repair. It can confidently be said, therefore, that unless the wiring harnesses are abused in any way, they are virtually foolproof.

3 Checks can be made with the aid of a wiring diagram. Unless there is an obvious major burn up which would need renewal of the complete section of harness anyway, attempts should not be made to renew or bypass individual sections of any one particular circuit. A mistake could lead to major damage in view of the delicate nature of the charging circuit.

23 Instrument voltage stabiliser unit

In order that all gauges shall read accurately, a voltage stabiliser unit is plugged into the instrument panel printed circuit. This overcomes fluctuations in battery voltage which could cause inaccurate readings.

If suspect, it can be unplugged and tested with a 12 volt battery and voltmeter.

Connect them as shown in Fig 10:25. The voltmeter needle should pulsate evenly on each side of 10 volts.

If there is no reading, or the needle remains static, the unit is faulty and must be renewed.

24 Flasher circuit - fault tracing and rectification

1 The flasher unit is clipped to the top edge of the vertical parcel shelf panel alongside the steering column. On models fitted with a hazard warning system a second unit is mounted alongside (nearer the driver) (Fig 10:26).

2 If, when operated to either side, the warning lamp on the dashboard stays on but does not flash and there is no audible warning, then one or both signal bulbs are not working.

3 If the warning bulb does not light and the frequency of flashing is slightly reduced, then the warning bulb is defective.

4 Before assuming that anything else is wrong (other than blown bulbs) make quite sure that the bulb caps and sockets are perfectly clean and corrosion free. Water, which may seep in due to a leaking seal or cracked lens, could cause this sort of trouble. Do not forget to check the fuse also.

5 Where there is total failure, the simplest check is to substitute a known good flasher unit. Finally check the switch if no other check reveals anything wrong.

6 The combined dip, horn, headlamp flasher and turn signal switch is mounted on the steering column with a clamp. Access to it can be gained after removing the column shroud. If the wires for the flasher circuit are detached and bridged, the fault can be isolated to either switch or wiring.

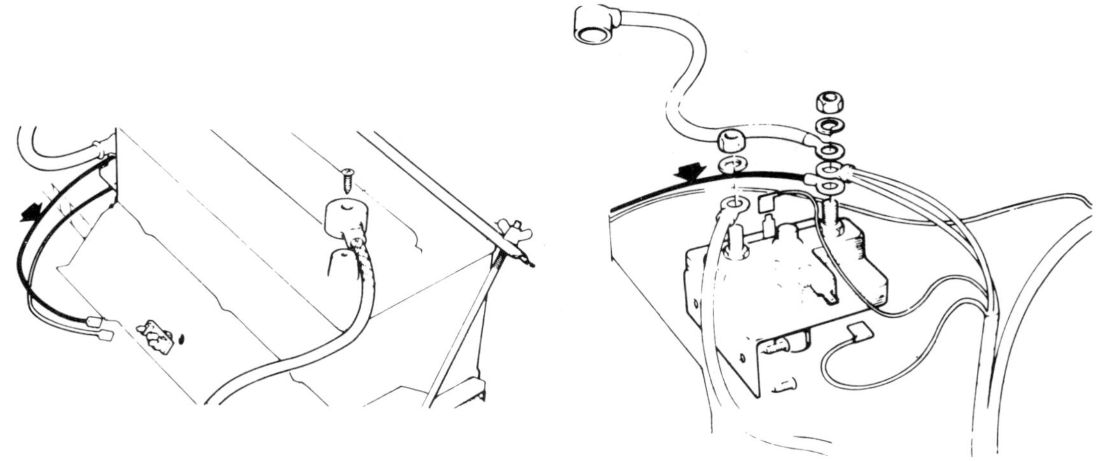

Fig 10:22 Fusible link (arrowed) showing connections to terminal block on battery tray and to positive terminal on starter solenoid switch

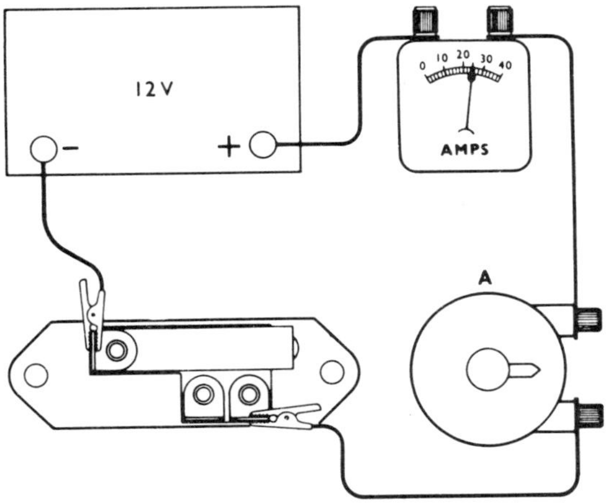

Fig 10:23 Thermal circuit breaker - testing circuit. 'A' is a variable resistance

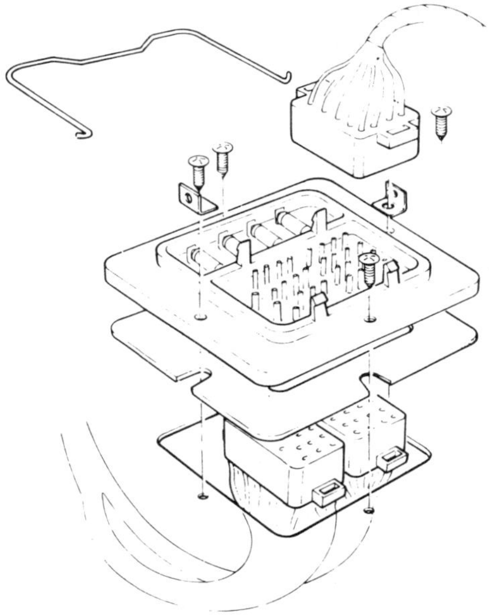

Fig 10:24 Wiring system - bulkhead multiple connector socket and combined fuse box

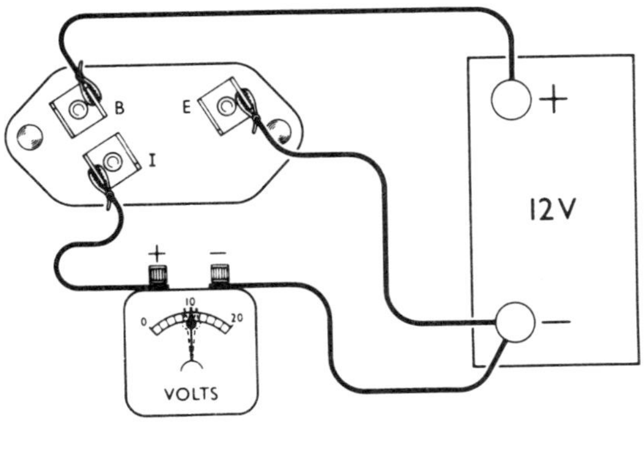

Fig 10:25 Instrument voltage stabiliser unit. Connections for testing

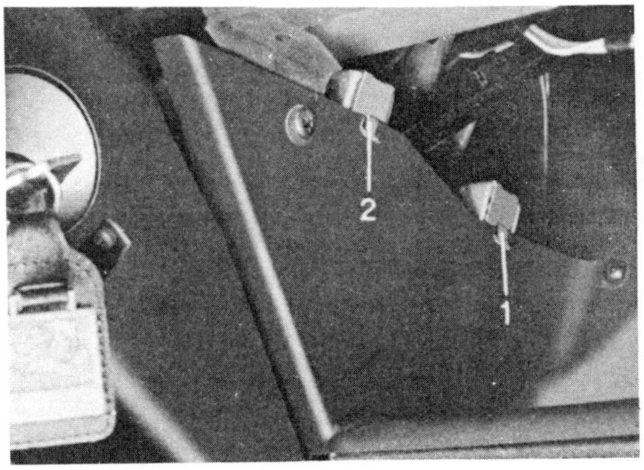

Fig 10:26 FLASHER CIRCUIT – POSITION OF FLASHER UNIT

1 Turn indicator unit 2 Emergency (all flashers) unit

25 Horn - fault tracing and rectification

1 If the horn works badly or fails completely, check the wiring leading to it for short circuits and loose connections. Check that the horn is firmly secured and that there is nothing lying on the horn body.
2 If the horn still does not work, or operates incorrectly, check the switch. This can be done by unscrewing the end cap on the combined switch. Bridge the two terminal blades and, if the horn works, check the contact piece inside the cap (Fig 10:28)
3 The horn adjusting screw is sealed on assembly to discourage tampering. If you do decide to tamper, it should be as a last resort only.
Turn it anticlockwise to obtain silence and clockwise to make noise to an acceptable level.

26 Headlamps and sidelamps - bulb and lens removal and replacement

1 The pre-focus headlamp units are fitted with detachable bulbs, and both side and headlamp bulbs are incorporated in the same housing.
2 To remove the bulb holder, raise the bonnet and pull the rubber cover off the back of the light unit. The lamp holder unit can then be detached and the bulb removed. The sidelamp bulb is a simple push fit into the unit below the headlamp bulb.
3 To remove the glass lens unit, it is first of all necessary to remove the radiator grille (see Chapter 12).
4 After the grille is removed, a spring clip at the lower edge of the glass is accessible. Remove this and the glass may be lifted out.
5 Replacements are a reversal of the removal procedures.
6 Headlamp beam adjustment is best done with proper equipment. As a guide, lights should be set when dipped and the light pattern should be as shown in the diagram. The horizontal level of the beam of light must be below the lamp centre line. The vertical division of the light area, where the horizontal joins the part which slopes away, must be to the left (right hand drive) of a vertical line between the two lamps. The adjuster screw is located on the inside edge of the headlamp surround panel under the bonnet. It has a small rectangular headed shank and care should be taken not to damage it when adjustments are made (Fig 10:29).

Twin headlamps

7 The four lamp system comprises either sealed beam or Unified European pre-focus type light units, according to the regulations of the country in which the car is marketed.
8 All four lamps are in operation on main beam, but when dipped, only the outer lamp operates on a dip or supplementary main beam filament.
9 Access to the lamp bulb on pre-focus units is from the engine compartment. The spring clip which secures the bulb can be released after squeezing the sides of the plastic cover and the bulb can be withdrawn from the lamp mounting ring.
10 To withdraw a light unit, remove the appropriate radiator grille insert (see Chapter 12), then slacken the three crosshead retaining screws and turn the unit anticlockwise.
11 The headlamps are secured to the front end panel by four screws and speed nuts; these are accessible with the radiator grille inserts removed.
12 Headlamp beam adjustment is best done with proper optical alignment equipment. However, if the lamps are seriously out, adjustment can be carried out on the two screws whose rear ends project into the engine compartment. Ideally, they are turned with a proper square section key.
13 The inner lamp should have a beam setting straight forward with $\frac{1}{2}^{o}$ downwards deflection. The outer lamps should be set when on dipped beams. They should then aim 2° down and 2° left (RHD) or 2° right (LHD).

27 Rear lights and flasher bulbs - removal and replacement

1 Access to the rear lamp bulbs is from inside the luggage boot.
2 Remove the cover and then pull out the appropriate bulb holder held in by spring clips round the edge of the holder.
3 Replace the bulb in the holder and then refit the holder so that the tongues locate in the spaces in the holder.

28 Windscreen wipers - fault finding

1 If the wipers do not work when they are switched on first check the No 3 fuse. If this is sound then there is either an open circuit in the wiring or switch, the wiper motor is faulty, or the pivot spindles or linkages may be binding.
2 If the wipers work intermittently then suspect a short circuit in the motor. Alternatively, the armature shaft end float adjustment may be too tight or the wiper linkage may be binding.
3 Should the wipers not stop when they are turned off there must be a short circuit in the switch or wiring.

29 Windscreen washer - fault finding

1 If the windscreen washers do not work when pumped, first check that there is water in the washer reservoir and that the jets in the discharge nozzles are clear (poke them with a pin).
2 Examine the water pipe connections at all junctions to ensure they are firmly fitted.
3 If there is still no jet from the screen nozzles, detach the pipes from the pump unit and remove the unit from the dashboard. The tube from the reservoir to the pump is coloured green and from the pump to the jets, yellow.
4 In a bowl of water, submerge the inlet of the pump, operate it, and water should come from the outlet only under water, when bubbles will come out. After a few strokes in this manner release the plunger and lift the outlet out of the water. If, on operating the pump again, some water comes from the unit then it means that the non-return valves inside are not functioning properly and the unit should be replaced. If the pump is satisfactory then the only possible faults can be in the suction and delivery pipes, unions or nozzles, all of which must be carefully examined for splits, kinks, blockages or leaking connections.
5 On models fitted with two speed wipers, the washer pump is electrically driven and is mounted on top of the water reservoir under the bonnet (Fig 10:32).
6 It is possible to renew a worn impeller by removing the end cover, seal, distance piece and washer. The procedure for this is straightforward and can be followed from the exploded illustration (Fig 10:34).

30 Windscreen wiper motor - removal and replacement

1 The windscreen wiper motor has to be removed complete with the wiper operating links and arms, which are held in a rigid frame and comprise the complete windscreen wiper assembly.
2 Take off the windscreen wiper arms.
3 From under the bonnet remove the panel shrouding the ventilator. The wiper motor and crank assembly can be seen and it is held by two bolts through rubber bushes in the bracket. To get at the bolts, position the links by running the wiper and switching off the ignition when they are in the proper place. If the motor is no use, remove the link arms by detaching the clips.
4 When the bracket mounting bolts are removed, the whole assembly can be eased down and out and the electrical connections detached from the motor.
5 Having unclipped the links, the motor may be removed from the mounting bracket.

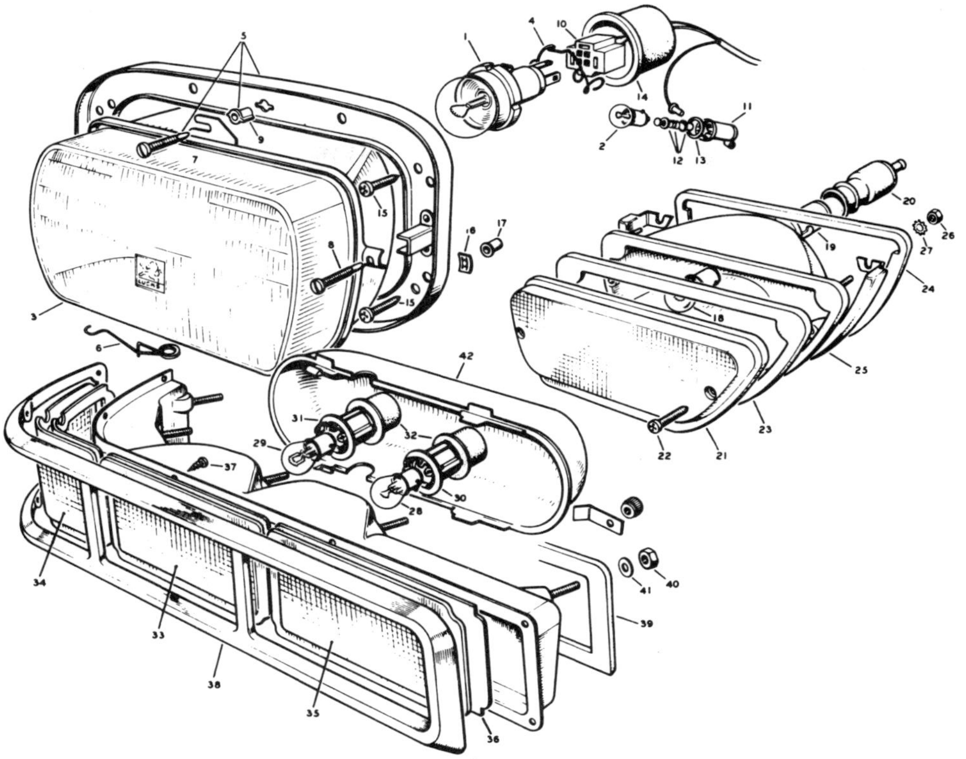

Fig 10:27a HEADLAMPS, FLASHERS AND REAR LAMP CLUSTERS – COMPONENTS – EXPLODED VIEW

1 Headlamp bulb
2 Sidelamp bulb
3 Lens unit
4 Bulb retaining clip
5 Mounting frame assembly
6 Retaining spring
7 Vertical adjuster
8 Horizontal adjuster
9 Retainer
10 Adaptor
11 Bulb holder
12 Bulb holder interior
13 Seating ring
14 Boot
15 Screw
16 Captive nut
17 Protector
18 Front flasher bulb
19 Bulb holder
20 Grommet
21 Amber lens
22 Lens screw
23 Seal
24 Seal
25 Moulding
26 Nut
27 Lockwasher
28 Flasher bulb
29 Rear lamp and stop bulb
30 Holder - flasher bulb
31 Holder - lamp/stop bulb
32 Dust cover
33 Flasher lens
34 Tail/stop lamp lens
35 Reflector lens
36 Lens seal
37 Screw
38 Bezel
39 Seal
40 Nut
41 Washer
42 Rear lamp cover

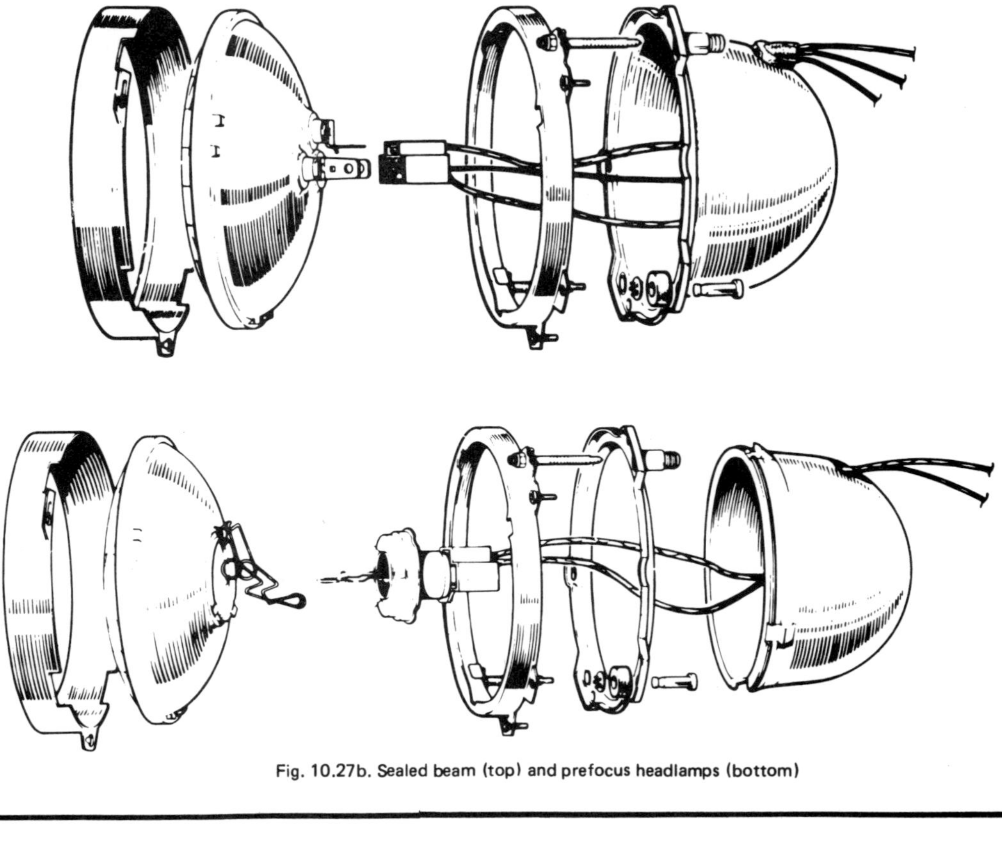

Fig. 10.27b. Sealed beam (top) and prefocus headlamps (bottom)

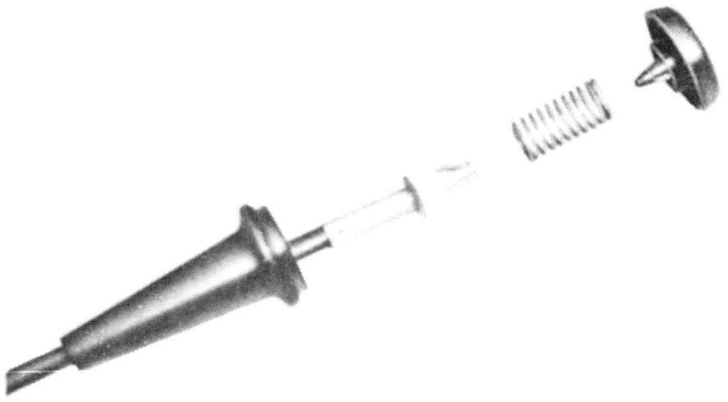

Fig 10:28 Horn switch - exploded illustration

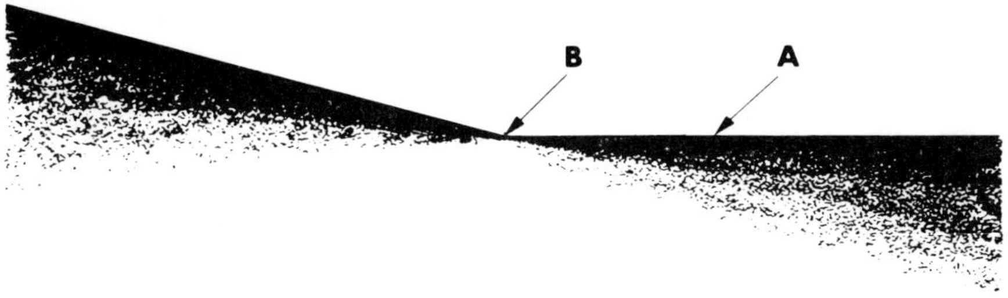

Fig 10:29 HEADLAMP BEAM ALIGNMENT (DIPPED)
A Horizontal light border
B Junction of horizontal and angled light

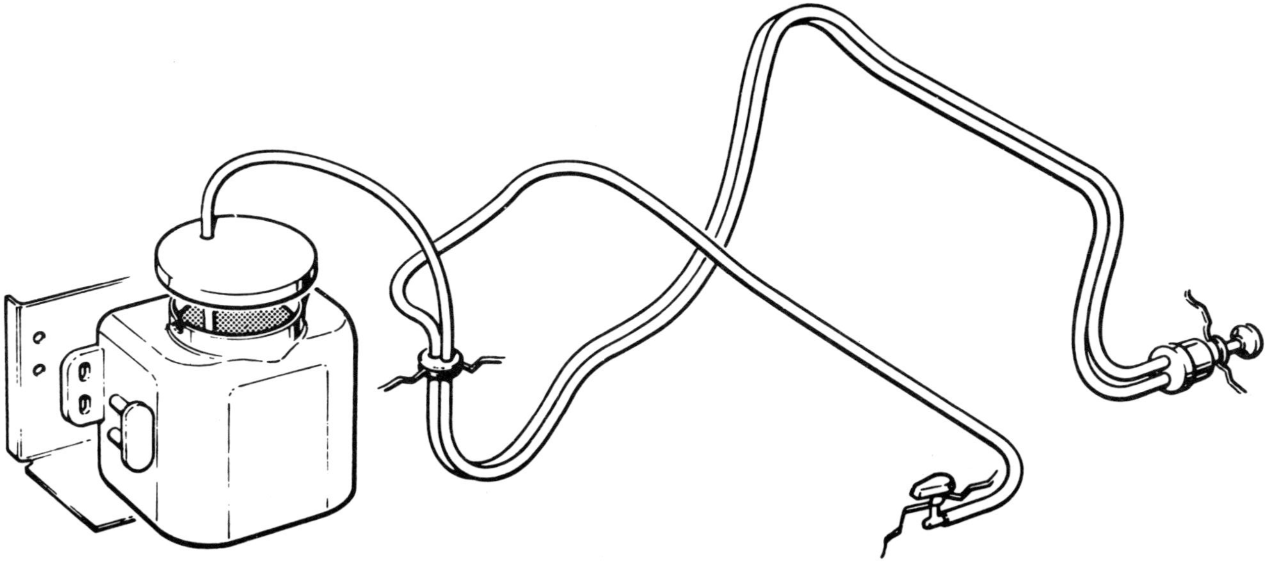

Fig 10:31 Windscreen washer - manually operated pump schematic layout

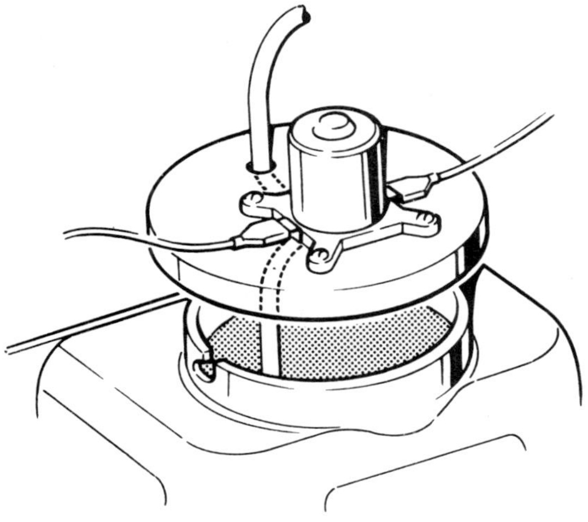

Fig 10:32 Windscreen washer - electric pump mounted on top the water reservoir

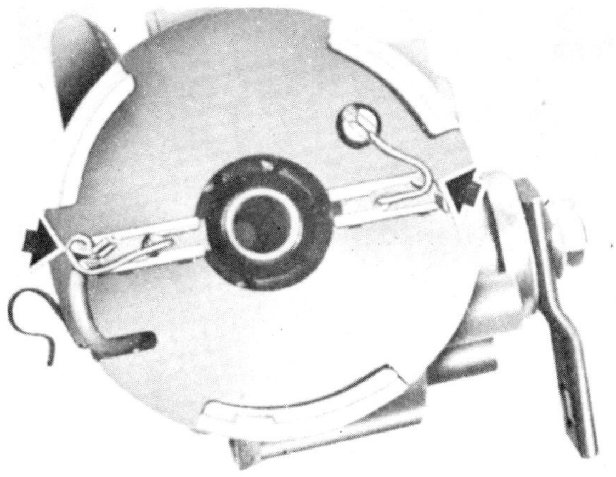

Fig 10:33 Windscreen wiper motor - terminal plate and carbon brushes (single speed motor)

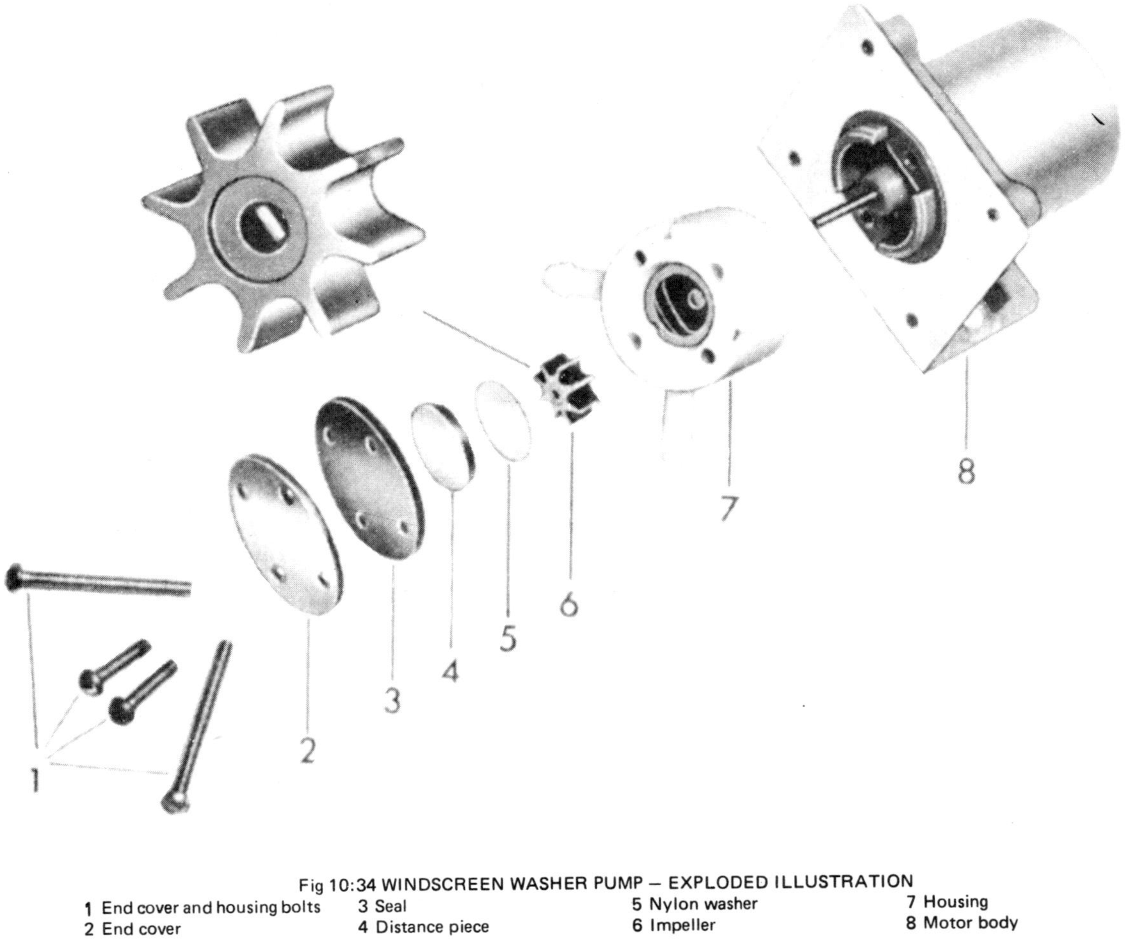

Fig 10:34 WINDSCREEN WASHER PUMP – EXPLODED ILLUSTRATION

1 End cover and housing bolts
2 End cover
3 Seal
4 Distance piece
5 Nylon washer
6 Impeller
7 Housing
8 Motor body

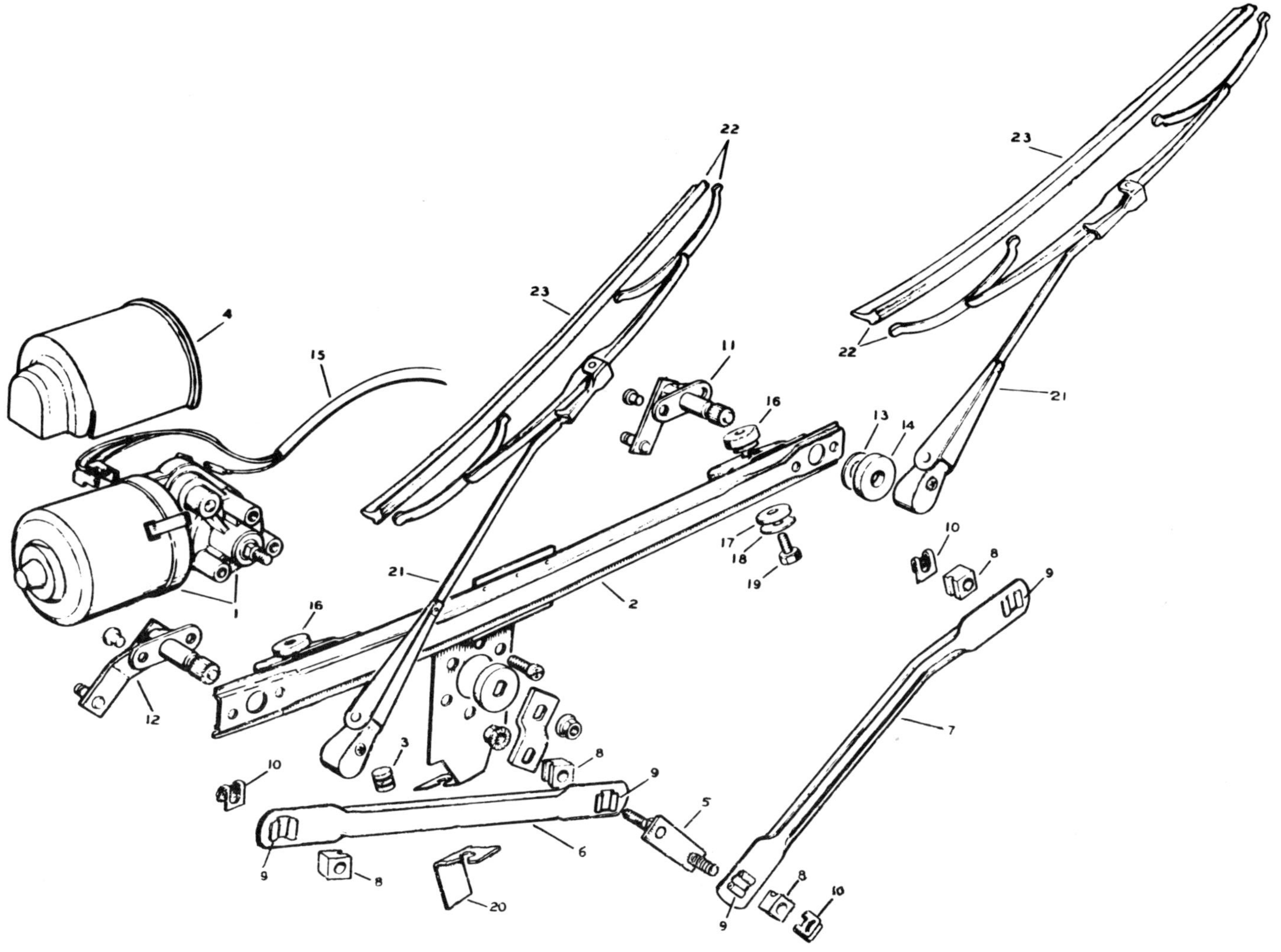

Fig 10:35 WINDSCREEN WIPER MECHANISM – COMPONENTS – EXPLODED VIEW

1 Motor assembly
2 Wiper mounting bracket
3 Retainer
4 Motor cover
5 Crank and pin assembly
6 Drive link
7 Cross link
8 Link fixing bush
9 Link fixing clip
10 Link fixing clip
11 Wiper pivot housing (driver side)
12 Wiper pivot housing (passenger side)
13 Wiper spindle washer
14 Wiper spindle seal
15 Wiper wiring harness
16 Frame insulator
17 Washer
18 Washer
19 Screw
20 Motor support bracket
21 Wiper arm
22 Wiper blade assembly
23 Squeegee

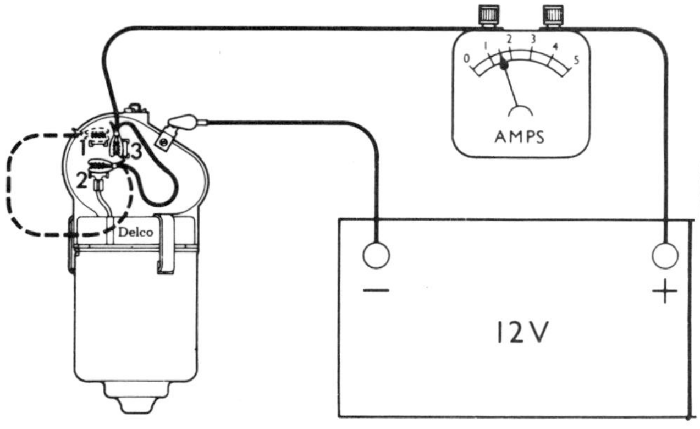

Fig 10:37 Windscreen wiper motor (single speed) - running test. Connection of battery and ammeter. Continuous line shows connection for normal running. Dotted line from terminal 2 to 1 shows additional lead connected to test self-park

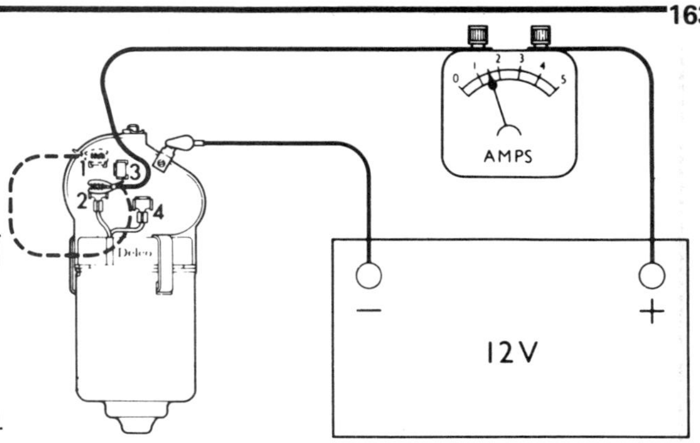

Fig 10:38 Windscreen wiper motor (2–speed) - running test. Connection of battery and ammeter. Continuous line shows connection for low speed running. For high speed running move lead from terminal from 2 to 4.
To check self-park move supply lead to terminal 3 and add another lead (dotted line) between terminals 1 and 2

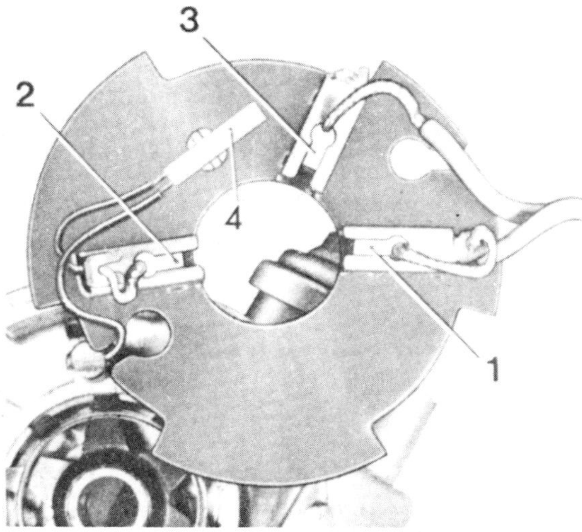

Fig 10:39 WINDSCREEN WIPER MOTOR (2–SPEED) ILLUSTRATION OF BRUSH PLATE
1 and 2 Slow speed brushes 4 Safety cut-out bi-metal strip
2 and 3 High speed brushes

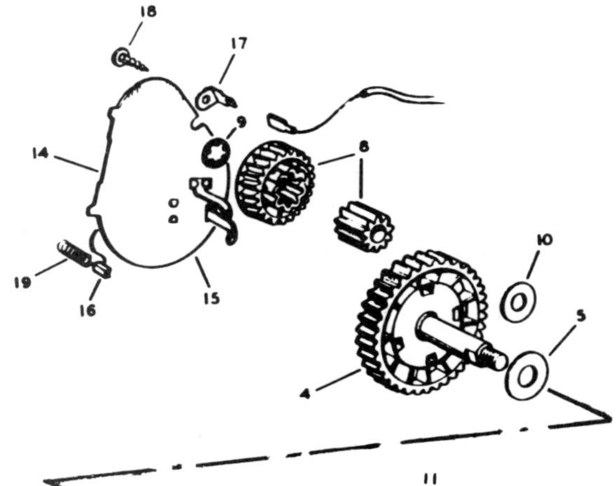

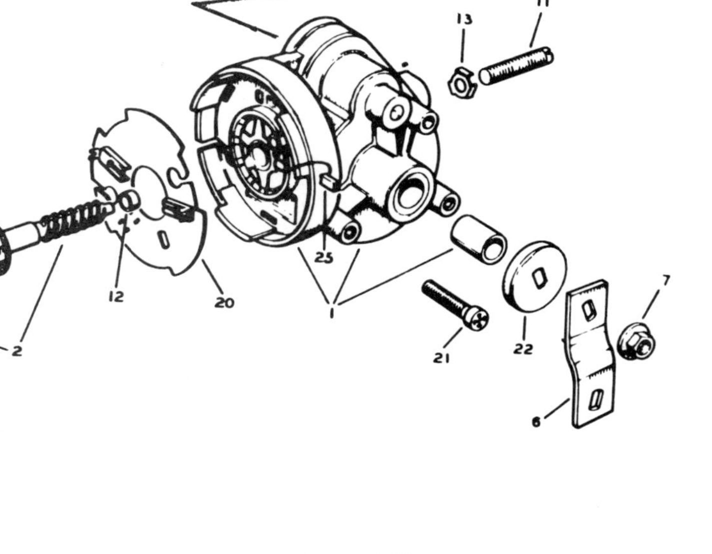

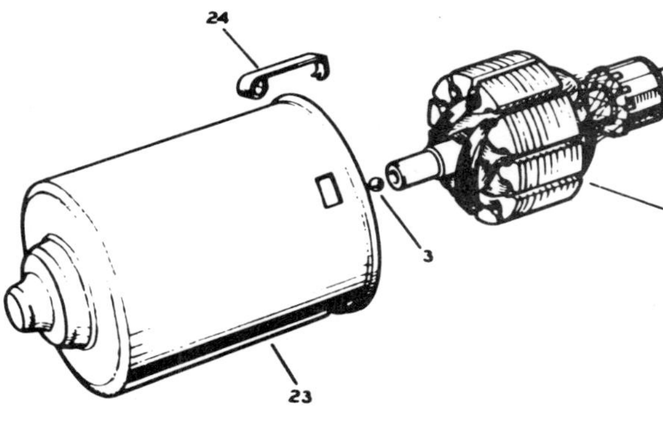

Fig. 10.36. WINDSCREEN WIPER MOTOR – COMPONENTS – EXPLODED VIEW

1 Housing and bearings assembly
2 Armature
3 Thrust ball
4 Cross shaft and gear
5 Thrust washer
6 Thrust washer
7 Crank locknut
8 Worn wheel and pinion
9 Worn wheel retainer
10 Thrust washer
11 Thrust screw
12 Thrust bearing
13 Locknut
14 Gear cover
15 Contact assembly
16 Brush and terminal assembly
17 Terminal
18 Terminal screw
19 Brush spring
20 Brush gear assembly
21 Cover screw
22 Water shield
23 Commutator end frame
24 End frame clip

31 Windscreen wiper motor - dismantling, inspection and reassembly

1 Other than for normal wear, the bearings and gears in the motor should not deteriorate and if for any reason the motor should cease to function altogether, it is probably due to the wiper mechanism jamming or seizing which has overloaded the motor and burnt it out. In such instances the purchasing of either armature or field coils, and probably brushes as well, is hardly comparable to buying an exchange unit. If the motor ceases to function for no immediately obvious reason proceed as follows:
2 Remove the end frame by levering out the two retaining clips.
3 Remove the screw securing the terminal end cover to expose the drive gears.
4 The end frame and armature can be withdrawn together, care being taken to prevent loss of the nylon thrust bearing at the end of the worm shaft.
5 The brush plate and gear cover are a complete assembly and can be withdrawn together.
6 If any of the spindle brushes are worn, the cover complete (in which they are fitted) will need renewal, as they are not serviced (by Vauxhall) separately.
7 New carbon brushes may be fitted. One of them has to be soldered to the existing lead, so when cutting the old lead leave sufficient length for soldering (Fig 10:33).
8 One of the most important features of reassembly is to make sure the end float in the armature and cross shafts is within specification. Excessive end float is what causes jamming and stiffness leading to overloading and a burnt out motor.
9 Make sure that the water shield and crank are assembled to the cross shaft so that the relative positions of the self-park segment and crank are correct as indicated in the illustration (Fig 10:40).
10 Before reassembly begins lubricate the armature end frame bush with engine oil and put anti-scuffing paste on the thrust ball recess in the shaft.
11 Hold back the carbon brushes by hooking the leads over the tags on the brush holders. Then they can be released through the clip holes in the end cover, after the end cover has been replaced.
12 When the motor is reassembled, test it with an ammeter in circuit if possible. Even though it runs, if the current used is above 1 amp then trouble may be expected because something is stiff and causing extra loading which the windup cannot take. While the motor is running, tap the end frame to help settle the bearings. If necessary adjust the end thrust screw but watch the loading on the ammeter whilst doing so to see that it does not increase more than .1 amp (Fig 10:37, 10:38).
13 Test that the self-park works also by re-arranging the wiring as mentioned in the connection illustration (Fig 10:40).
14 If any of the nylon bushes in the pivots and link arms are worn, they can be removed and renewed by simply snapping them out of position. The pivot assembly can be renewed also. Drill out the rivets and re-rivet new units in position. Refit the motor as shown in the illustration (Fig 10:41).
15 Two speed wipers are similar in construction - the difference being in the terminal plate and armature. The terminal plate also carries a thermal overload cut-out which protects the motor if the mechanism jams. Test connections for the two speed version are illustrated (Fig 10:39).

32 Instrument panel and switches - removal and replacement

1 As mentioned in the general description - the circuiting in the panel is printed and a multi-socket connector plugs into it, to provide all the necessary connections.
2 Disconnect the battery. Remove the panel cover which is held by six nuts, two studs and six clips. To get at the nuts, the cover panels and parcel shelves must be first removed. The illustration shows location and detail of the fixings. In addition, the switches on the cover (below the instrument panel) must be removed. This is simply done by prising them out and disconnecting the wires from each one (Fig 10:44 and 10:45).
3 Remove the three screws holding the instrument panel and disconnect the speedometer cable from behind by access from underneath (the aperture to the left of the steering column) (Fig 10:46).
4 The multi-socket connector is held in position in the back of the panel by a catch at each end. If these are squeezed inwards the unit can be drawn out. The instrument panel may then be lifted (Fig 10:47).
5 Replacement of the panel assembly is a reversal of the removal procedure.

33 Speedometer, fuel gauge, temperature gauge and ignition and oil warning lamps

1 Remove the instrument panel assembly as described in the preceding Section.
2 Before assuming that either the fuel gauge or temperature gauge is faulty, make sure that the checks as described in Chapters 2 and 3 are carried out.
3 When handling the individual instrument gauges, be very careful and avoid bending or straining the pointers which might upset their delicate balance.
4 Remove all bulb holders, voltage stabiliser and the four nuts and washers, and the printed circuit sheet can be lifted off. Note that the black bulb holders are for 3 watt bulbs for ignition warning and illumination lamps (Fig 10:48).
5 Prise off the six spring clips round the edge of the case to release the lens.
6 Remove the two screws holding the speedometer, and then push the top of the instrument face away from the case by carefully inserting a screwdriver through one of the lamp apertures in the back of the case. This will move the face of the sealer strip which surrounds the warning lamp windows.
7 The instruments and face can then be lifted out of the case. The speedometer unit is simply renewed after undoing two more screws and sliding it out of the support channels (Fig 10:49).
8 Both the fuel and temperature gauges are held to the face in channels with spring clips and can be removed by simply easing them out (Fig 10:50).
9 Reassembly of the components is a reversal of this procedure. Take great care when handling instruments that the indicator needles are not bent. When the speedometer has been replaced, made sure that the needle swings clear across the dial before replacing the lens. Bulb holders should be fitted so that the contact strips touch the metal of the printed circuit (Fig 10:51).

34 Stop lamp switch

1 The stop lamp switch is mounted on the brake pedal support bracket adjacent to the brake pedal. It is operated mechanically on depression of the brake pedal.
2 The switch should operate when the brake pedal is depressed. It can be adjusted by slackening the retaining nut, and moving it in relation to the brake pedal. This should be so that when the pedal is at rest, the plunger is pressed in only just enough to operate the switch. If set too near the pedal, it could restrict the return movement with serious consequences to braking.
3 If the switch is suspected of not working, first check that the circuit is working, by bridging the two terminals of the switch and noting whether the stop lights come on (with the ignition switch on). If they do, the switch is at fault. If not, check the stop lamp circuit wiring.

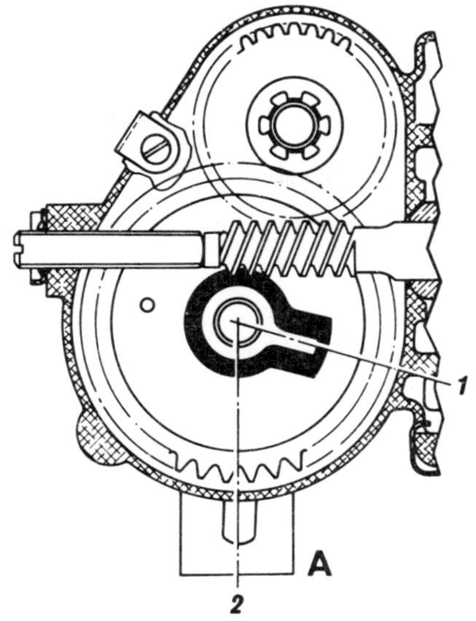

Fig 10:40 WINDSCREEN WIPER MOTOR – REASSEMBLY SHOWING CORRECT POSITION OF PARKING SEGMENT (1) RELATIVE TO THE CRANK (2)

A Left hand drive B Right hand drive

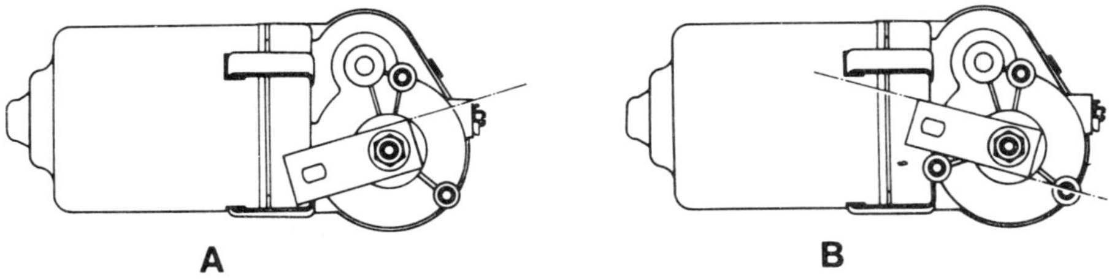

Fig 10:41 WINDSCREEN WIPER MOTOR. POSITION OF CRANK AT SELF-PARK

A Left hand drive B Right hand drive

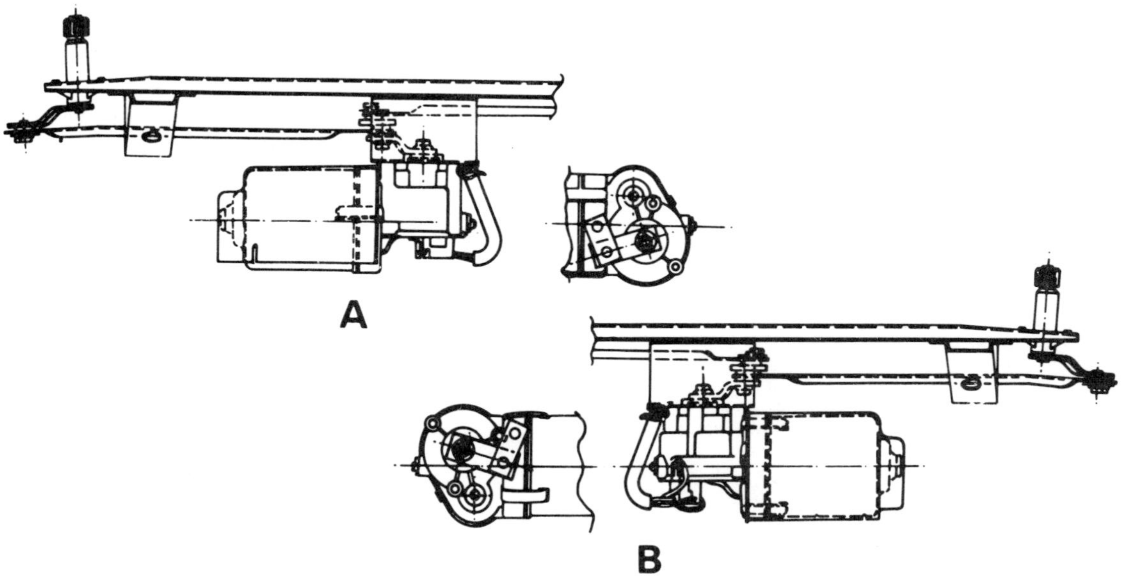

Fig 10:42 WINDSCREEN WIPER MECHANISM ASSEMBLY. POSITION OF MOTOR RELATIVE TO MOUNTING BRACKET

A Left hand drive B Right hand drive

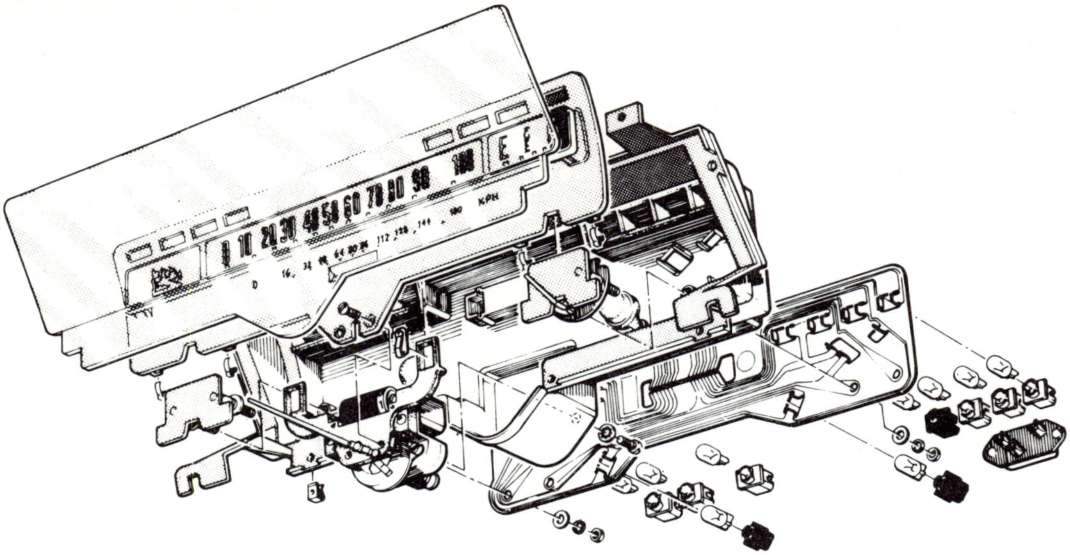

Fig 10:43 Instrument panel assembly - exploded illustration

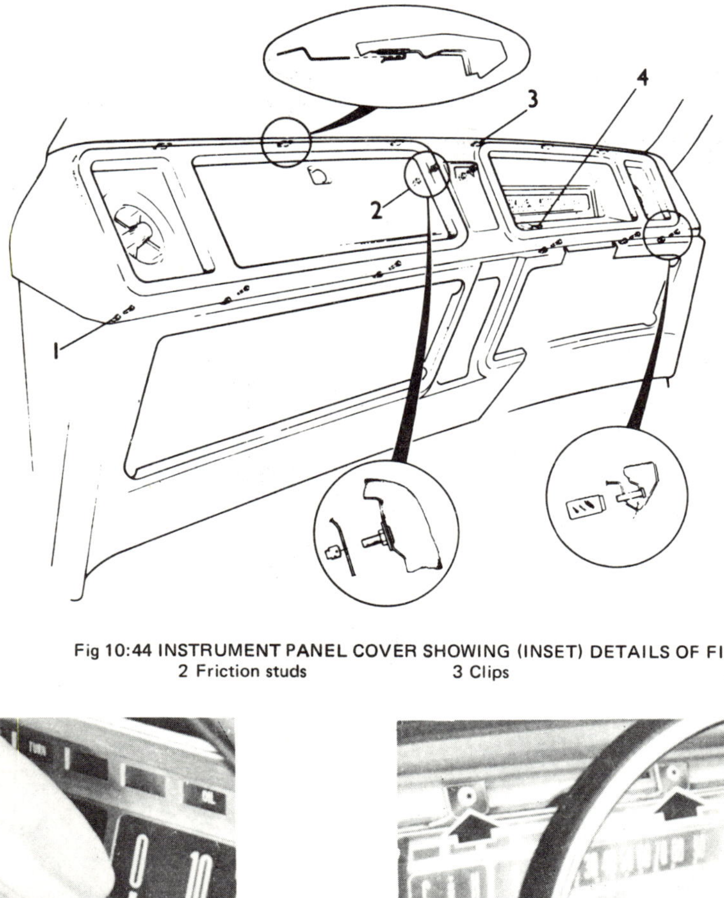

Fig 10:44 INSTRUMENT PANEL COVER SHOWING (INSET) DETAILS OF FIXINGS
1 Nuts 2 Friction studs 3 Clips 4 Switches

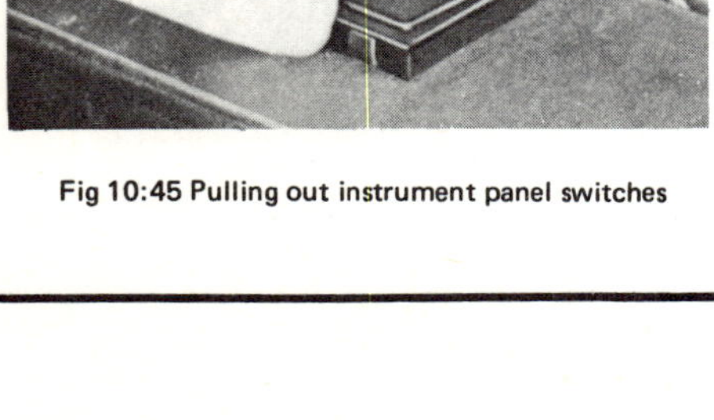

Fig 10:45 Pulling out instrument panel switches

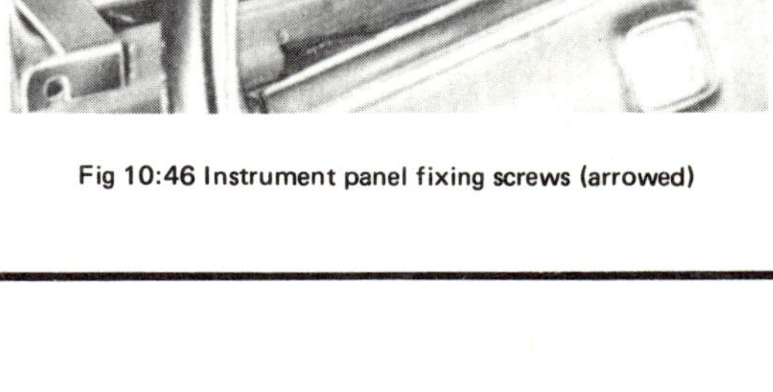

Fig 10:46 Instrument panel fixing screws (arrowed)

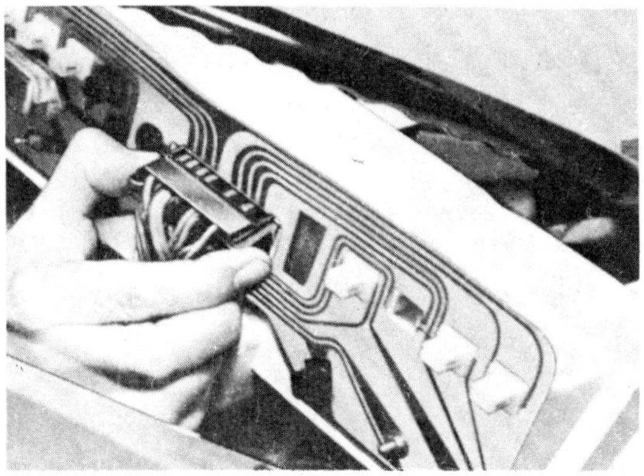

Fig 10:47 Instrument panel - removing multi-connector from printed circuit socket

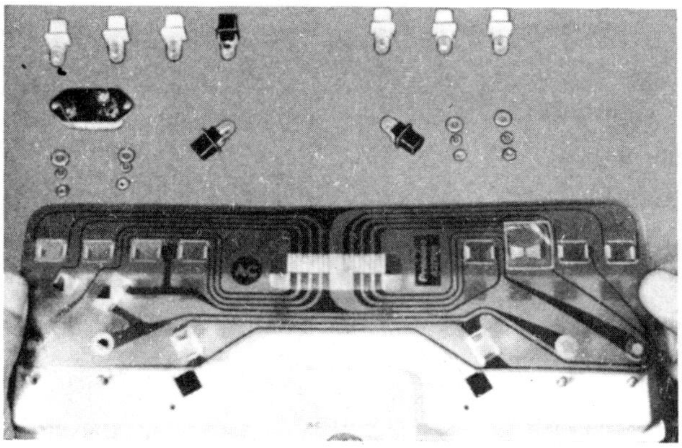

Fig. 10.48. Instrument panel. View of printed circuit showing bulb holders, voltage stabiliser and retaining nuts and washers

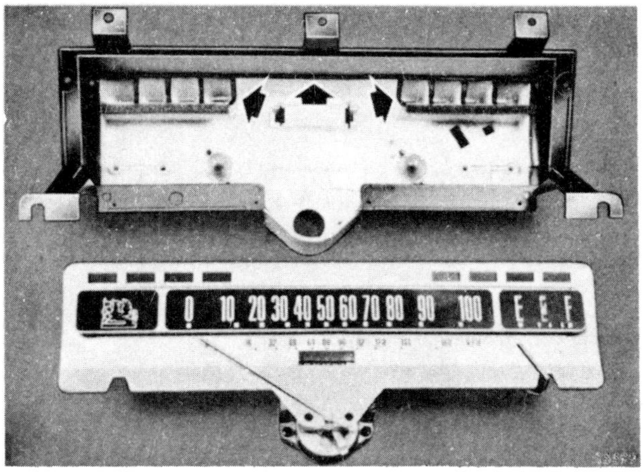

Fig. 10.49. Instrument panel - speedometer instruments and scale detached from housing. Note seals (arrowed)

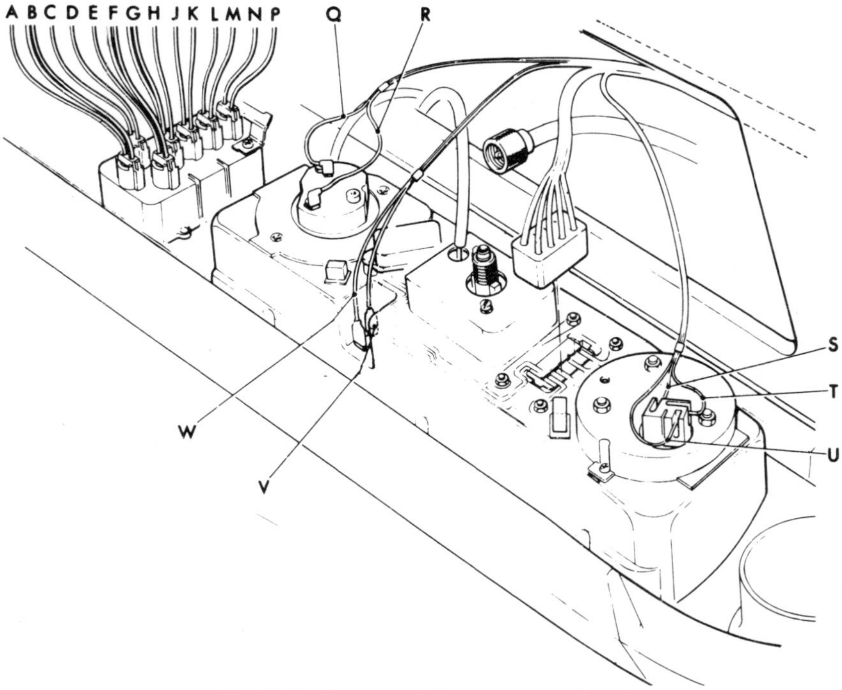

Fig. 10.49a. The seven dial instrument assembly wiring

A Blue/white	G Green	M Green	S Green
B Black	H Brown/yellow	N Green/yellow	T White/black
C Black	J Green	P Black	U Black
D Green/red	K White/brown	Q Purple	V Black
E Green/white	L Green/brown	R Black	W Green
F Black			

35 Reverse lamp switch

1 Where a reverse lamp is operated by the gearchange lever, the switch is screwed into the rear end of the remote control housing on the gearbox.

Early models

2 If a new switch is to be fitted, a lead washer is necessary and jointing compound is used on the threads. The clearance between the front shoulder of the switch and the housing cover should be 0.20 in (5 mm) (Fig 10:52).

Later models

3 On later models, no adjustment is possible. When fitting a new switch ensure that the sealing washer is sound and that the O-ring is in position on the housing.

36 Ignition switch

1 The steering column lock and barrel operated by key, also operate the ignition switch.

To remove the switch is a simple matter and does not need to disturb the lock and key.

2 Remove the screws securing the lower shroud round the steering column and then remove the two screws holding the switch. The socket connector can be pulled out of the switch and a new switch fitted (Fig 10:53).

37 Seven dial instrument panel - lamp, switch and voltage stabiliser renewal

1 The bulb holders in the inner end of the instrument assembly and in the warning lamp panel can be removed after detaching the filler panel adjacent to the steering column. The voltage stabiliser can also be removed through this aperture. Testing of the stabiliser is described in Section 18.

2 The bulb holders at the outer end of the instrument assembly can be reached from above the driver's parcel shelf. (Fig 10:49a).

38 Seven dial instrument panel - removal, servicing and replacement

1 To remove the complete instrument assembly, release the two screws at each end and lift the assembly away together with the cover. If the printed circuit is removed, ensure that the instrument assembly is not held face down, or the gauge may drop through and damage the movement when the nuts are removed.

2 The clock and tachometer can be withdrawn from the rear of the instrument case after removing the two screws. The speedometer, battery condition meter, oil pressure, fuel and water temperature gauges can be removed from the front of the case after removing the five screws securing the facia.

3 The warning lamp panel can be withdrawn, when the instrument assembly is removed, if the two retaining nuts and one screw are released. (Fig 10:49a).

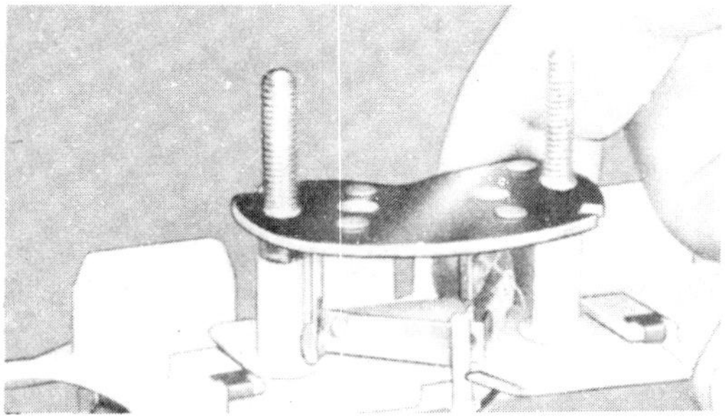

Fig 10:50 Instrument panel - removal of fuel and water temperature gauges

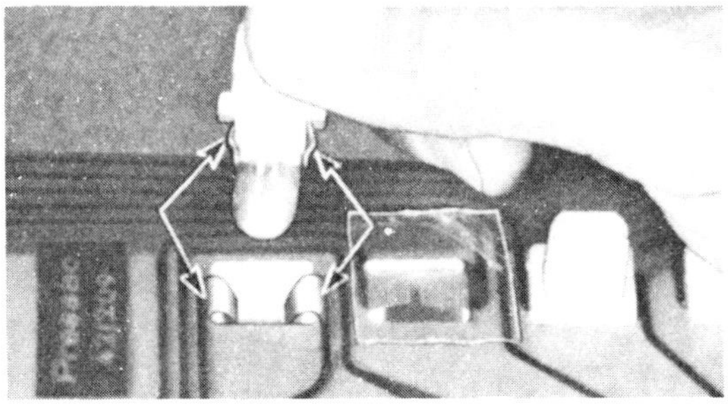

Fig 10:51 Instrument panel showing positioning of bulb holder contacts into printed circuit

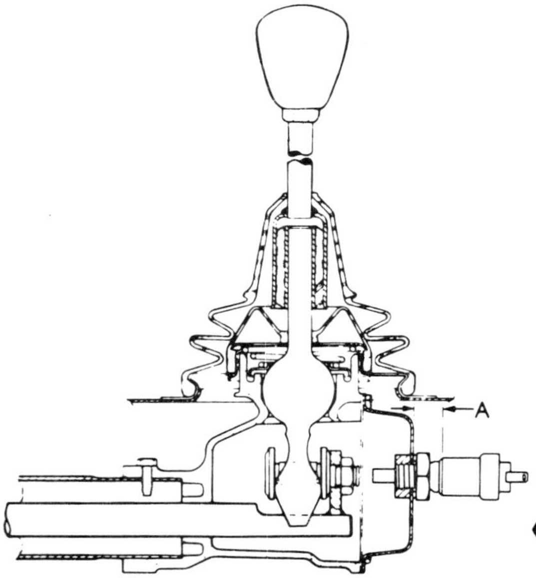

Fig 10:52 Reverse lamp switch location in gearbox. Dimension 'A' is 0.2 inch (5 mm)

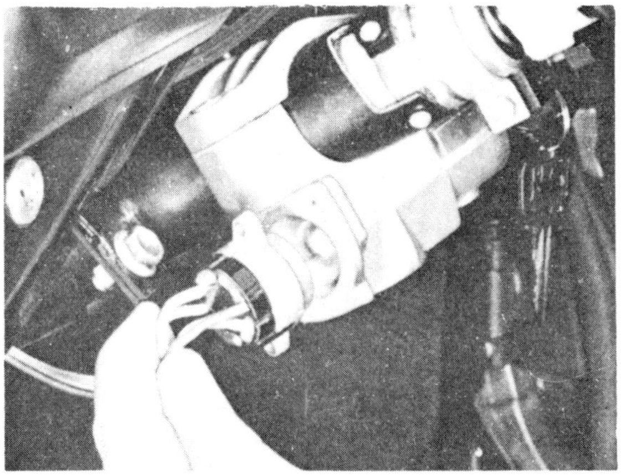

Fig 10:53 Ignition switch - removal

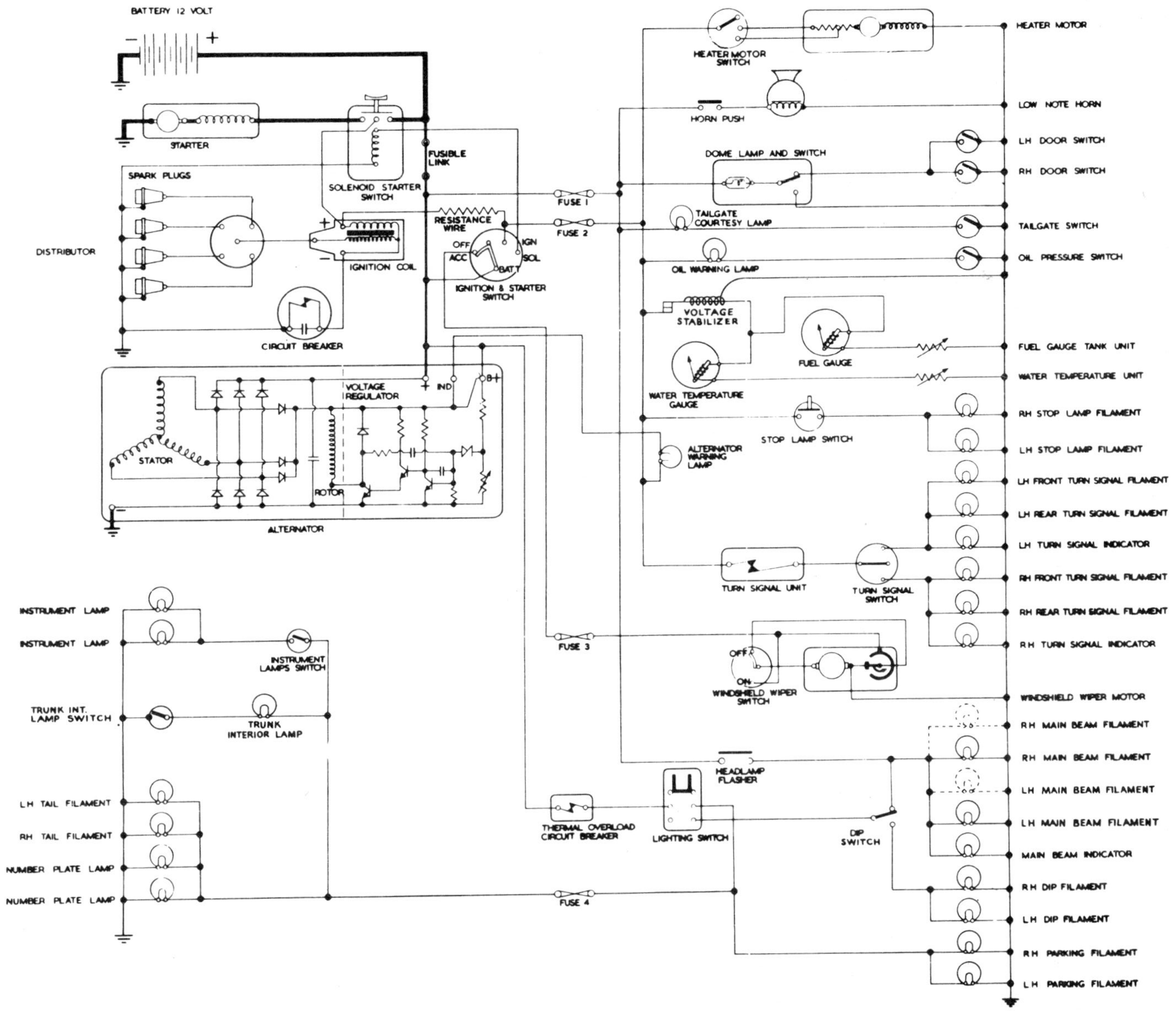

WD 1 WIRING DIAGRAM – SCHEMATIC

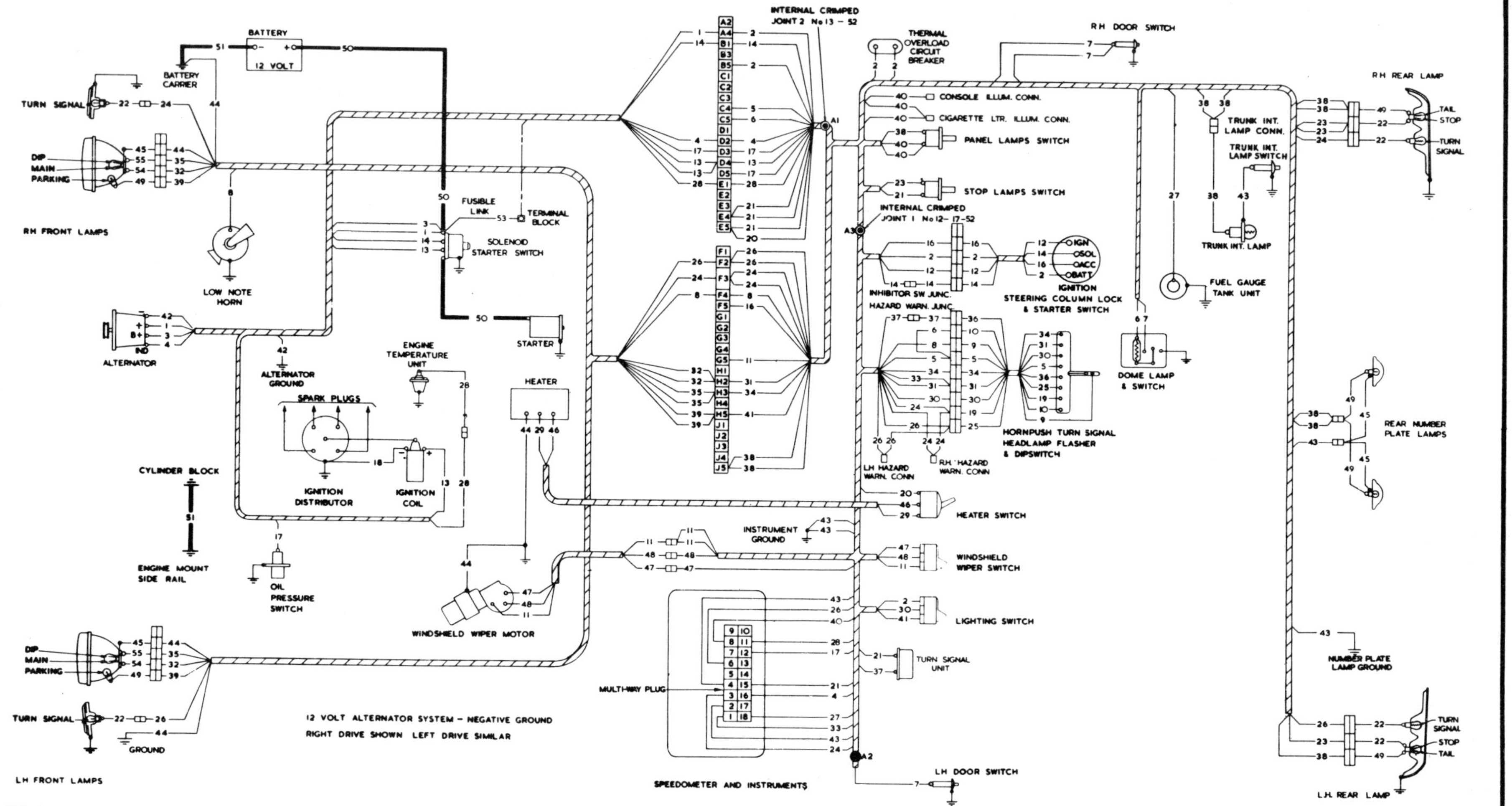

No	COLOUR	SIZE	No	COLOUR	SIZE	No	COLOUR	SIZE	No	COLOUR	SIZE	No	COLOUR	SIZE
1	BROWN	65/O12	12	WHITE	35/O12	23	GREEN/PURPLE	9/O12	34	BLUE/RED	28/O12	45	BLACK	14/O10
2	BROWN	35/O12	13	WHITE	14/O12	24	GREEN/WHITE	9/O12	35	BLUE/RED	14/O12	46	BLACK/BLUE	14/O12
3	BROWN	14/O12	14	WHITE/RED	28/O12	25	GREEN/RED	14/O10	36	LT GREEN/BROWN	14/O10	47	YELLOW/LT GREEN	14/O12
4	BROWN/YELLOW	9/O12	15	WHITE/RED	9/O12	26	GREEN/RED	9/O12	37	LT GREEN/BROWN	9/O12	48	RED/LT GREEN	14/O12
5	PURPLE	28/O12	16	WHITE/BLUE	28/O12	27	GREEN/BLACK	9/O12	38	RED	9/O12	49	RED	14/O10
6	PURPLE	9/O12	17	WHITE/BROWN	9/O12	28	GREEN/BLUE	9/O12	39	RED/GREEN	9/O12	50	———	37/O28
7	PURPLE/WHITE	9/O12	18	WHITE/BLACK	7/16/O04	29	GREEN/YELLOW	14/O12	40	RED/WHITE	9/O12	51	———	16/16/O12
8	PURPLE/BLACK	9/O12	19	GREEN/WHITE	14/O10	30	BLUE	28/O12	41	RED/BROWN	28/O12	52	RESISTANCE WIRE	◉
9	PURPLE/BLACK	14/O10	20	GREEN	14/O12	31	BLUE/WHITE	28/O12	42	BLACK	65/O12	53	BROWN	35/O12
10	PURPLE/BROWN	14/O10	21	GREEN	9/O12	32	BLUE/WHITE	14/O12	43	BLACK	9/O12	54	BLUE/WHITE	14/O10
11	WHITE/LT GREEN	14/O12	22	GREEN	14/O10	33	BLUE/WHITE	9/O12	44	BLACK	14/O12	55	BLUE/RED	14/O10

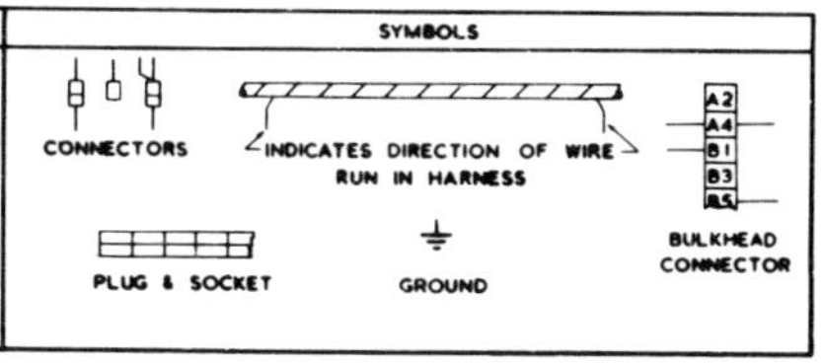

◉ DENOTES SINGLE STRAND RESISTANCE WIRE GIVING A TOTAL OF 2Ω ± 20Ω ROUTED BETWEEN POINTS A1 A2 – A3 RESISTANCE WIRE 1/O28 OR 22SWG RESISTANCE PER INCH ·O312Ω ± 5%.

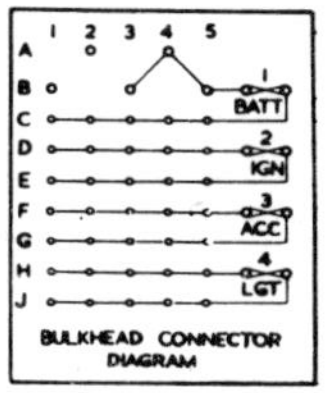

WD 2 WIRING DIAGRAM – ACTUAL CONNECTIONS

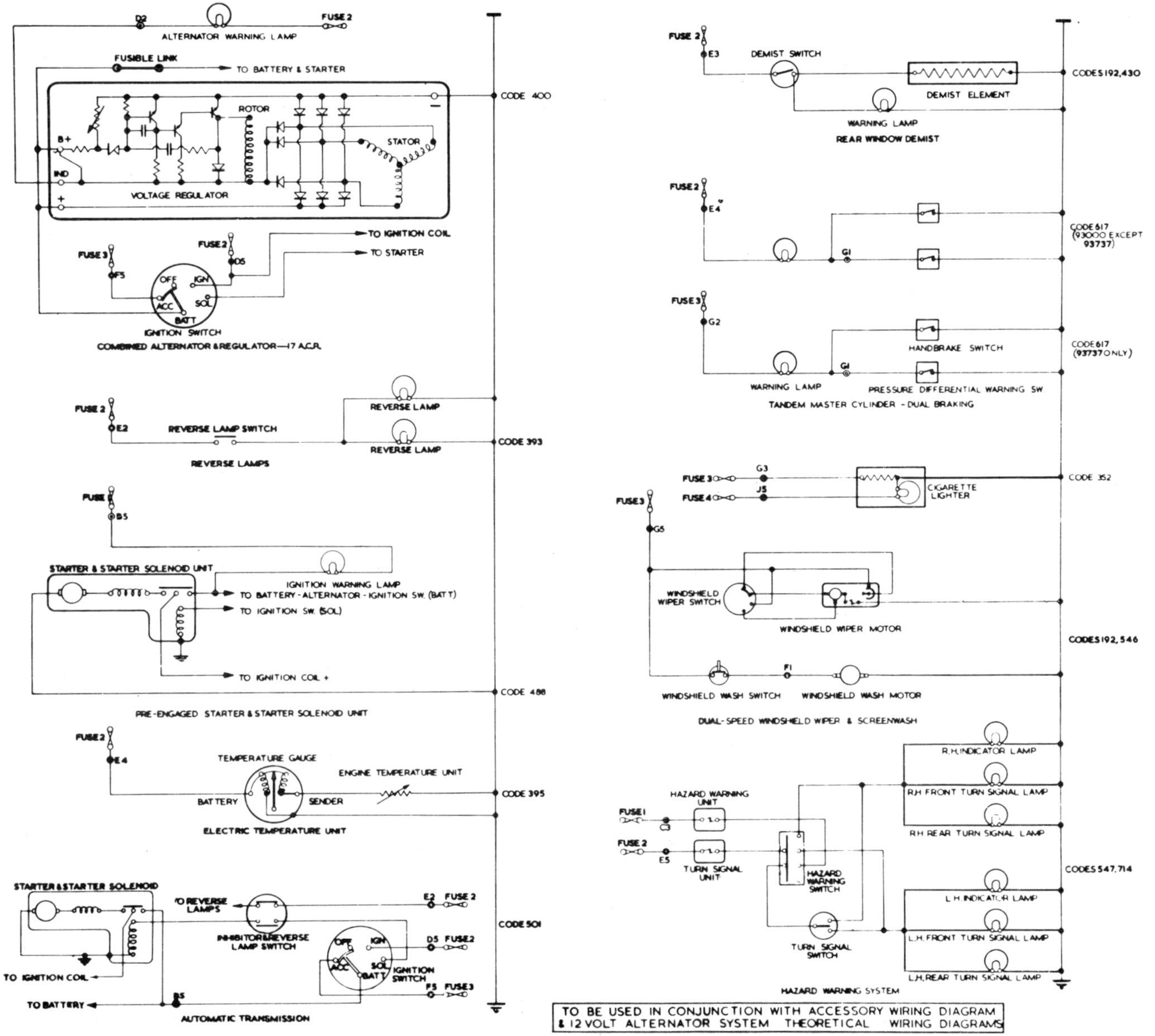

TO BE USED IN CONJUNCTION WITH ACCESSORY WIRING DIAGRAM & 12 VOLT ALTERNATOR SYSTEM THEORETICAL WIRING DIAGRAMS

WD 3 WIRING DIAGRAM – OPTIONAL EQUIPMENT

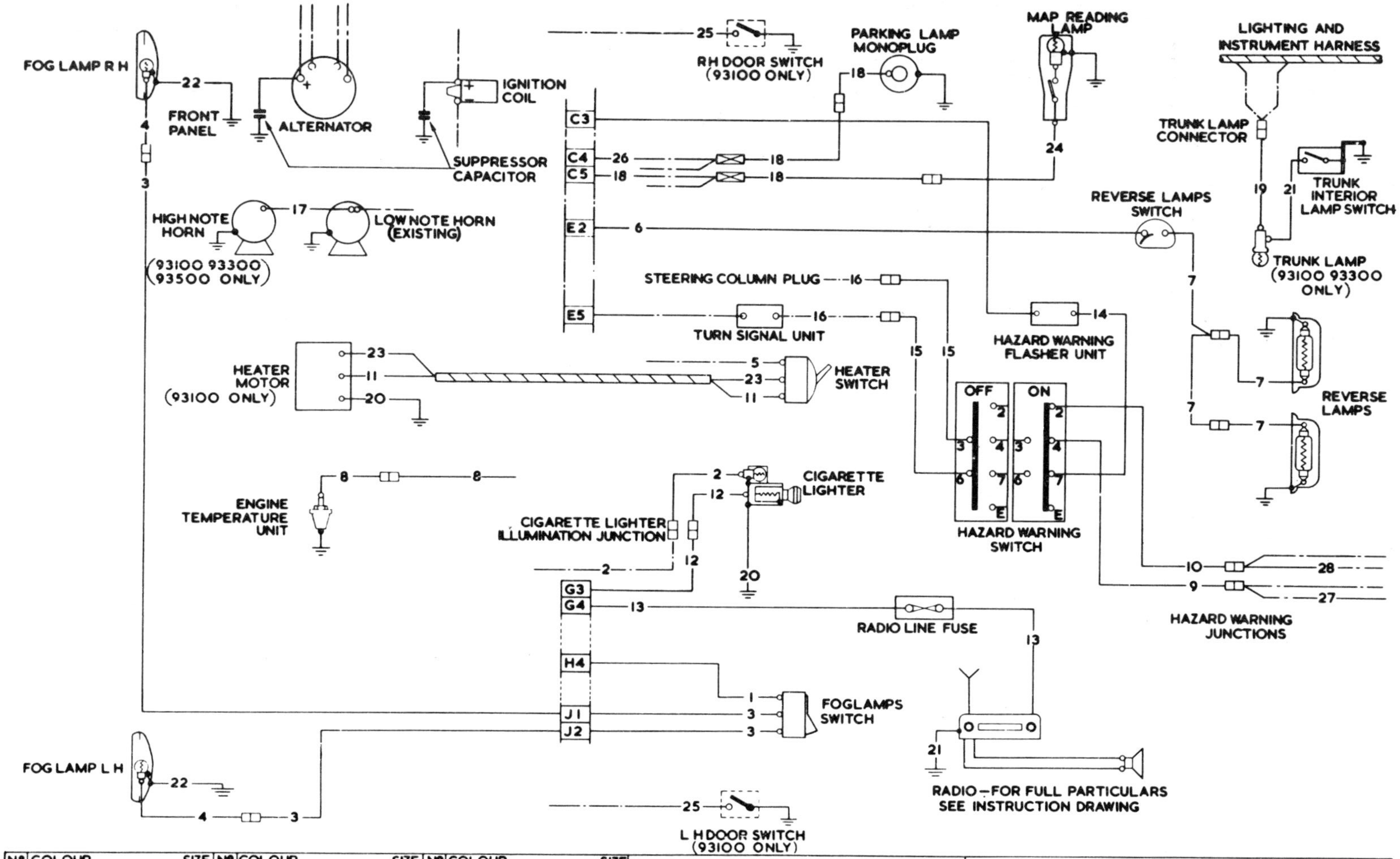

Nº	COLOUR	SIZE	Nº	COLOUR	SIZE	Nº	COLOUR	SIZE
1	RED	28/·012	11	GREEN/YELLOW	14/·012	21	BLACK	9/·012
2	RED/WHITE	9/·012	12	WHITE/GREEN	14/·012	22	BLACK	14/·010
3	RED/YELLOW	14/·012	13	WHITE/GREEN	9/·012	23	BLACK/BLUE	14/·012
4	RED/YELLOW	14/·010	14	LT GREEN/PURPLE	14/·010	24	GREY	14/·010
5	GREEN	14/·012	15	LT GREEN/BROWN	14/·010	25	PURPLE/WHITE	9/·012
6	GREEN	9/·012	16	LT GREEN/BROWN	9/·012	26	PURPLE	28/·012
7	GREEN/BROWN	9/·012	17	PURPLE	14/·012	27	GREEN/RED	9/·012
8	GREEN/BLUE	9/·012	18	PURPLE	9/·012	28	GREEN/WHITE	9/·012
9	GREEN/RED	14/·010	19	PURPLE/BLACK	14/·012			
10	GREEN/WHITE	14/·010	20	BLACK	14/·012			

TO BE USED IN CONJUNCTION WITH MAIN AND THEORETICAL WIRING DIAGRAMS

SYMBOLS

BULKHEAD CONNECTOR (C4, C5) — EXISTING CABLES — SINGLE CONNECTOR — TAP CONNECTOR — GROUND

WD 4 WIRING DIAGRAM – ACCESSORIES

Chapter 11 Suspension, dampers and steering

Contents

Specifications

Front Suspension

Type	Independent, coil springs with single lower arms and wishbone upper arms

Rear Suspension

Type	Coil spring, four arm location

Steering

Type	Rack and pinion
Make	Burman or Cam Gears
Oil capacity	¼ pint (Imperial)

Steering Geometry

(Up to Chassis Nos OE339002 and OV252006)
(Maximum limits given)

Toe-in measured on rims at height of wheel centres	0.045 in toe-in to 0.045 in toe-out
Camber angle	0–2° 30′ positive
Steering pivot (king pin) inclination	5° 38′ to 8° 38′
Castor angle	2° to 2° 30′
Toe-out on turns	Outer wheel 18° 45′ from straight with inner wheel at 20°

(Chassis Nos OE339003 and OV252007 on)
(Maximum limits given)

Toe-in measured on rims at height of wheel centre	0.045 in toe-in to 0.045 in toe-out
Camber angle	2° negative to 0°
Steering pivot (king pin) inclination	7° 55′ to 10° 25′
Castor angle	2° 30′ to 4°
Toe-out on turns	Outer wheel 18° 30′ from straight with inner wheel at 20°

Dampers

Type	Telescopic double acting front and rear

Wheels and Tyres

Type	Steel disc - stud fixing
Tyre sizes and type	5.20 x 13 4 ply. Standard suspension 6.25 – 13 4 ply low profile. Heavy duty suspension and estate cars

Tyre Pressures

Saloons	24 lb/in^2 front and rear	
Estate cars	22 lb/in^2 front and rear	
Saloons at GVW (2740 lbs)	24 lb/in^2 front	30 lb/in^2 rear
Estate cars at GVW (2910 lbs)	24 lb/in^2 front	30 lb/in^2 rear
SL models, saloon and estate	22 lb/in^2 front and rear	
SL saloons at GVW (2850 lbs)	22 lb/in^2 front	26 lb/in^2 rear
SL estates at GVW (2940 lbs)	24 lb/in^2 front	30 lb/in^2 rear

NOTE: De luxe saloons fitted with optional 6.25 – 13 tyres are as for SL models.

Torque Wrench Settings	lb/ft
Front crossmember upper mounting to body nuts	19
Front crossmember rear mounting to body bolts	32
Spring upper mounting to crossmember bolts	32
Upper wishbone pivot bolt	55
Lower arm pivot bolt	42
Upper ball joint securing bolts (replacement)	22
Steering knuckle to ball joint nuts	33
Steering arm to steering knuckle	25
Track rod end to steering arm nut	24
Damper top mounting bolt	32
Damper lower mounting bolt	57
Control rod to lower arm bolts	32
Control rod rear end nut	47
Rear suspension arms to hangars and axle mounting bolts	38
Rear spring to suspension arm mounting nut	19
Steering wheel nut	45
Steering gear to crossmember bolts	19
Steering column upper support bracket bolts	17

1 General description

The Viva suspension system is based on four coil springs - one at each wheel. At the front, the wheels are suspended independently on an upper wishbone and lower arm, with the spring fixed to the lower arm and passing through the wishbone to an upper anchorage on the ends of the crossmember pillars.

The live rear axle is located by four arms, two being longitudinal pivoted at the forward end on plates welded to the frame side rails, and at the rear fixed to flanges on the outer ends of the axle tubes.

The other two arms also pivot at their front ends on brackets welded to the frame side rails (behind the longitudinal arms) and run at an angle of approximately 45° to mounting lugs integral with the differential casing.

The rear coil springs are located between the rear ends of the longitudinal arms and the side frame rails which curve up at that point. All pivoting is done on steel bushed rubber mountings which require no maintenance.

Double acting telescopic hydraulic dampers are used front and rear, those at the front being positioned inside the coil spring and mounted at the same points as the spring. Those at the rear are attached to the flange on the outer end of the axle tube at the bottom, and the upper end to a special support bracket fixed to the side rail.

The steering is rack and pinion of conventional design and the products of either of two manufacturers is fitted, namely Burman or Cam Gears. These can be identified by the manufacturer's name cast on to the bottom of the gear housing. The assembly, comprising a housing, rack and pinion, is supported in rubber mountings on the front of the axle crossmember. The rack is mounted in one end of the housing by a bush, and at the other by a spring loaded adjustable yoke which also maintains engagement with the pinion. The pinion is mounted between ball thrust bearings, the preloading of which is also adjustable. The inner ends of the steering tie rods are attached to the rack by adjustable ball joints. The outer ends are fixed to the steering knuckle by sealed ball joints.

The steering column is the safety type with a collapsible 'lattice work' section in the column which will crumple on impact. The shaft itself is able to telescope and it is retained in its normal position by injections of plastic material which will shear on impact.

A steering column lock is fitted as standard.

2 Springs and dampers - inspection

1 With the tyre pressures correct, fuel tank full and the car standing on level smooth ground, bounce it up and down a few times and let it settle. Then measure the distance from the lower arm fulcrum bolt centre to the ground, and from the rear longitudinal arm front bolt centre to the ground.

If there is a difference of more than ¼ inch (6 mm) side for side at front and rear, then further investigation for either a broken spring or faulty damper is necessary. If no consistent measurements can be obtained after several 'bouncings' it usually indicates that a damper is binding and should be renewed (Fig 11:4).

2 Dampers may be checked by bouncing the car at each corner. Generally speaking the body will return to its normal position and stop after being depressed. If it rises and returns on a rebound, the damper should be suspect. Examine also the damper mounting bushes for any sign of looseness and the cylinders themselves for traces of hydraulic fluid leaks. If there is any sign of the latter, the unit must be renewed. Static tests of dampers are not entirely conclusive and further indications of damper failure are: noticeable pitching (bonnet going up and down when the car is braked and stopped sharply); excessive rolling on fast bends; and a definite feeling of insecurity on corners, particularly if the road surface is uneven. If you are in doubt it is a good idea to drive over a roughish road and have someone follow you to watch how the wheels behave. Excessive up and down 'patter' of any wheel is usually quite obvious, and denotes a defective damper.

3 Front dampers - removal and replacement

All figures in text refer to Fig 11:8

1 Removal of the dampers is made easier with the special Vauxhall tool to compress the spring, but is not essential.

2 Jack up the car so that the front wheel is clear of the ground by a few inches.

3 Undo the three nuts (39, 57) which hold the lower damper mounting plates to the lower suspension arm and also the nut (54) on the lower damper mounting bolt (51) (Fig 11:1).

4 Place a block under the front wheel and then lower the car so that the brackets and lower damper mounting come clear of the arm. Remove the lower mounting pin and detach the brackets.

5 Remove the upper mounting bolt and nut (49, 50) and the damper may be withdrawn from below.

6 The lower damper mounting bushes may be renewed separately if required but the top bush is part of the damper and is not supplied separately.

7 Reassembly is a reversal of the removal procedure but, for convenience, refit the lower mounting bushes and brackets loosely before reconnecting the top. Tighten all nuts to the specified torques.

4 Front springs - removal and replacement

All figures in text refer to Fig 11:8.

1 Proceed as for front damper removal as described in Section 3, as far as paragraph 4 inclusive. Slacken also the nut (36) on the lower arm fulcrum pin (35). The coil spring is now held in compression by the lower suspension arm which is secured only by the lower steering joint (24) to the steering arm. Obviously the joint cannot be separated from the arm without taking measures to control the expansion of the spring when the lower arm is released. With a spring compressor this is no problem. A spring compressor can be made using short lengths of iron rod with the ends bent over to form hooks. Three of these will hold the coils of the spring sufficiently to enable the lower arm to be disengaged.

2 Alternatively, one may proceed as follows: Jack up the lower suspension arm and then support the car under the front crossmember in the centre.

3 Place a block of wood between the spring upper mounting and the upper wishbone to hold the wishbone up in position.

4 Disconnect the lower suspension arm ball joint as described in Section 5.

5 Lower the jack under the lower suspension arm until the spring is relaxed. The spring may then be detached.

6 Replacement is a reversal of this procedure. Make sure that the spring locates correctly in the lower arm seat and that the tapers of the ball joint pin and steering knuckle are perfectly clean and dry before reconnection.

7 When the weight of the car is finally resting on the suspension, retighten the lower arm fulcrum bolt to the recommended torque of 32 lb/ft.

5 Front suspension arm ball joints - removal and replacement

1 The front suspension ball joints will need renewal if they are worn beyond the acceptable limits. This wear is one of the items checked when MOT tests become necessary. The lower joint can be checked by jacking the car up so that the wheel hangs free and then placing another jack under the lower suspension arm and raising the arm so that movement in the joint can be detected. If there is any movement the joint must be renewed.

2 To remove the joint, first take off the self-locking nut (32) (Fig 11:8) from the pin located in the steering knuckle.

3 Separate the joint from the knuckle. This can only be done with surety by using a claw clamp. However, it is possible to tap out the taper pin but only if the knuckle is firmly supported. The joint

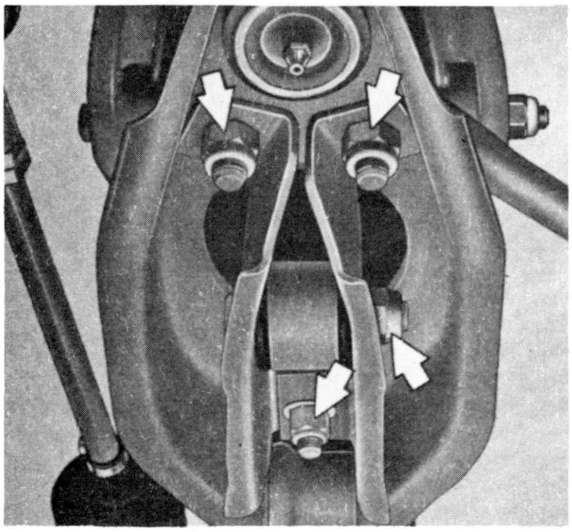

Fig 11:1 Front suspension lower arm. Nuts (arrowed) to be removed when removing damper lower mounting

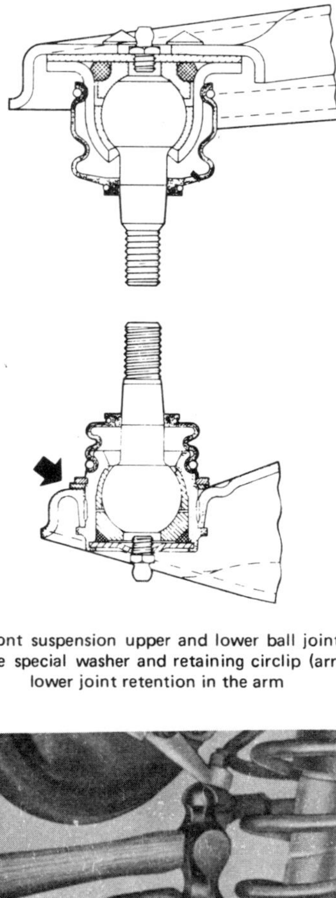

Fig 11:2 Front suspension upper and lower ball joints – Cross section. Note special washer and retaining circlip (arrowed) for lower joint retention in the arm

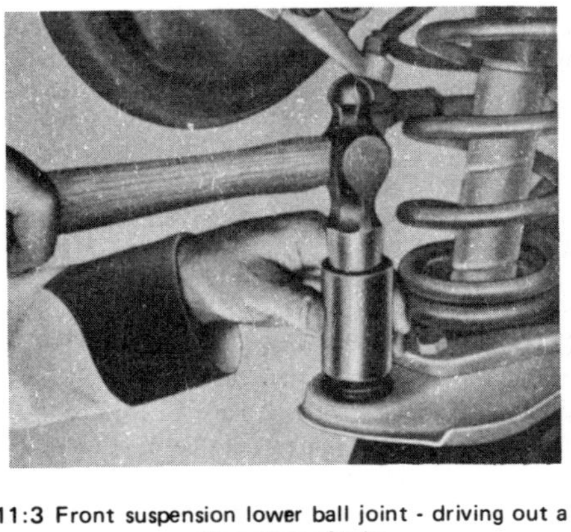

Fig 11:3 Front suspension lower ball joint - driving out a joint with a piece of tube

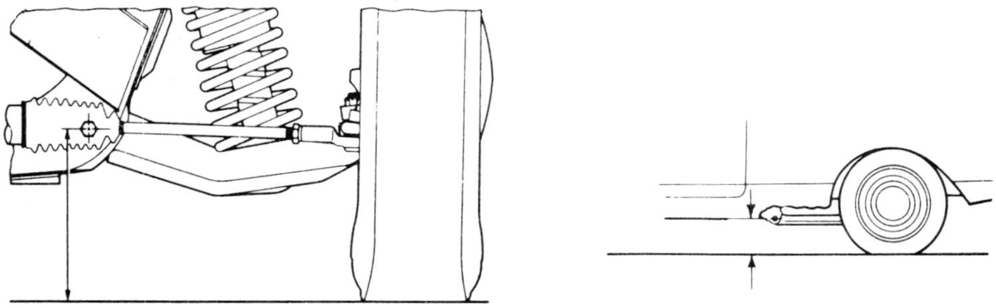

Fig 11:4 Measurement points for front and rear body height

Fig 11:5 Front suspension lower ball joint - correct position of flat (arrowed) on special assembly washer

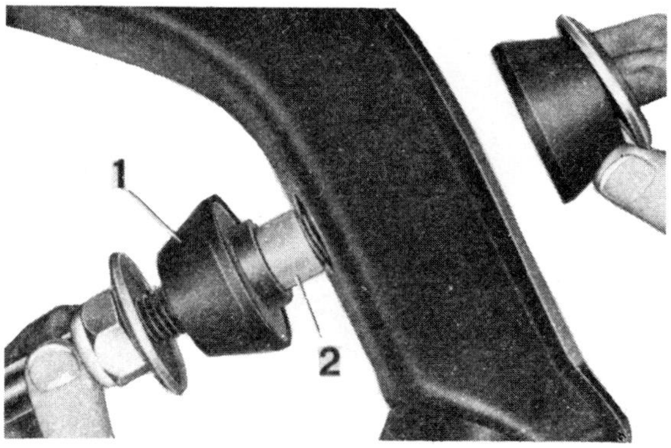

Fig 11:6 CORRECT FITMENT OF FRONT SUSPENSION CONTROL ROD BUSHES
1 Rubber bush with collar 2 Spacer

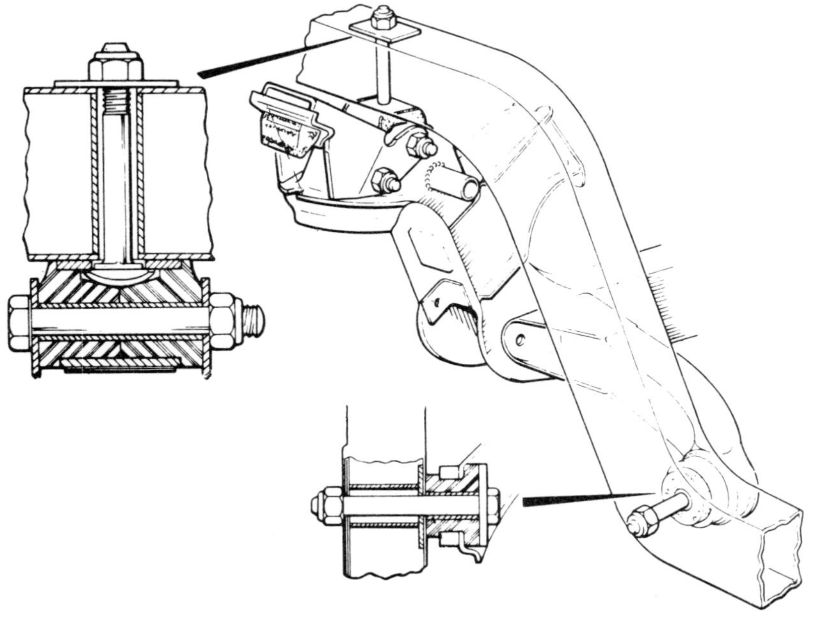

Fig 11:7 Front suspension crossmember. Detail of body frame mountings

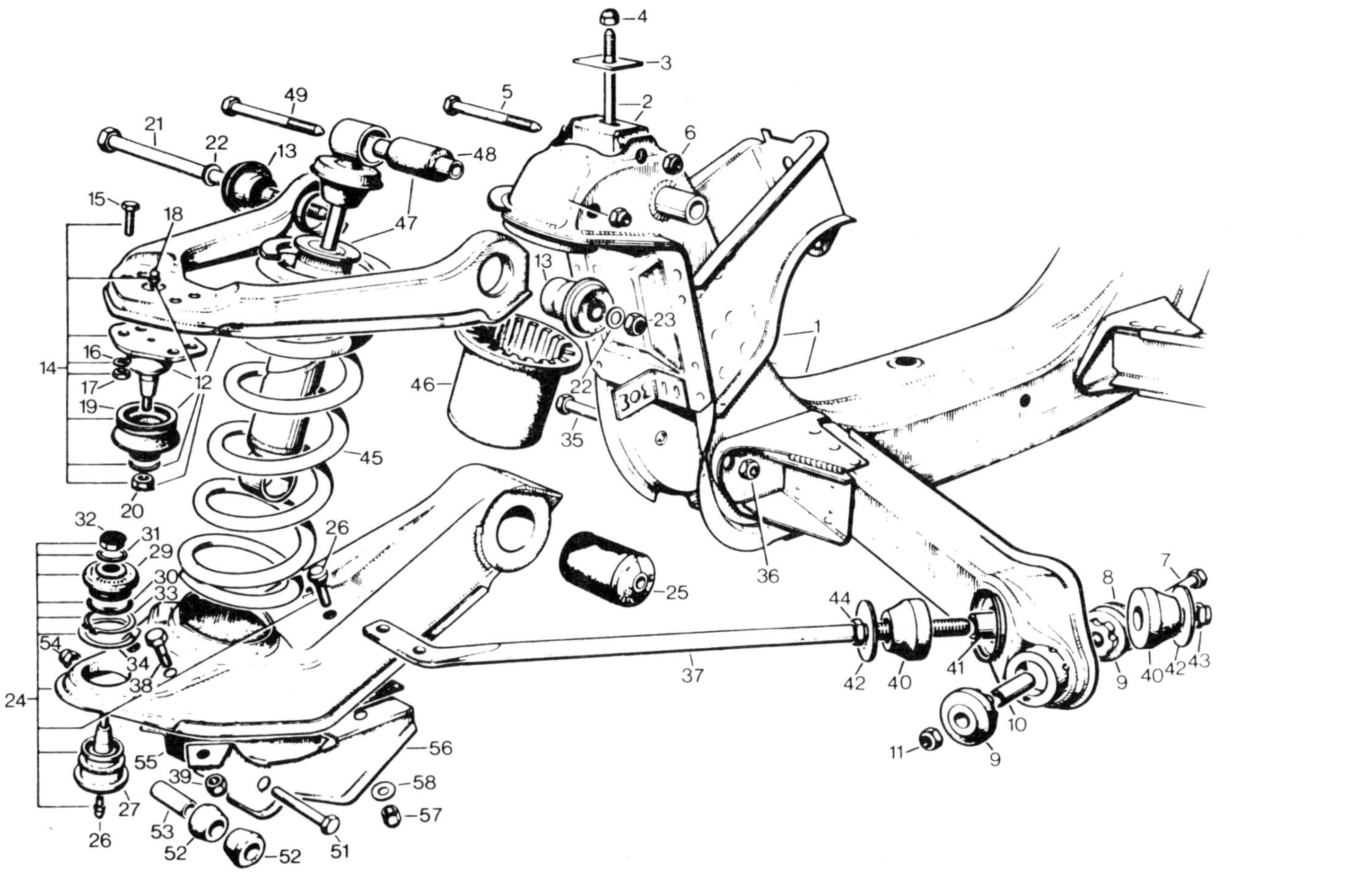

Fig 11:8 FRONT SUSPENSION AND CROSSMEMBER – EXPLODED ILLUSTRATION

1 Main crossmember
2 Crossmember mourfting
3 Plate
4 Nut
5 Mounting bolt
6 Nut
7 Brace mounting bolt
8 Washer
9 Insulator
10 Spacer
11 Nut
12 Upper arm assembly
13 Upper arm bushes
14 Upper arm ball joint
15 Bolt
16 Washer
17 Nut
18 Grease nipple
19 Boot
20 Nut
21 Upper arm bolt
22 Washer
23 Nut - self-locking
24 Lower arm assembly
25 Lower arm bush
26 Bolt
27 Lower arm ball joint
28 Grease nipple
29 Boot
30 Spring garter
31 Boot ring
32 Nut
33 Circlip
34 Washer
35 Lower arm bolt
36 Nut
37 Control rod
38 Control rod securing bolt
39 Nut
40 Insulator
41 Spacer
42 Washer
43 Nut
44 Nut
45 Spring
46 Spring insulator
47 Damper assembly
48 Damper bush
49 Damper mounting bolt
50 Nut
51 Damper mounting bolt
52 Bush
53 Spacer
54 Nut
55 Anchor bracket
56 Anchor bracket
57 Nut
58 Washer

will almost certainly be damaged in the process. Another method is to strike the side of the knuckle where the pin goes through whilst holding the head of another hammer on the opposite side. This has a squeezing out effect on the tapered pin. The lower suspension arm will only move a limited way as it is held by the front damper.
4 The car should then be jacked up under the front crossmember, a block placed under the wheel, and the car lowered again until the lower ball joint is clear of the knuckle and can be got at for removal from the arm.
5 To remove the joint from the arm, first remove the circlip and washer.
6 Using a piece of tube, drive the joint out of the arm. It may be necessary to put a block of wood under the arm to provide a firm support for this (Fig 11:3).
7 Drive in a new joint so that the splines engage in the arm. Then fit the special washer with the flat lined up and the concave side upwards, and replace the circlip (Fig 11:5).
8 Reconnect the pin to the steering knuckle ensuring that the mating surfaces of the taper are clean. Replace the nut.
9 The upper arm ball joint can be checked by jacking the car up and rocking the wheel whilst holding the joint to detect any play. If there is any play the joint must be renewed. This necessitates removal of the upper wishbone (see Section 6).
10 With the upper wishbone removed, the four rivets holding the joint must be drilled out without damaging the holes in the arm.
11 The holes in the arm must then be drilled out to $^5/_{16}$ inch. The new joint will be supplied with the necessary mounting bolts. Install these with the heads uppermost and tighten the nuts to the specified torque of 22 lb/ft.
12 Replace the arm as described in Section 6.
13 Apply a grease gun to the ball joint nipples.

6 Front suspension arms - removal and replacement

Figures in the text refer to Fig 11:8.
1 The suspension arms will need to be removed if the bushes are worn. Also the upper suspension arm needs to be removed in order to renew the ball joint.
2 To remove the upper arm, jack up the suspension under the lower arm and then remove the upper ball joint from the steering knuckle. This entails removal of the unit. The joint can then be removed as described in Section 5.
3 Next, remove the long fulcrum bolt (21) and nut (23) which hold the wishbone to the crossmember. If necessary, it is permissible to bend back the lashing eye (used for securing the car when carried on a transporter) which may get in the way when withdrawing the bolt. The arm can then be taken off.
4 To renew the bushes (13) calls for care as the arms of the wishbone must not be distorted during the course of removing and replacing the bushes. It is best to get the old bushes out by cutting through them.
5 New bushes should be lubricated with soapy water and drawn in. Use a long nut and bolt together with a tubular spacer (on the inside of the arm) and large washers to ensure the bushes are drawn in square. Under no circumstances should attempts be made to drive the bushes in with a hammer.
6 Replace the arm in the reverse order of removal but do not tighten the nut on the fulcrum bolt to the full torque until the weight of the car is resting on the suspension.
7 To remove the lower suspension arm, proceed as for removal of the front spring as described in Section 4. Then remove the fulcrum bolt, and the arm can be drawn away from the mounting brackets.
8 The single large bush needs careful treatment when removed, otherwise the arm may be distorted. Do not try and drive it out. Use a bolt, spacer and washers and draw it out. Note that there is a steel sleeve on the outside of the bush which must not be left behind.
9 The new bush should be oiled on the steel sleeve to assist fitting. Make sure it is drawn in from the side that will not let it foul the lip inside the arm. When installed, the centre sleeve should project equally on each side.
10 Replacement of the arm is in the reverse order of removal. Do not tighten the fulcrum bolt nut until the weight of the car is resting on the suspension.

7 Front suspension control rods - removal and replacement

All numbers refer to Fig 11:8.
1 The front suspension lower arms are stabilised fore and aft by a rod (37) which is bolted to their outer ends, and located in rubber bushes (40) at the other end into the front crossmember support stays. The length of the rod is adjustable to achieve the correct degree of castor angle on the front wheels.
2 If the control rod bushes need renewal it will be necessary to remove the rod first.
3 Jack up the suspension under the lower arm and remove the two nuts and bolts (38 and 39) securing the forward end to the suspension arm.
4 Slacken the two nuts (43 and 44) noting the position of the inner one first in relation to the thread. Then remove the end one (43) followed by the washer (42) and half bush. The rod may then be drawn out.
5 Return the inner nut to its original position before fitting new bushes and then replace the arm in the reverse order of removal. Fit the bushes correctly (Fig 11:6).
6 Tighten the rear nut to 32 lb/ft.
7 It is advisable to have the steering geometry checked at a garage, with suitable testing equipment, after removal and replacement of these control rods.

8 Front crossmember - removal, checking and replacement

1 Under certain circumstances it may be advantageous to remove the front suspension, as a complete assembly, from the car. It is attached to the underbody side members at two points on each side, one at the top and the other at the rear of the bracing arm at each side as shown in Fig 11:7.
2 With the car jacked up and supported on stands under the body, disconnect the steering column from the steering gear (see Section 15) and disconnect the hydraulic brake line unions.
3 Support the axle beam in the centre and then undo the nuts from the upper and lower mounting bolts - four in all. Then remove the lower bolts, and the axle assembly can be lowered from the car.
4 Renew the axle mountings, if necessary, before replacing the axle assembly.

9 Rear dampers - removal and replacement

1 The rear dampers will need removal if their mounting bushes are worn or if indications are that the unit is no longer performing properly.
2 It is not necessary to raise the car to remove the dampers but, if it is raised, the axle should be supported as well.
3 To detach the top mounting, remove the rubber plug from the wheel arch inside the boot when the two upper nuts will be accessible. The slotted spindle will need holding firmly whilst the nuts are removed.
4 The lower mounting eye may be disconnected by removing the nut on the mounting pin and pulling it out.
5 Replacement is a reversal of the removal procedure. All bushes are renewable and care should be taken to arrange the bushes and washers correctly as shown in Fig 11:10.
6 Tighten the top securing nuts down to the bottom of the stud threads.

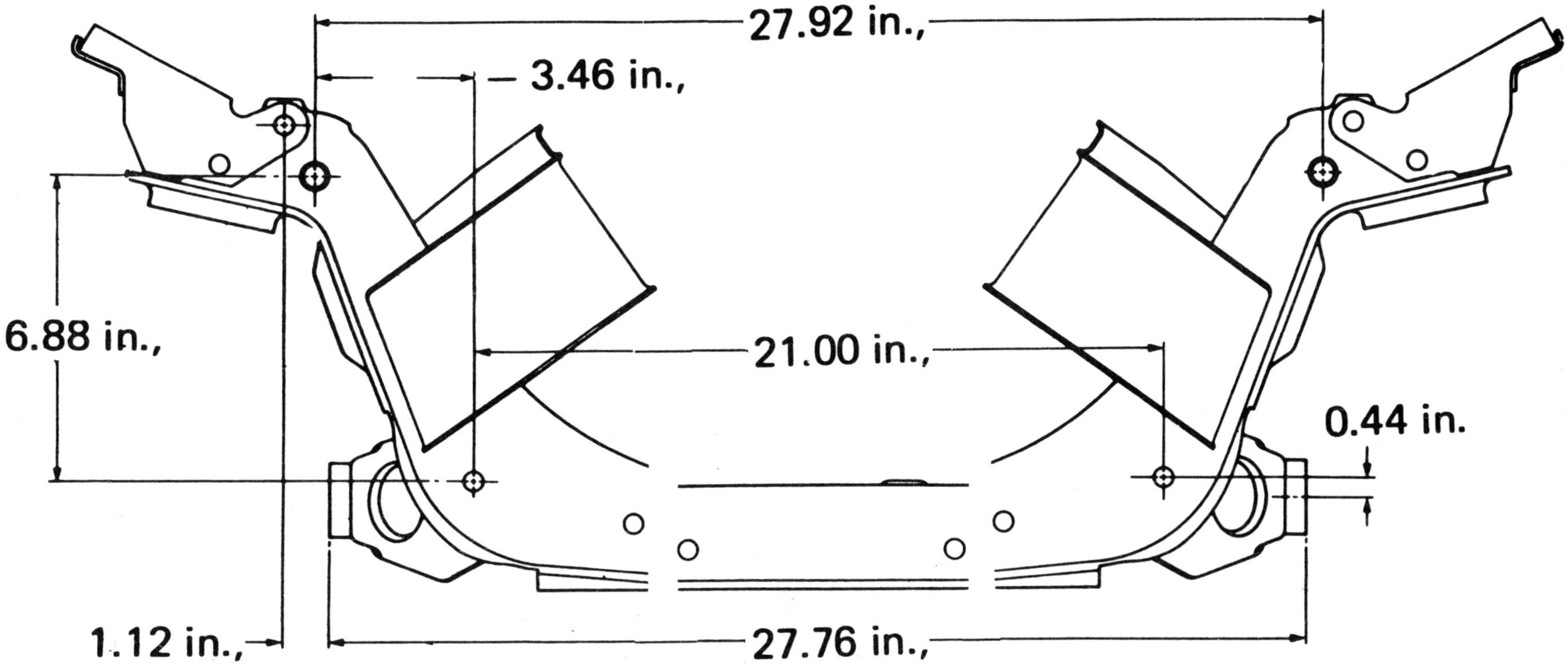

Fig 11:9 Front suspension crossmember. Dimension details

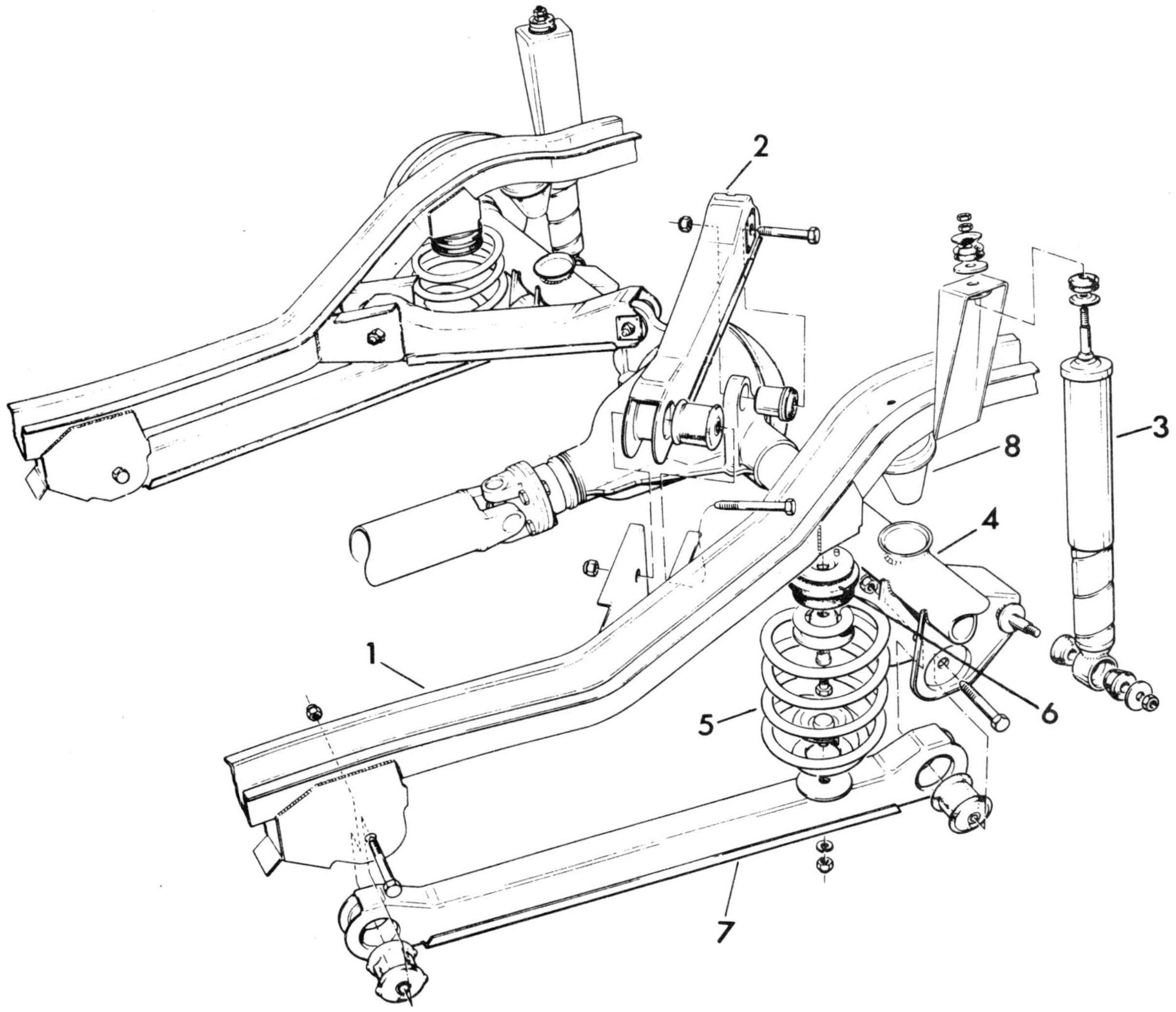

Fig 11:10 REAR SUSPENSION – DETAIL OF COMPONENTS – EXPLODED VIEW

1 Body frame side rail
2 Upper radius arm
3 Telescopic damper
4 Rear axle tube
5 Spring
6 Upper spring seat
7 Lower longitudinal arm
8 Rebound rubber

10 Rear springs - removal and replacement

1 Jack up the car and support the body under the rear frame members on stands. Then support the axle on a jack.
2 Undo the lower spring mounting stud nut which is underneath the longitudinal suspension arm.
3 Lower the jack under the axle until the suspension arm is clear of the spring.
4 Using a socket wrench, remove the upper bolt which is in the centre of the upper spring seat and fits into a captive nut in the side member.
5 Lift out the spring complete with the upper mounting seat.
6 New springs, spring seats, rubber insulators and retainers are supplied individually. They are all assembled and held together by the centre bush which is peėned over on to the upper seat. Fig 11:11 shows a cross section of the assembly with a special Vauxhall tool used to draw the new bush into position and peen over the top. If other improvised tools are used to carry out this job, the main thing to remember is that the upper seat spring retainer must be held tightly together when the bush is being peėned over.
7 Note that the upper seat has a dowel peg which locates in a corresponding hole in the side member. This must be correctly positioned when refitting the spring assembly which is otherwise a straightforward reversal of the removal procedure.

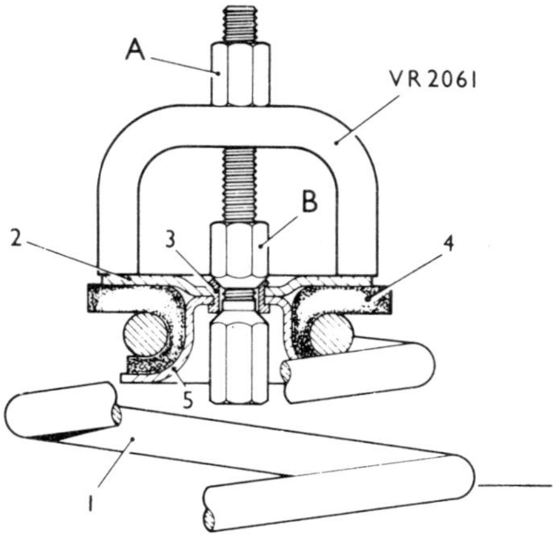

Fig 11:11 REAR SUSPENSION – SPRING UPPER SEAT DETAIL CROSS SECTION SHOWING SPECIAL TOOL BEING USED TO PEEN THE CENTRE BUSH
A,B Peening tool
1 Spring
2 Upper seat
3 Bush
4 Insulator
5 Retainer

11 Rear suspension arms - removal and replacement

1 If the rubber mounting bushes at each end of any of the arms are worn, it will be necessary to remove the arms to replace them.
2 The upper (radius) arms are bushed at the front end and bolted between brackets on the rear body frame member. The rear end locates over a lug on the rear axle, which is rubber bushed and is secured by a bolt and nut.
3 Simply by removing the nuts and bolts at each end, the arm can be removed.
4 The front bush should be removed and replaced in exactly the same manner as described for the front suspension lower arm in Section 6.
5 The rear bush will need drawing out of the lug on the axle using a long nut and bolt with a tubular spacer and large flat washers. The new bush should be drawn in, in the same manner after it has been well lubricated with soapy water.
6 Replace the arm and bolts but before tightening the nuts to the correct torque (38 lb/ft) let the weight of the car settle on the springs.
7 The lower (longitudinal) arms can be removed after first raising and supporting the car on stands under the body side frame members and placing a jack under the axle.
8 Remove the nut securing the lower spring mounting plate to the arm and then lower the jack until the arm is clear of the spring.
9 Remove the mounting bolts and nuts at each end and the arm can then be detached.
10 The same precautions for removal and replacement of the bushes apply as for the upper arms. In addition, the front bush has two locating ridges which must be set correctly into the arm as shown in Fig 11:13.
11 When refitting the arm, replace the mounting bolts but do not tighten the nuts to their correct torque (38 lb/ft) until the weight of the car is resting on the springs.

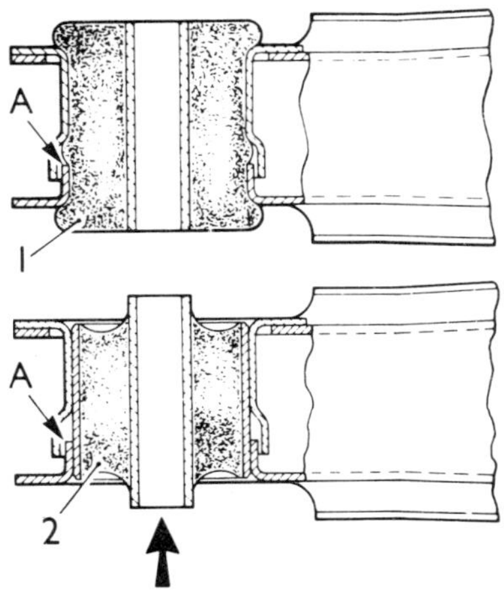

Fig 11:12 REAR SUSPENSION ARM BUSHES
1 Plain bush
2 Steel jacketed bush (heavy duty)
A Lip in arm fabrication requiring bush to be pressed in, in direction of arrow

12 Front wheel hub bearings - inspection and adjustment

1 The steering qualities of the car will deteriorate if the front wheel bearings are maladjusted or worn.
2 To check the bearings, first jack up the car so that the wheel is clear of the ground. Check that the wheel spins freely with the brakes off.

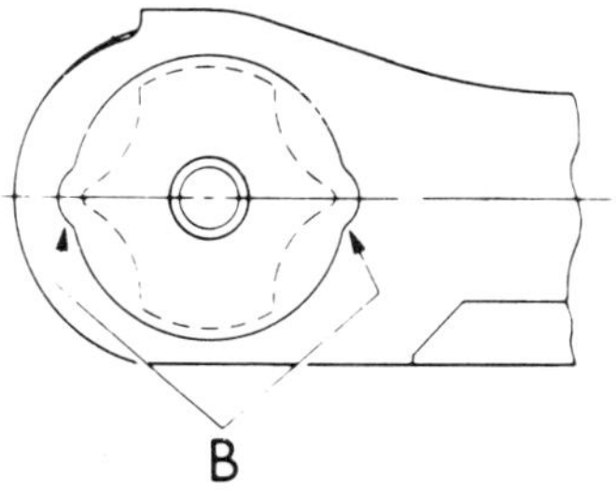

Fig 11:13 REAR SUSPENSION – LONGITUDINAL ARM FRONT BUSH SHOWING
Special locating lugs (B) to fit in arm preventing rotation of bush

3 Then grip the edge of the tyre at top and bottom and try and rock it in a vertical plane. If movement can be felt it is normally due to looseness in the bearing but at the same time it should be noted whether there is any sign of lateral movement in either the upper or lower suspension arm ball joints. (If there is, they must be renewed as described in Section 5).

4 There should be no detectable movement in the wheel bearings and if there is, they should be adjusted. Fig 11:14 gives an exploded view of a front hub with either disc or drum brakes.

5 Remove the hub cap from the wheel and then tap the dust cover out of the centre of the hub.

6 Remove the split pin and, using a tubular spanner, tighten the nut whilst continuing to revolve the wheel. Then slacken the nut off and retighten it using only the tubular spanner without a tommy bar. This will provide the maximum permissible loading on the bearing.

7 The wheel should now spin freely with no indications of movement when rocked vertically. If any roughness is felt on spinning the wheel the bearings should be removed for further examination.

8 Provided the adjustment is satisfactory, replace the nut retainer so that two of its slots line up with the pin hole, and fit a new split pin.

Where a dial gauge is available, the bearing end float must be adjusted to between 0.002 and 0.005 in.

13 Front wheel hubs and bearings - removal, inspection and replacement

1 Jack up the car and remove the road wheel.

2 On drum brake models slacken off the brake adjusters as described in Chapter 9. On disc brake models remove the brake caliper as described in Chapter 9.

3 Remove the grease cap and the split pin locking the hub nut.

4 Remove the hub nut and the washer behind it which is keyed to the shaft.

5 Withdraw the hub together with the brake drum, or disc, as appropriate.

6 The ball races will have come off with the hub, and the inner race of the outer bearing will be loose so that it can be taken out.

7 The inner bearing will need to be driven out with a drift from the inside of the hub. Locate the drift against the outside race. The oil seal will come out with the bearing. The outside race of the outer bearing should come out easily, but may need tapping from the inside with a drift.

8 Thoroughly clean the bearings and examine the rollers and races for signs of wear. If in doubt, renew them. Wear can be detected by running perfectly clean, lightly oiled bearings in their races and feeling for traces of roughness. Blue discolouration indicates overheating, but brown discolouration will only be lubricant stain and is not to be taken as an adverse indication.

9 Reassembly of the outer bearing races into the hub is a reversal of the removal procedure. Make sure that the open ends of the tapers of the outer races face outwards from the centre of the hub. Pack the inner bearing inner race and rollers with grease, and place it in position in the hub.

10 With the inner bearing fit a new seal, which on drum brake models fits over the hub boss and, on disc brake models, recesses into the hub behind the bearing. The lip of the seal should face to the centre of the hub on the latter.

11 Refit the brake drum or disc to the hub, (if removed) and on disc brakes the bolts should be renewed (they incorporate nylon thread inserts for self-locking purposes). The bolt threads should be treated with 'Loctite' Grade C.

12 Pack the outer bearing inner race and rollers with the recommended grease, place it in position and refit the complete hub assembly to the spindle. Replace the washer and castellated nut and then adjust the bearings as described in Section 12.

13 Do not pack the hub itself with grease in the space between the bearings.

14 Replace the disc caliper if appropriate as described in Chapter 9.

15 Adjust drum brakes as described in chapter 9, having first replaced the road wheel.

14 Steering mechanism - inspection

1 The steering mechanism on the Viva is uncomplicated and easy to check. The owner can save himself a lot of trouble by regular examination, apart from, of course, keeping a check for his own safety.

2 Assuming that the suspension joints and bushes and front wheel bearings have been checked and found in order, the steering check involves tracing the amount of lost motion between the rim of the steering wheel and road wheels. If the rim of the steering wheel can be moved more than 1 to 2 inches at its periphery with no sign of movement at either, or both, of the front wheels it may be assumed that there is wear at some point. If there are signs of lost motion, jack up the car at the front and support it under the front crossmember so that both wheels hang free.

3 Grip each wheel in turn and rock it in the direction it would move when steering. It will be possible to feel any play. Check first for any sign of lateral play in the ball joints which connect the tie rods from the steering gear to the steering arms on the wheel knuckles. This is the more common area for wear to occur and, if any is apparent, the ball joint/s must be renewed.

4 Having checked the ball joints, next grip the tie rod and get someone to move the steering wheel. Do this with the bonnet open and if there is any play still apparent, look first to see whether the coupling in the steering column shaft is causing the trouble. If it is, it should be renewed.

5 Finally, if play still exists, it must be in the steering gear itself. This is more serious (and expensive!). If either of the rubber boots at each end of the gear housing is damaged, resulting in loss of oil from the unit, then various bearings and teeth on the rack and pinion may have been severely worn. In such cases renewal of the complete steering gear assembly may be necessary. Certainly adjustments will be required.

15 Steering gear - examination, adjustment, removal and replacement

1 Assuming that all ball joints and front wheel bearings are in order, it may be necessary to remove and replace, or renovate, the steering gear if there is excessive play between the steering shaft and the steering tie rods. This can be checked by gripping the inner end of both the rods in turn near the rubber boot, and getting someone to rock the steering wheel. If the wheel moves more than 1/16th of a revolution (11¼° in either direction) without moving the steering tie rod, then the wear is sufficient to justify overhaul. If the rubber boots have leaked oil they will also need renewal and, in order to do this and effectively refill the unit with the proper oil, it is easiest in the long run to remove the assembly from the car.

2 To remove the steering gear from the car, first slacken the upper half of the coupling from the steering column by undoing the pinch bolt. Then disconnect both tie rod outer ball joints from the steering arms as described in Section 16. The three mounting bolts holding the assembly to the front crossmember may then be removed and the unit taken off. If the coupling tends to stick on the splines of the steering column do not strike it, but ease it off by moving the steering wheel from side to side.

3 If it is necessary to replace only the rubber boots and refill the assembly with lubricant, remove both outer ball joints from the tie rods together with the locknuts, having noted their original position carefully. Slacken off the boot retaining clips noting their position in relation to the assembly housing. If the steering arms are dirty, clean them thoroughly and slide off the old boots.

4 Refit the clips to new boots and slide them onto the rods. Tighten the clips in position on one boot only. Stand the unit on end, refill the housing with ¼ pint (no more) of Castrol 'Hipress' (or equivalent EP SAE 90 oil), and then refit the other boot and tighten the clips.

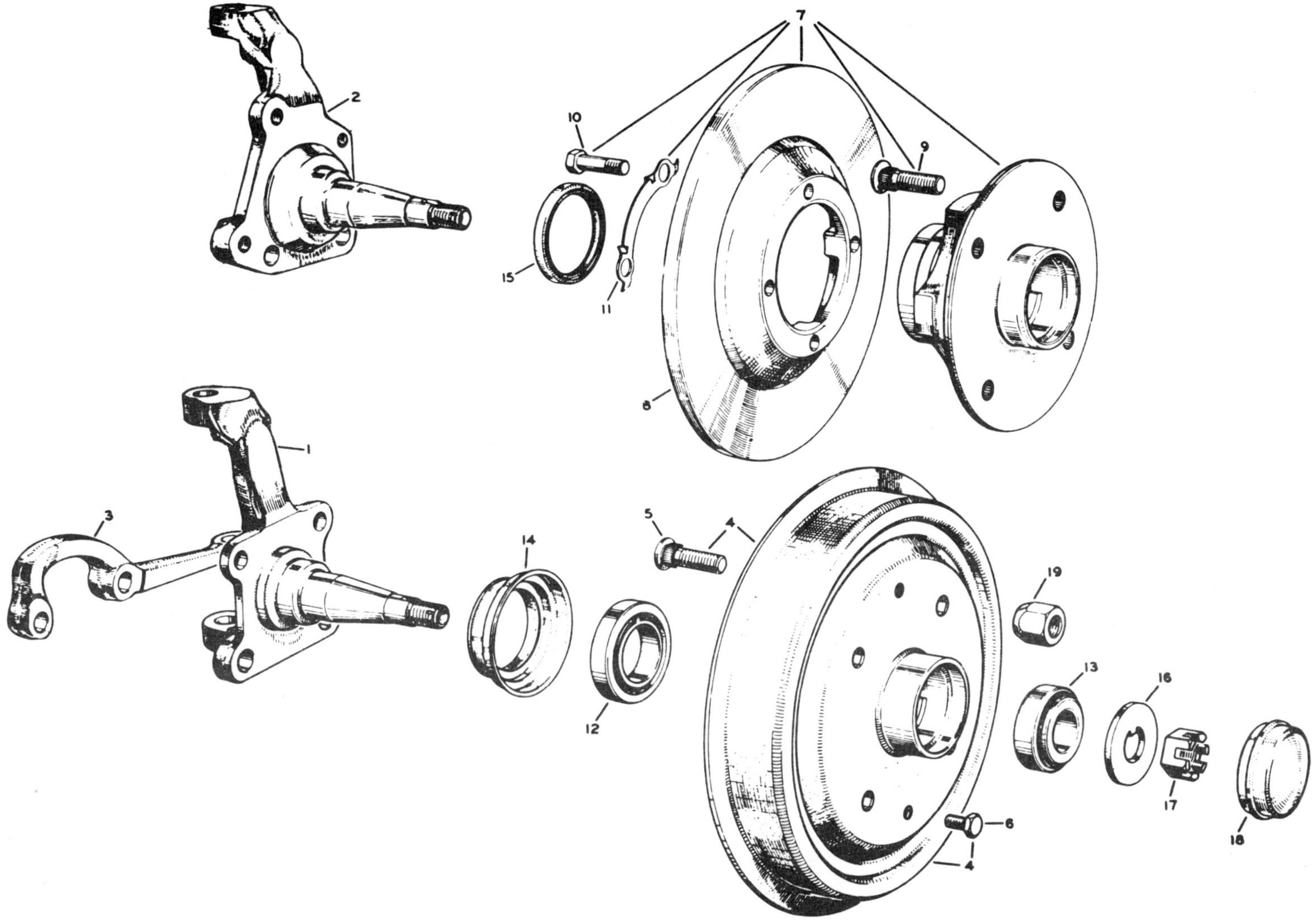

Fig 11:14 FRONT WHEEL HUB ASSEMBLIES – EXPLODED ILLUSTRATION

1 Steering knuckle – drum brakes
2 Steering knuckle – disc brakes
3 Steering arm
4 Drum and hub assembly
5 Wheel stud
6 Drum to hub locating bolt
7 Disc and hub assembly
8 Disc
9 Wheel stud
10 Bolt - disc to hub
11 Tab washer
12 Inner wheel bearing
13 Outer wheel bearing
14 Oil seal - drum brakes
15 Oil - disc brakes
16 Thrust washer
17 Bearing nut
18 Grease retaining cap
19 Wheel nut

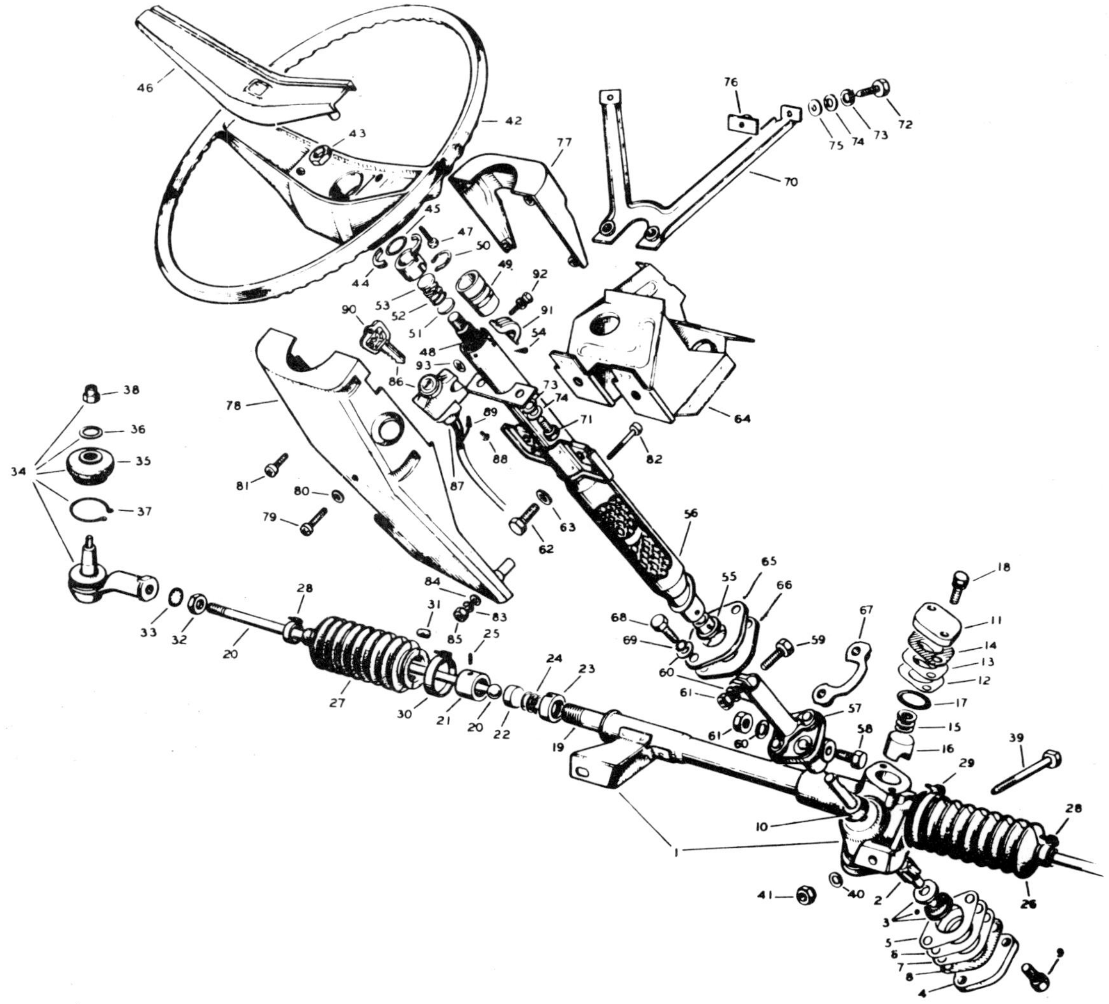

Fig 11:15 STEERING COLUMN AND STEERING GEAR – COMPONENTS – EXPLODED VIEW

1 Steering gear housing
2 Pinion
3 Pinion bearing
4 End cover
5 Shim
6 Shim
7 Shim
8 Gasket
9 Cover bolt
10 Shaft seal
11 Yoke cover
12 Shim
13 Shim
14 Gasket
15 Spring
16 Yoke
17 'O' ring
18 Bolt and washer
19 Rack
20 Tie rod
21 Tie rod bellhousing
22 Tie rod ball seat
23 Nut - self-locking
24 Rack spring
25 Ball housing pin
26 Rack boot
27 Rack boot
28 Clip
29 Clip
30 Clip
31 End cap
32 Locknut
33 Lockwasher
34 Ball joint assembly
35 Boot
36 Retaining ring
37 Circlip
38 Locknut
39 Steering gear mounting bolt
40 Washer
41 Nut, self-locking
42 Steering wheel
43 Steering wheel nut
44 Thrust collar
45 Retaining band
46 Spoke pad
47 Pad screw
48 Steering shaft
49 Upper bearing and housing
50 Circlip
51 Washer
52 Spring
53 Washer
54 Bearing screw
55 Lower bearing
56 Cover mesh
57 Coupling assembly
58 Bolt
59 Bolt
60 Washer
61 Nut
62 Bolt
63 Washer
64 Column front support
65 Column lower support bracket
66 Lower support insulator
67 Lower support fixing plate
68 Bolt
69 Lockwasher
70 Column brace
71 Bolt
72 Bolt
73 Lockwasher
74 Washer
75 Washer (special)
76 Clamp plate
77 Column canopy (upper)
78 Column canopy (lower)
79 Screw
80 Washer
81 Screw
82 Screw
83 Washer
84 Special washer
85 Nut
86 Steering column lock
87 Ignition/starter switch
88 Fixing screw
89 Cable connector plug
90 Key
91 Lock clamp
92 Bolt (break-head)
93 Washer

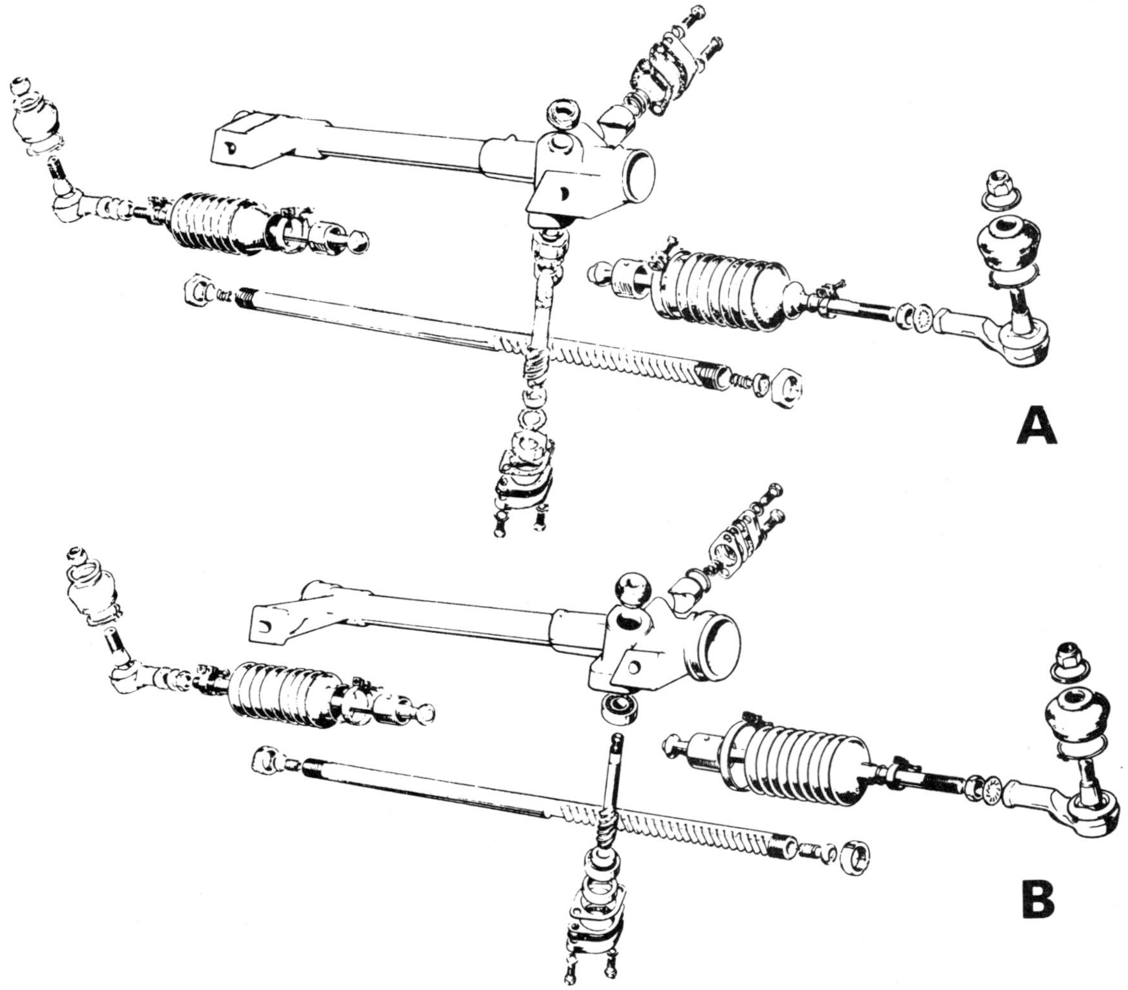

Fig 11:16 Steering gear - Exploded illustrations showing detailed differences between Burman A and Cam Gears B types

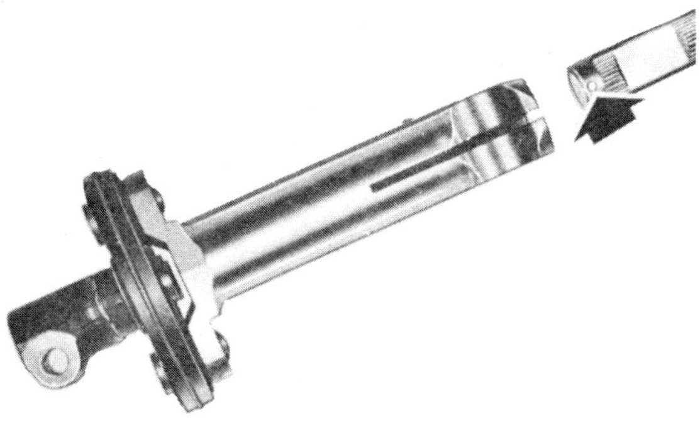

Fig 11:17 Steering shaft coupling showing peg (arrowed) on shaft which engages in slot in coupling

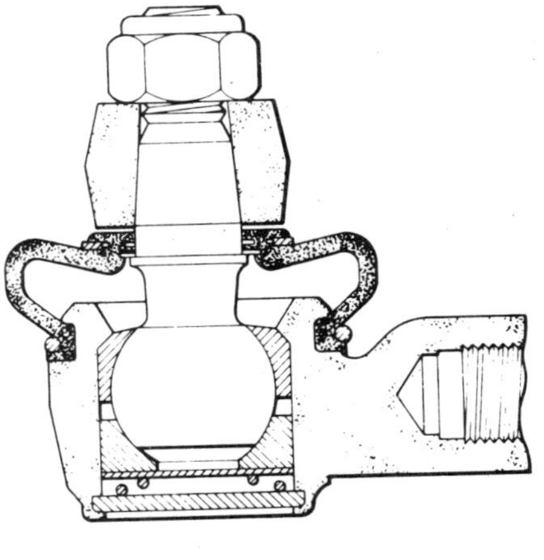

Fig 11:18 Steering track rod outer end ball joint cross section

5 It is possible to alleviate some of the play in the gear (between rack and pinion) by checking that the yoke pre-load is correct.
6 Remove the yoke cover plate and remove the shims and spring followed by the yoke. On Burman units a paper gasket is fitted each side of the shim pack. On Cam Gears only one is fitted between pack and cover.
7 Replace the yoke and cover without the spring or any shims and lightly tighten the bolts.
8 Measure the gap between the cover and the housing with a feeler gauge. The thickness of the shims should be the gap measurement PLUS .0005 - .0030 inch on Burman units, or PLUS .0005 - .0060 inch on Cam Gear units. Make up the shim packs accordingly and reassemble. Adjust the shims required to give a turning torque on the pinion of 12 lb/in. Take into account the paper gaskets.
9 Similarly, any sign of end float and slackness in the pinion shaft may be taken up by removing the cover opposite the pinion extension and reducing the thickness of the shims behind the cover accordingly. In this case the shims should be .001–.003 inch LESS than the measured clearance between the cover and housing.
10 It must be emphasised that the adjustments mentioned in this paragraph are not sufficient to compensate for extreme wear. Before making them, therefore, it must be decided whether the wear apparent is beyond adjustment, or sufficient to warrant adjustment anyway.
11 Any play in the tie rod INNER ball joints may be adjusted but involves drilling and re-pinning the joint and this calls for precision work.
12 Replacement of the assembly is a reversal of the procedure as described in paragraph 2. Make sure that the steering rack is in its central position and, if the coupling has been detached from the pinion, line up the bolt hole in the lower part of the coupling with the flat on the pinion. Replace but leave the bolt slack. When fitting the upper coupling to the column line up the peg with the coupling slot. When everything is assembled tighten first the steering gear mounting bolts, then the lower coupling pinch bolt followed by the upper pinch bolt. Before tightening the cotter pin nut securing the coupling flange to the pinion shaft make sure that the hub of the steering wheel is not rubbing on the column shroud below it. This may be caused by the steering column dropping down a little when the steering gear unit was disconnected. Lift the steering wheel before tightening the cotter clamp nut. Tighten all nuts and bolts to the correct torques as specified (Fig 11:15).
13 The front wheel toe-in should then be checked at a garage with the proper equipment.

16 Steering track rod outer ball joints - removal and replacement

1 The removal of the ball joints is necessary if they are to be renewed, or if the rubber boots on the steering gear are being renewed.
2 It is not necessary to jack the cap up but the increase in height above ground level may make it more convenient to do so.
3 Slacken the self-locking nut, completely remove it to clear the threads, and replace it after oiling the threads until the head of the nut is level with the end of the stud. This will protect the threads in subsequent operations if the same joint is being replaced.
4 If a claw clamp is being used to 'break' the taper of the joint pin from the steering arm, the joint may be disconnected without further ado.
5 If no claw clamp is available and it is necessary to strike the pin out, it is essential to provide a really firm support under the steering arm first. In the photo it can be seen how a socket on the top of a jack was used to achieve this. A firm tap with a normal weight hammer is all that is then necessary to move the pin out of the steering arm. Another method is to strike one side of the arm whilst holding the head of another hammer against the opposite side. This tends to 'squeeze' the taper pin out.
6 If the nut now turns the pin when trying to remove it, (despite the precaution taken in paragraph 3) jam the pin back into the arm with the jack to hold it whilst the nut is removed. If difficulty is experienced with a joint being renewed then cut it off.
7 Once the ball joint is clear of the arm, slacken the locknut but leave it at its original position. The joint may then be removed and a new one fitted by screwing it up as far as the locknut. The pin should point upwards and then be fitted into the steering arm (photos).
8 As the nut is self-locking it will be necessary to prevent the pin turning whilst tightening it. This can be done by putting a jack under the joint so that the weight of the wheel rests on the taper (photo).
9 Tighten the locknut on the tie rod, at the same time holding the track rod end still by applying a spanner to its flats.
10 It is advisable to have the front wheel alignment checked as soon as possible.

17 Steering knuckle and steering arm

1 Neither the steering knuckle (or stub axle as it is sometimes called) nor steering arm, normally need any attention. It is possible, however, in the case of severe shock or damage to the front suspension and steering, that either or both of them could be bent or distorted. If it is necessary to remove them for checking or renewal proceed as follows:
2 Remove the front hub as described in Section 12.
3 Detach the upper and lower wishbone ball joints as described in Section 5.
4 Detach the steering arm outer ball joint as described in Section 16.
5 Disconnect the hydraulic brake pipe from the wheel cylinder mounted on the brake backplate (details in Chapter 9).
6 Remove the brake backplate and separate the steering arm from the knuckle by undoing the bolts and nuts joining them together.
7 Reassembly and replacement is a reversal of the procedure. Bleed the brake system when reassembly is complete (see Chapter 9).

18 Steering geometry - checking and adjustment

1 Unless the front axle and suspension has been damaged, the castor angle, camber angle and steering pivot angles will not alter, provided, of course, that the suspension ball joints and wishbone fulcrum pin bushes are not worn in any way.
2 The toe-in of the front wheels is a measurement which may vary more frequently and could pass unnoticed if, for example, a steering tie rod was bent. When fitting new tie rod ball joints, for example, it will always be necessary to reset the toe-in. Similarly, the control rods running back from the lower suspension arm, if bent, would affect the castor angle.
3 Indications of incorrect wheel alignment (toe-in) are uneven tyre wear on the front tyres and erratic steering particularly when turning. To check toe-in accurately needs optical aligning equipment, so get a garage to do it. Ensure that they examine the tie rods for straightness and all ball joints and wheel bearings at the same time, if you have not done so yourself.

19 Steering wheel - removal and replacement

1 The steering wheel is located on splines to the column shaft, seats on tapered split collars and is secured by a nut.
2 First remove the centre medallion by undoing the two screws on the underside of the spoke.
3 Undo the nut with a tubular spanner and then mark the relative position of the wheel to the shaft by making two marks with a centre punch. Then pull the wheel off. If the wheel is stuck tight, do not jar it off as this may wreck the in-built safety feature of the column, use a suitable extractor.
4 Replacement is a reversal of the removal procedure noting the following points: Check that the road wheels point straight ahead and the wheel spoke is horizontal if you have not made a mark. The

16.5

Steering track rod outer ball joint. Supporting steering arm on jack and socket to assist removal of joint pin from arm

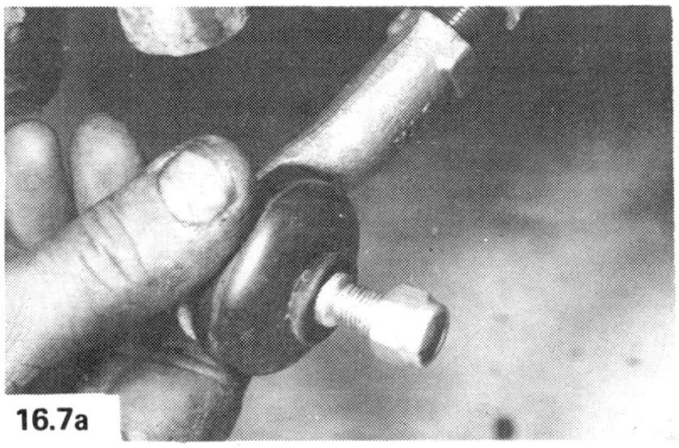

16.7a

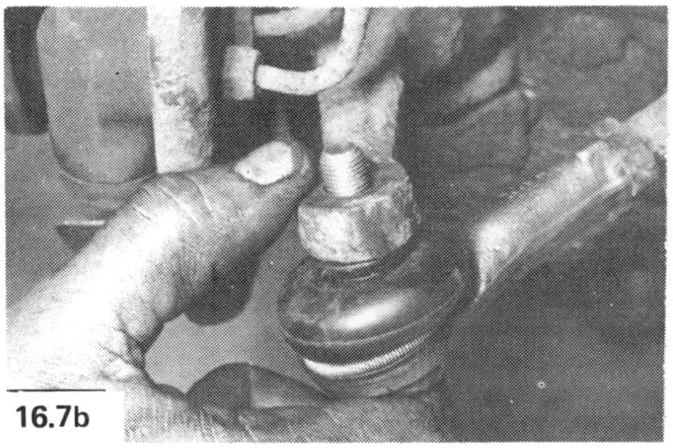

16.7b

Steering track rod outer ball joint - fitting a new joint to the rod

16.8

Steering track rod outer ball joint - tightening the nut

Fig 11:19 STEERING WHEEL. POSITION OF:
1 Indicator cancelling sleeve
2 Lugs
3 Tapered split collars and retaining band

Fig 11:20 Steering column lock showing break-head bolts (arrowed)

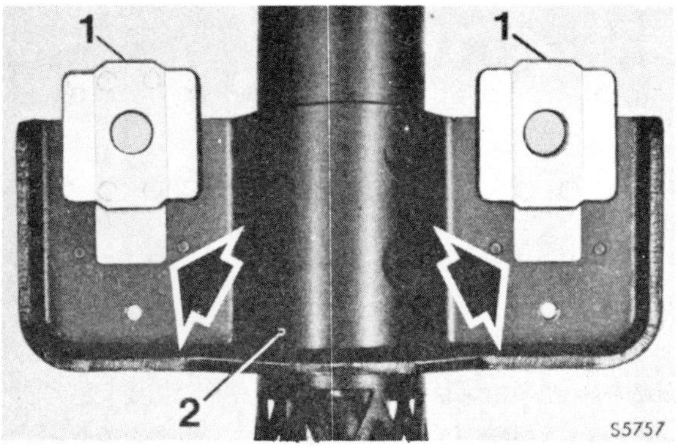

Fig 11:21 STEERING COLUMN UPPER MOUNTING BRACKET
1 Shear type mountings 2 Column Arrows point to gaps indicating collapse of column

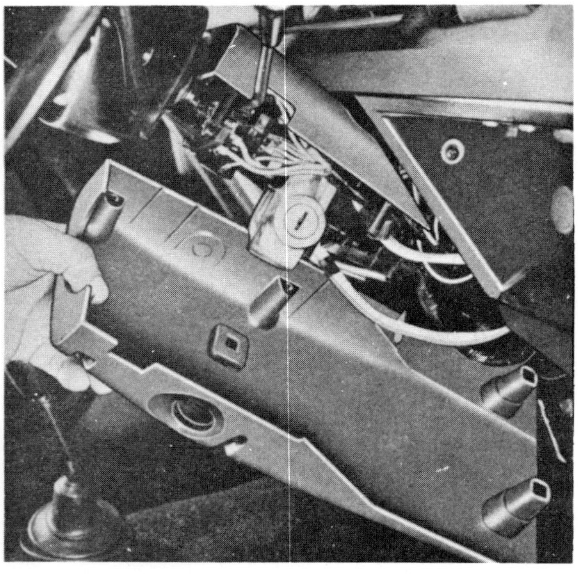

Fig 11:22 Steering column shroud. Shroud being removed after taking out three screws

split collars should be held in position on the column with a rubber band to aid assembly. The direction indicator cancelling sleeve should be positioned so that it lines up with the lugs on the wheel hub boss.
5 Tight the wheel nut to the specified torque.

20 Steering shaft coupling - removal and replacement

1 The coupling is supplied complete and a new one should be fitted if any signs of wear are apparent.
2 To remove the coupling, the steering gear mounting bolts should be slackened to allow it to drop down sufficiently for the coupling to be drawn off. Once the pinch bolts are slackened on the upper and lower halves it can be drawn off. Any tendency to stick must be overcome by rocking from side to side. Impacts from hitting it with a hammer could render useless the shock absorbing material of the steering column.
3 When replacing the coupling, follow the procedure as described in paragraph 12 of Section 15.

21 Steering column lock

1 **The ignition and starter switch key also operates a lock on the steering column. Great care must be taken to ensure that if the car is moved or towed without the engine running that the key is not in the lock position. Otherwise the steering will be inoperative.**
2 In order to remove the lock for any reason, the steering column assembly must be removed first. The lock is clamped to the column and two break-head bolts are used which must be drilled out with $^1/_8$ inch drill and removed with a screw extractor (Fig 11:20).
3 When fitting the lock use only break-head bolts. Tighten them sufficiently to hold the lock in position for testing and finally tighten until the heads break off.

22 Steering column, steering shaft and bearings - removal, inspection and replacement

1 The steering column assembly is designed to collapse under impact to lessen the risk of injury to the driver. The column has a central section of expanded metal and the steering shaft is telescopic. The telescopic halves are maintained in their normal operating position by a plastic substance injected between them which shears on impact.

It is most important that the assembly and its components are handled with care. Any damage causing them to bend or lose their correct dimensions means that they will have to be renewed.
2 The steering column and mounting bracket may be inspected in situ. The upper mounting bracket (do not confuse it with the upper column brace bracket) under the dash panel is mounted on shear type pads which will also break if an excessive load is imparted on the steering wheel.

Examine the bracket to see whether there is any gap between the pads and the metal part of the bracket. If there is, it means that the bracket must have moved forward - indicating that the column has collapsed to a certain extent. The column must, therefore, be renewed (Fig 11:21).
3 To remove the column assembly, remove the steering wheel (Section 19) and slacken the pinch bolt securing the upper half of the coupling flange assembly (Section 20).

Remove the shroud surrounding the upper part of the column by taking out the screws (Fig 11:22).

Disconnect the leads from the combined horn/dipper/flasher switch and from the ignition switch (Fig 11:23).
4 Undo the bolts securing the upper support brace and the upper mounting bracket.
5 Slacken the bolts holding the plate at the bottom of the column and lift the whole assembly out.
6 The upper bearing and shaft are removed from the column together after the two screws securing the bearing to the column have been taken out. The bearing is a bayonet fit and after turning it anticlock-

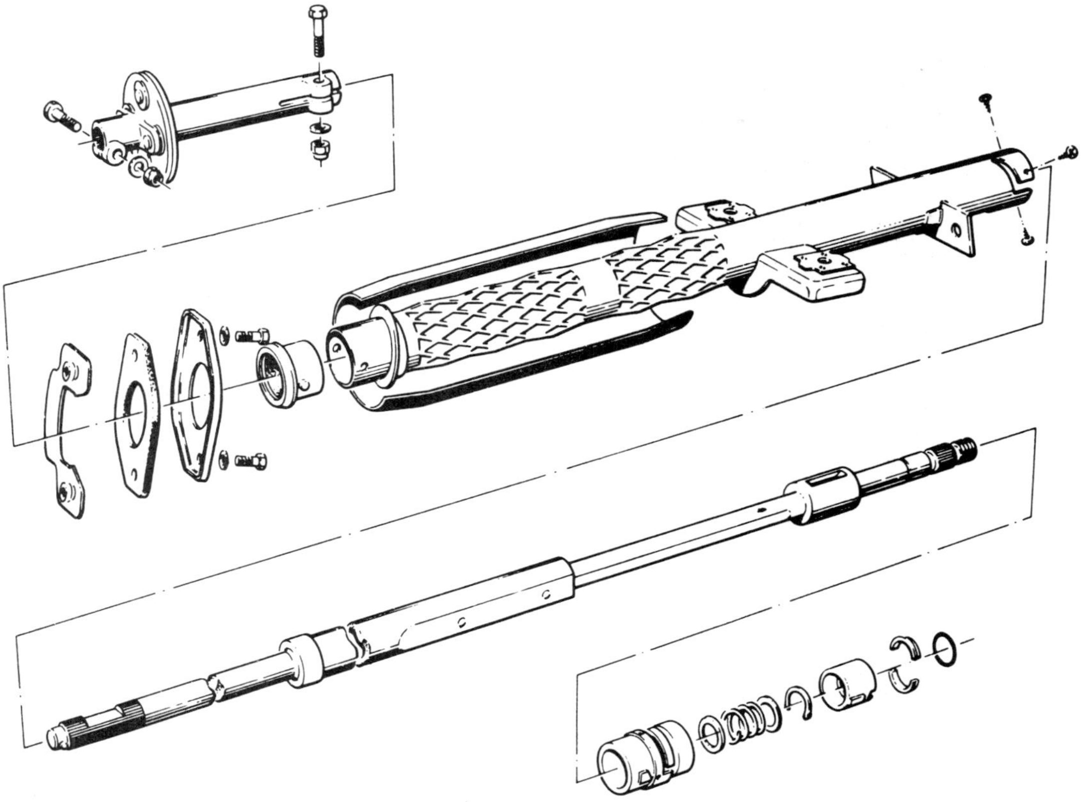

Fig 11:23 Steering column and shaft - detailed illustration

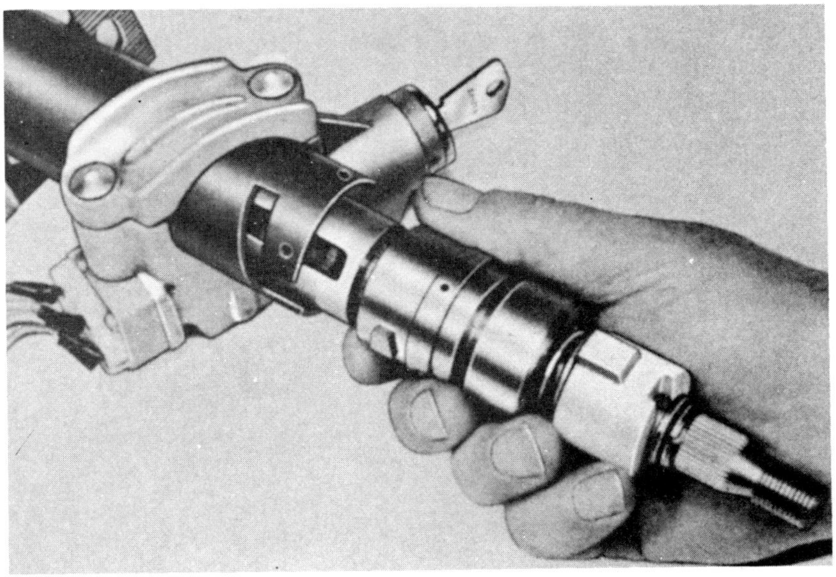

Fig 11:24 Steering column and shaft. Withdrawing the shaft and bearing together from the column

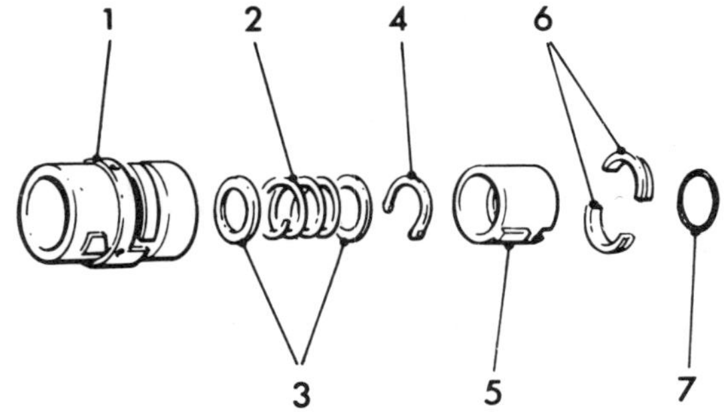

Fig 11:25 STEERING SHAFT BEARING DETAILS
1 Bearing assembly
2 Spring
3 Retainer washers
4 Circlip
5 Indicator cancelling sleeve
6 Split collars
7 Collar retaining band

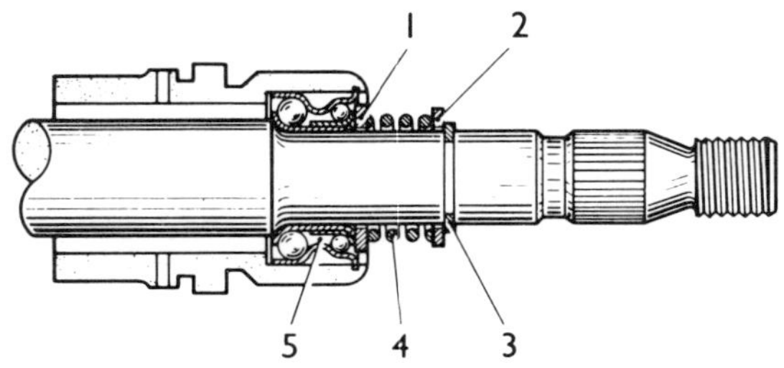

Fig 11:26 STEERING COLUMN AND SHAFT UPPER BEARING ASSEMBLY – CROSS SECTION
1 Thick washer
2 Thin washer
3 Circlip
4 Spring
5 Bearing

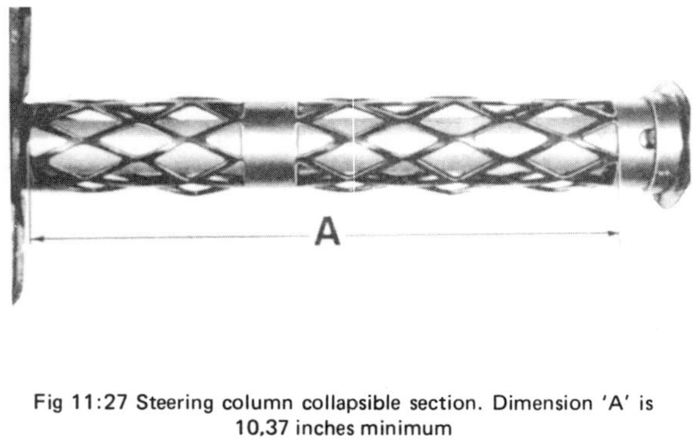

Fig 11:27 Steering column collapsible section. Dimension 'A' is 10.37 inches minimum

wise it can be drawn out (Figs 11:24 and 11:25).

7 If the bearing is worn out it is best to replace the complete unit. Remove the retaining circlip, split collars and indicator cancelling sleeve to take it off the shaft (Fig 11:26).

8 The shaft is checked by measuring its length which should be within 35.96–35.88 inch. If outside these dimensions it should be renewed.

9 The lattice work Section of the column should not be less than 10.37 inch in length. The lattice work is covered by a plastic shroud which can be prised off in two halves. The two halves can be refitted by tacking them with a hot iron in five or six spots along the seam (Fig 11:27).

10 The bearing at the bottom of the column is a nylon bush. Two small lugs on it engage in a snap fit into two holes in the column. To renew it prise it out and push in another (Fig 11:28).

11 When fitting a new upper bearing to the shaft it will be necessary to drill two holes in the outer sleeve with a No 31 drill to accept the retaining screws. Do this when the bearing has been fitted into the column.

12 If the lower support bracket has inadvertently come apart due to undoing the screws too much, note that the nut plate (on the front side of the panel) should be replaced with the flat side against the panel and the open U facing the centre line of the car. The insulation pad should be in good condition and, to assist pushing the column back into it, smear the inner edges with soap.

13 Reassembly and replacement of the column and shaft is a straightforward reversal of the removal procedure. If the lower support bracket bolts were not slackened when the column was lifted out slacken them now and first introduce the bottom end of the column through it. Make sure that the peg on the end of the shaft engages the slot in the coupling. Reassemble all the bracket bolts and tighten them in the following order: 1 Upper support bracket; 2 Lower bracket to dash panel; 3 Upper column brace bracket; 4 Coupling flange pinch bolt. Finally tighten the upper support bracket bolts to the specified torque of 17 lb ft (Fig 11:29).

14 Replace the steering wheel and column shroud.

23 Wheels and tyres

Crossply tyres

1 To provide equal, and obtain maximum wear from all the tyres, they should be rotated on the car at intervals of 6000 miles to the following pattern, provided the wheels have been balanced off the car.

Spare to offside rear;
Offside rear to nearside front;
Nearside front to nearside rear;
Nearside rear to offside front;
Offside front to spare.

Wheels which were balanced on the car, should be re-balanced when this is done. However, some owners baulk at the prospect of having to buy five new tyres all at once and tend to let two run on and replace a pair only. In this case the new pair should always be fitted to the front wheels, as these are the most important from the safety aspect of steering and braking.

Radial tyres

Rotation is not recommended but if essential, move between front and rear on the same side of the car only.

2 Never mix tyres of a radial and crossply construction on the same car, as the basic design differences can cause unusual and, in certain conditions, very dangerous handling and braking characteristics. If an emergency should force the use of two different types, make sure the radials are on the rear wheels and drive particularly carefully. If three of the five wheels are fitted with radial tyres then make sure that no more than two radials are in use on the car (and those at the rear). Rationalise the tyres at the earliest possible opportunity.

3 Wheels are normally not subject to servicing problems, but when tyres are renewed or changed the wheels should be balanced to reduce

Fig 11:28 Steering column lower shaft bearing. Prising out bush from column

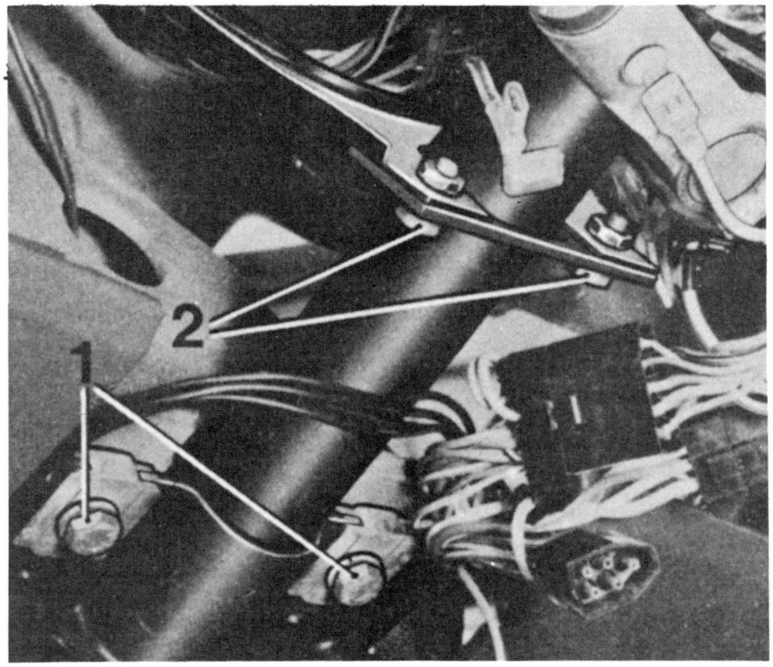

Fig 11:29 Steering column. Upper attaching bolts at: 1 mounting bracket and 2 upper brace

vibration and wear. If a wheel is suspected of damage - caused by hitting a kerb or pot hole which could distort it out of true, change it and have it checked for balance and true running at the earliest opportunity.

4 When fitting wheels do not overtighten the nuts. The maximum possible manual torque applied by the manufacturer's wheel brace is adequate. It also prevents a struggle when the same wheel brace has to be used in emergency to remove the wheels. Overtightening may also distort the stud holes in the wheel causing it to run off centre and off balance.

Fault Finding Chart –Suspension, Dampers and Steering

Before diagnosing faults from the following chart, check that any irregularities are not caused by:

1 Binding brakes.
2 Incorrect 'mix' of radial and cross-ply tyres.
3 Incorrect tyre pressures.
4 Misalignment of the body frame or rear axle.

Symptom	Reason/s	Remedy
Steering wheel can be moved considerably before any sign of movement of the wheels is apparent	Wear in the steering linkage, gear and column coupling	Check movement in all joints and steering gear and overhaul and renew as required.
Vehicle difficult to steer in a consistent straight line - wandering	As above Wheel alignment incorrect (indicated by excessive or uneven tyre wear)	As above Check wheel alignment.
	Front wheel hub bearings loose or worn	Adjust or renew as necessary.
	Worn ball joints, track rods or suspension arms	Renew as necessary.
Steering stiff and heavy	Incorrect wheel alignment (indicated by excessive or uneven tyre wear).	Check wheel alignment.
	Excessive wear or seizure in one or more of the joints in the steering linkage or suspension arm ball joints	Renew as necessary or grease the suspension unit ball joints.
	Excessive wear in the steering gear unit	Adjust if possible or renew.
Wheel wobble and vibration	Road wheels out of balance	Balance wheels.
	Road wheels buckled	Check for damage.
	Wheel alignment incorrect	Check wheel alignment
	Wear in the steering linkage, suspension arm ball joints or suspension arm pivot bushes	Check and renew as necessary.
	Broken front spring	Check and renew as necessary.
Excessive pitching and rolling on corners and during braking	Defective dampers and/or broken spring	Check and renew as necessary.

Chapter 12 Bodywork and underframe

Contents

1 General description

The combined body shell and underframe is an all welded unitary structure of sheet steel. Openings in it provide for the engine compartment, luggage compartment, doors and front and rear windows. The rear axle is attached to the body by arms bolted directly to it on rubber bushes, and a detachable crossmember across the bottom of the engine compartment provides support for the engine and front suspension. A second detachable item is a central crossmember bridging the transmission tunnel which is the rear support of the engine/gearbox unit.

Two door models are available in Standard, de luxe and 'SL' versions whereas the four door model is available only in de luxe and 'SL' versions. The estate model has two single side doors and a full width single rear door hinged at the top. This comes in de luxe and 'SL' versions only.

2 Maintenance - bodywork and underframe

1 The general condition of a car's bodywork is the one thing that significantly affects its value. Maintenance is easy but needs to be regular and particular. Neglect, particularly after minor damage, can lead quickly to a further deterioration and costly repair bills. It is important also to keep watch on those parts of the car not immediately visible, for instance the underside, inside all the wheel arches and the lower part of the engine compartment. If your car is not fitted with mud flaps at the front, it is strongly recommended that they are installed. Vauxhall agents will supply them made to measure for the car at a very fair price. These protect the door undersills which are otherwise soon stripped to the bare metal by the water/grit slurry thrown up by the front wheels in wet weather.

2 The basic maintenance routine for the bodywork is washing preferably with a lot of water, from a hose. This will remove all the loose solids which may have stuck to the car. It is important to flush these off in such a way as to prevent grit from scratching the finish. The wheel arches and underbody need washing in the same way to remove any accumulated mud which will retain moisture and tend to encourage rust. Paradoxically enough, the best time to clean the underbody and wheel arches is in wet weather when the mud is thoroughly wet and soft. In very wet weather the underbody is usually cleaned of large accumulations automatically and this is a good time for inspection.

3 Periodically, it is a good idea to have the whole of the underside of the car steam cleaned, engine compartment included, so that a thorough inspection can be carried out to see what minor repairs and renovations are necessary. Steam cleaning is available at many garages and is necessary for removal of accumulations of oily grime which sometimes cakes thick in certain areas near the engine, gearbox and back axle. If steam facilities are not available, there are one or two excellent grease solvents available which can be brush applied. The dirt can then be simply hosed off.

4 After washing paintwork, wipe it off with a chamois leather to give an upspotted clear finish. A coat of clear protective wax polish will give added protection against chemical pollutants in the air. If the paintwork sheen has dulled or oxidised, use a cleaner/polisher combination to restore the brilliance of the shine. This requires a little more effort, but the condition is usually caused because regular washing has been neglected. Always check that door and ventilator opening drain holes and pipes are completely clear so that water can drain out. Bright work should be treated the same way as paintwork. Windscreens and windows can be kept clear of the smeary film which often appears if a little ammonia is added to the water. If they are scratched, a good rub with a proprietary metal polish will often clear them. Never use any form of wax or chromium polish on glass.

3 Maintenance - interior

1 Mats and carpets should be brushed or vacuum cleaned regularly to keep them free of grit. If they are badly stained remove them from the car for scrubbing or sponging and make quite sure they are dry before replacement. Seats and interior trim panels can be kept clean by a wipe over with a damp cloth. If they do become stained

(which can be more apparent on light coloured upholstery) use a little liquid detergent and a soft nail brush to scour the grime out of the grain of the material. Do not forget to keep the head lining clean in the same way as the upholstery. When using liquid cleaners inside the car do not over wet the surfaces being cleaned. Excessive damp could get into the seams and padded interior causing stains, offensive odours or even rot. If the inside of the car gets wet accidentally it is worthwhile taking some trouble to dry it out properly, particularly where carpets are involved. Do NOT leave oil or electric heaters inside the car for this purpose.

4 Minor repairs to bodywork

1 A car which does not suffer some minor damage to the bodywork from time to time is the exception rather than the rule. Even presuming the gatepost is never scraped or the door opened against a wall or high kerb, there is always the likelihood of gravel and grit being thrown up and chipping the surface, particularly at the lower edges of the doors and sills.
2 If the damage is merely a paint scrape which has not reached the metal base, delay is not critical, but where bare metal is exposed action must be taken immediately before rust sets in.
3 The average owner will normally keep the following 'first aid' materials available which can give a professional finish for minor jobs:

a) A resin based filler paste;
b) Matched paint either in an aerosol can or 'touch up' tin;
c) Fine cutting paste;
d) Medium and fine grade wet and dry abrasive paper.

4 Where the damage is superficial (ie not down to the bare metal and not dented), fill the scratch or chip with sufficient filler to smooth the area, rub down with paper and apply the matching paint.
5 Where the bodywork is scratched down to the metal, but not dented, clean the metal surface thoroughly and apply a suitable metal primer first, such as red lead. Fill up the scratch as necessary with filler and rub down with wet and dry paper. Apply the matching colour paint.
6 If more than one coat of colour is required rub down each coat with cutting paste before applying the next.
7 If the bodywork is dented, first beat out the dent as near as possible to conform with the original contour. Avoid using steel hammers - use hardwood mallets or similar and always support the back of the panel being beaten with a hardwood or metal 'dolly'. In areas where severe creasing and buckling has occurred it will be virtually impossible to reform the metal to the original shape. In such instances a decision should be made whether or not to cut out the damaged piece or attempt to re-contour over it with filler paste. In large areas where the metal panel is seriously damaged or rusted, the repair is to be considered major and it is often better to replace a panel or sill section with the appropriate part supplied as a spare. When using filler paste in largish quantities, make sure the directions are carefully followed. It is false economy to try and rush the job, as the correct hardening time must be allowed between stages or before finishing. With thick application the filler usually has to be applied in layers - allowing time for each layer to harden. Sometimes the original paint colour will have faded and it will be difficult to obtain an exact colour match. In such instances it is a good scheme to select a complete panel - such as a door, or boot lid, and spray the whole panel. Differences will be less apparent where there are obvious divisions between the original and resprayed areas.

5 Major repairs to bodywork

1 Where serious damage has occurred or large areas need renewal due to neglect, it means certainly that completely new sections or panels will need welding in and this is best left to professionals. If the damage is due to impact it will also be necessary to completely check the alignment of the body shell structure. Due to the principle of construction, the strength and shape of the whole can be affected by damage to a part. In such instances the services of a Vauxhall agent with specialist checking jigs are essential. If a body is left mis-aligned, it is first of all dangerous as the car will not handle properly - and secondly, uneven stresses will be imposed on the steering, engine and transmission, causing abnormal wear or complete failure. Tyre wear will also be excessive.

6 Front and rear bumpers - removal and replacement

1 The front and rear bumpers are mounted similarly - that is on brackets which in turn are bolted to the longitudinal side rails at the front and to the end panel at the back. The ends of the bumper are attached to the body panels also.

When removing bumpers, remove the bolts securing the brackets to the car so that the whole lot can be lifted off together.
2 On estate cars the cargo floor must be lifted up to get at the bolts.
3 When replacing the rear bumper, the bolt holes must have sealing putty applied before the nuts and bolts are tightened. Otherwise water can get into the boot.

7 Windscreen rear window and rear quarter fixed glass - removal and replacement

1 Unless the glass has been broken it is assumed that it is being removed because the sealing strip is leaking. If you are buying a secondhand screen from a breaker's yard, ask them to remove it for you before paying for it, but if the screen is already removed check the edges very carefully for signs of chipping. The screen should be smoothly ground all round the edge and any chip is a potential starter for a future crack.
2 Check whether the screen is made of toughened or laminated glass.
3 Remove the windscreen wiper arms by slackening the wedge screw, tapping it to loosen the wedge and lifting the wiper off. Disconnect the battery and remove the interior mirror.
4 Toughened glass screens can be removed by bumping the glass from inside with the flat of the hand. Wear stout gloves as a precaution. If moderate bumping fails, use foot pressure with pads under the feet to distribute pressure.
5 With laminated glass remove the glazing channel insert strip, where fitted, and cut away the lip of the glazing channel on the outside of the glass. Apply firm steady pressure from inside. Do not bump the glass or it may crack.
6 If a broken screen is being removed, cover up the scuttle ventilation grill to prevent pieces falling into the heater or ventilator.
7 To replace the glass, first clean all old sealing compound off the frame and sealing strip if the sealing strip can be re-used. If the screen is being replaced because of vehicle damage, make sure the frame is not distorted in any way. This can be checked by carefully holding the new screen in position to see that its contour and shape is reasonably well matched. Take care not to chip the glass edges.
8 Fit the glazing channel to the screen with the securing lip towards the inner (concave) side. Fig 12:3 gives a cross-section showing chrome inserts fitted on SL models.
9 Fit a piece of thin, strong cord into the inner groove so that a loop is left in the top centre and the ends come out at the bottom centre. Make sure that the cord crosses over in the channel at the loop and ends (otherwise the centre pieces of the glazing channel cannot be pulled over the flange with the cord). Identify each end and the halves of the loop so that the running direction of each piece of cord is known.
10 Using a suitable container fitted with a fine nozzle, apply sealer (Bostik No 6) to the bottom of the corner of the body frame flange and also round the front edge of the glass between the glass and the glazing strip.

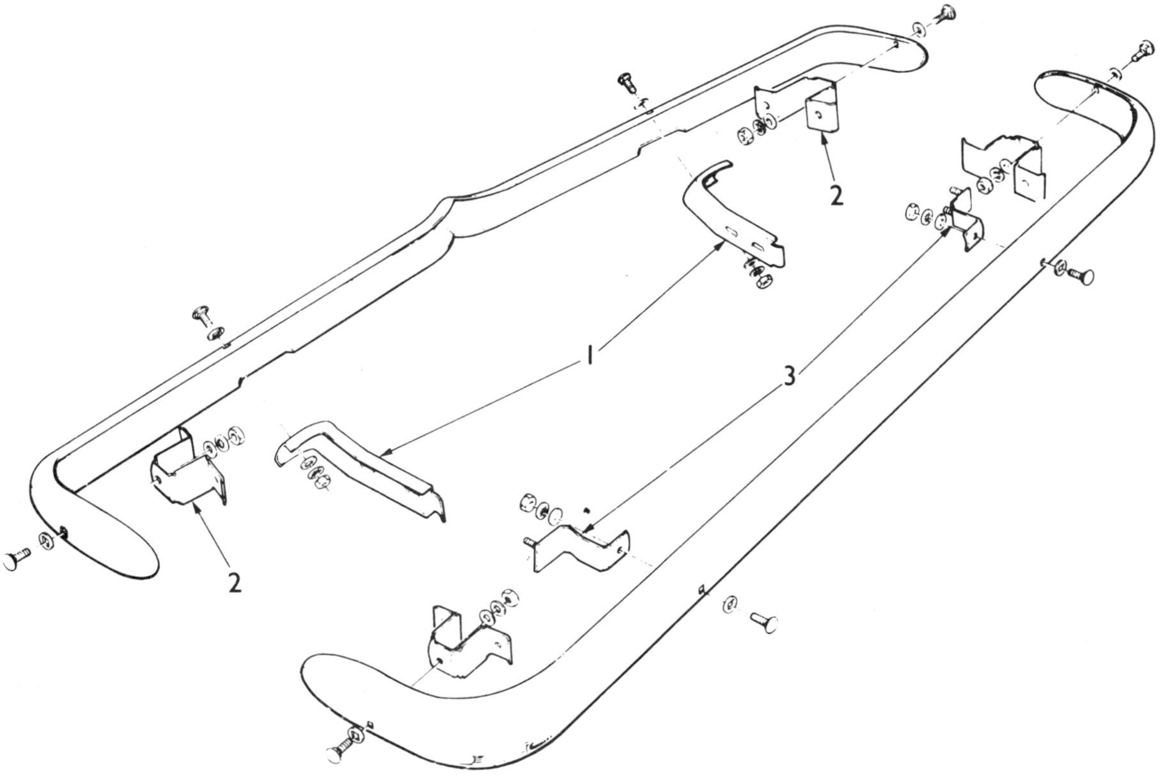

Fig 12:1 FRONT AND REAR BUMPERS AND BRACKETS – DETAILS
1 Front bumper mounting brackets 2 Front bumper end brackets 3 Rear bumper brackets

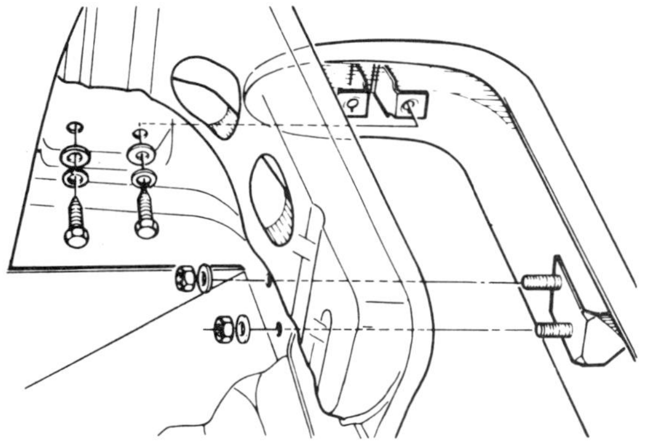

Fig 12:2 Rear bumper end fixings

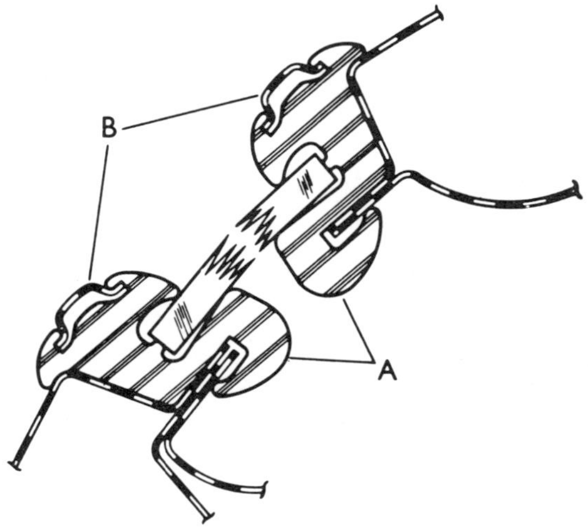

Fig 12:3 WINDSCREEN GLAZING STRIP – CROSS SECTION
A Glazing channel B Moulding insert(SL models)

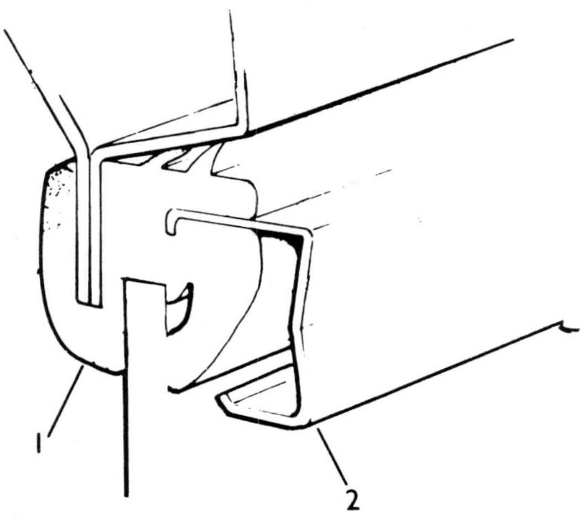

Fig 12:4 REAR FIXED QUARTER LIGHT GLASS
1 Glazing strip 2 Moulding

Fig 12:5 REAR FIXED QUARTER LIGHT GLASS
Pop rivets (arrowed) securing moulding to body panel

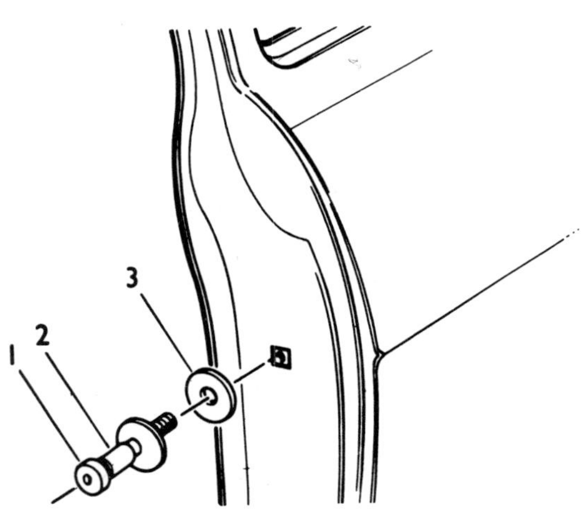

Fig 12:6 DOOR LATCH STRIKER PIN FIXING
1 Striker 3 Packing washer
2 Silencing bush

11 Place the screen in position, pressing lightly from the outside, and pull the strings from the bottom edge so that the glazing channel comes over the edge of the bottom flange up to within six inches of each bottom corner. Make sure the glass is kept central and repeat the procedure along the top edge followed last of all by the sides. Check that the screen is properly seated both inside and out, and clear away surplus sealing compound.

12 Where a heated rear window is installed, ensure that the wires are disconnected from the terminals and take great care that the printed circuit on the inner surface is not scratched or abraded. Minor repairs to the printed circuit can be repaired using a high conductivity paint.

13 The rear quarter fixed windows may be removed in the same manner but first of all three pop rivets must be drilled out of the forward edge of the reveal moulding (Fig 12:4 and 12:5).

8 Doors - tracing of rattles and setting of latch striker pin

1 Check first that the door is not loose at the hinges and that the latch is holding it firmly in position. Check also that the door lines up with the aperture in the body.

2 If the hinges are loose, or the door is out of alignment, the hinge securing bolts should be tightened up first. If necessary, new hinge pins should be fitted. If the door is still out of alignment then the problem is more serious and the hinges may need renewal. As they are welded to the door this needs specialist attention.

3 If the latch is holding the door correctly it should be possible to press the door inwards fractionally against the rubber weatherstrip. If not, adjust the striker pin which is screwed into a plate behind the door pillar either inwards or outwards as required. Use an Allen key to undo the striker sufficiently to move it as needed. If the striker should need moving forward or backwards (relative to the front and rear of the car) packing washers can be fitted as necessary so that the lock fork engages over the pin where the rubber silencing bush is fitted (Fig 12:6).

9 Door trim panel and interior handles - removal and replacement

1 Remove the window regulator (winder) handle by first prising out the trim cover and undoing the screw. Do not lose the wearing washer fitted behind (Fig 12:7).

2 Remove the arm rest or door pull by undoing the mounting screws. Remove the escutcheon round the latch handle - it is a press fit and will lever off.

3 Slide a thin stiff blade (such as a putty knife) behind the edge of the trim, run it round next to each fixing clip in turn and prise the clip out of the hole in the door. Do not prise anywhere except next to a clip, or the clip will probably tear out of the trim panel. On SL models the trim pad is thicker and is hung over the door sill at the top. Although the arm rest is part of the trim pad, do not overlook the two screws which hold the centre of the pad to the door panel. Also the inside lock button must be pulled off (Fig 12:8).

4 On estate car rear doors the pad is held by clips which are prised off in the same manner.

5 The remote control latch handle is fixed to the door and the trim pad can be taken off over it.

6 Under the trim pad there is a polythene sheet stuck to the lower half of the door panel as a water deflector (to keep water that gets inside the door from soaking the trim pad before it runs out of the bottom of the door). This should be securely fixed - particularly along the bottom edge.

7 Replacement is a straightforward reversal of the removal procedure.

10 Door glasses - removal and replacement

1 Remove the trim pad as described in the previous Section and the sealing strips along the door aperture.

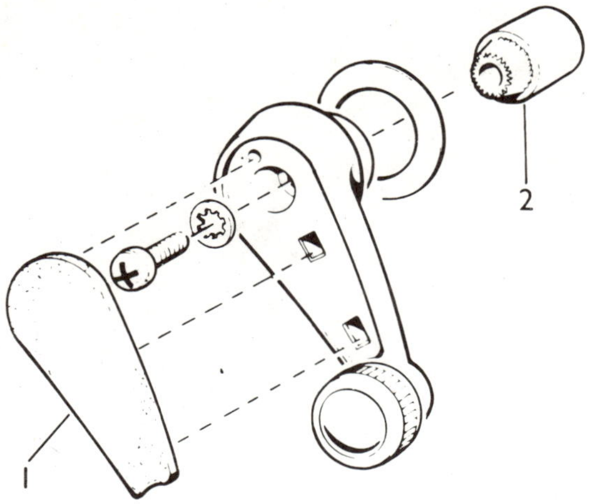

Fig 12:7 WINDOW WINDING HANDLE DETAILS
1 Spring fit insert
2 Extension piece for thicker trim pad on SL models

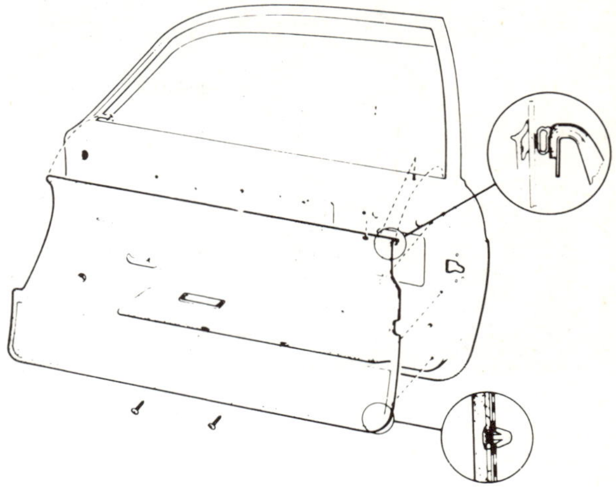

Fig 12:8 Door trim pad - SL models - shown (inset) fixing over window sill and clip fixing

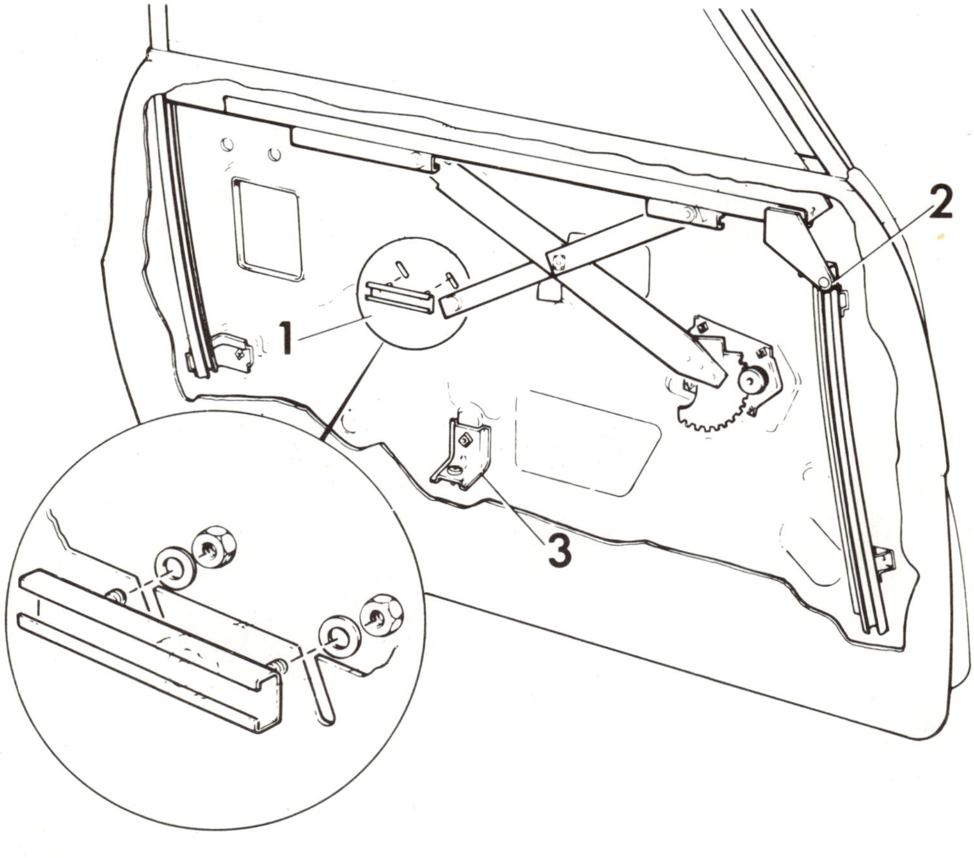

Fig 12:9 WINDOW WINDING MECHANISM – FRONT DOORS INSET – DETAIL OF HORIZONTAL CHANNEL
1 Support channel 2 Fixed roller in guide channel 3 Stop buffer bracket

This sequence of photographs deals with the repair of the dent and scratch (above rear lamp) shown in this photo. The procedure will be similar for the repair of a hole. It should be noted that the procedures given here are simplified - more explicit instructions will be found in the text

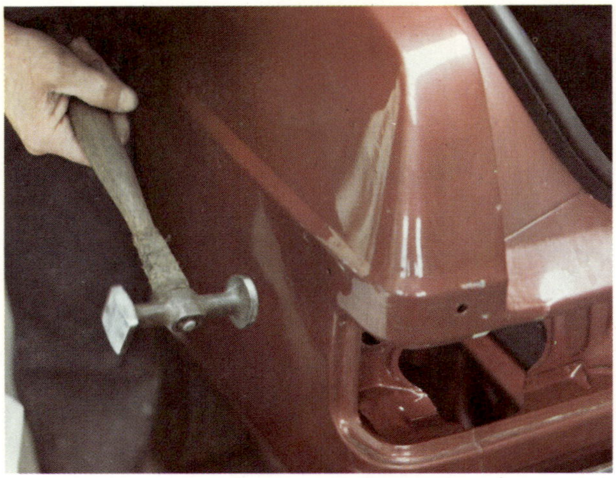

In the case of a dent the first job - after removing surrounding trim - is to hammer out the dent where access is possible. This will minimise filling. Here, the large dent having been hammered out, the damaged area is being made slightly concave

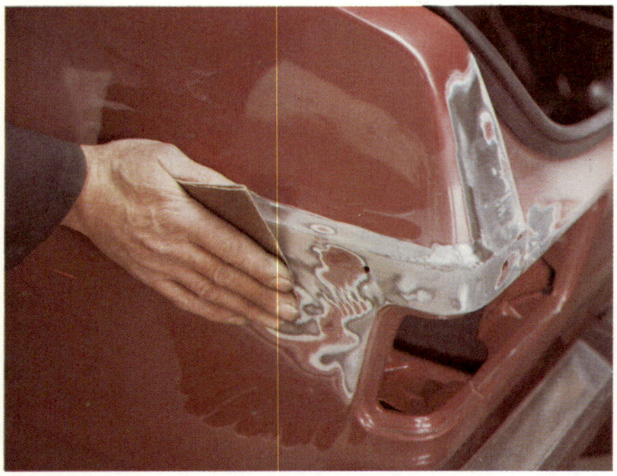

Now all paint must be removed from the damaged area, by rubbing with coarse abrasive paper. Alternatively, a wire brush or abrasive pad can be used in a power drill. Where the repair area meets good paintwork, the edge pf the paintwork should be 'feathered', using a finer grade of abrasive paper

In the case of a hole caused by rusting, all damaged sheet-metal should be cut away before proceeding to this stage. Here, the damaged area is being treated with rust remover and inhibitor before being filled

Mix the body filler according to its manufacturer's instructions. In the case of corrosion damage, it will be necessary to block off any large holes before filling - this can be done with zinc gauze or aluminium tape. Make sure the area is absolutely clean before ...

... applying the filler. Filler should be applied with a flexible applicator, as shown, for best results: the wooden spatula being used for confined areas. Apply thin layers of filler at 20-minute intervals, until the surface of the filler is slightly proud of the surrounding bodywork

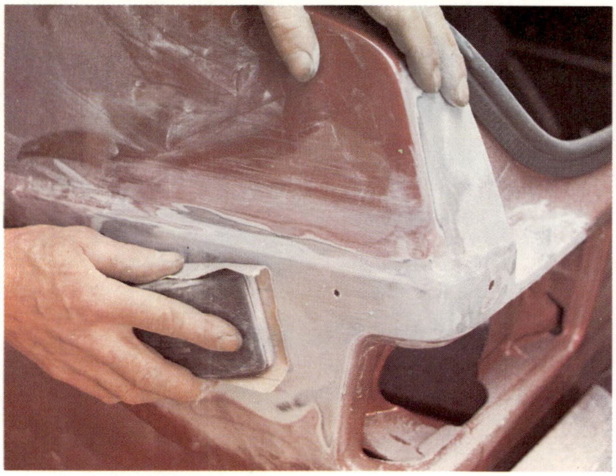

Initial shaping can be done with a Surform plane or Dreadnought file. Then, using progressively finer grades of wet-and-dry paper, wrapped around a sanding block, and copious amounts of clean water, rub-down the filler until really smooth and flat. Again, feather the edges of adjoining paintwork

The whole repair area can now be sprayed or brush-painted with primer. If spraying, ensure adjoining areas are protected from over-spray. Note that at least one-inch of the surrounding sound paintwork should be coated with primer. Primer has a 'thick' consistency, so will fill small imperfections

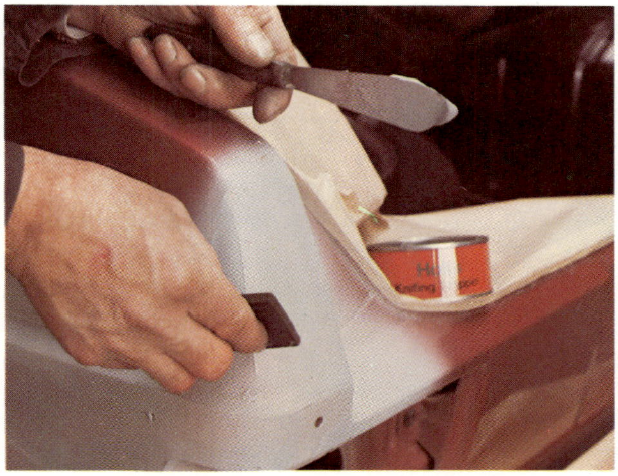

Again, using plenty of water, rub down the primer with a fine grade of wet-and-dry paper (400 grade is probably best) until it is really smooth and well blended into the surrounding paintwork. Any remaining imperfections can now be filled by carefully applied knifing stopper paste

When the stopper has hardened, rub-down the repair area again before applying the final coat of primer. Before rubbing-down this last coat of primer, ensure the repair area is blemish-free - use more stopper if necessary. To ensure that the surface of the primer is really smooth use some finishing compound

The top coat can now be applied. When working out of doors, pick a dry, warm and wind-free day. Ensure surrounding areas are protected from over-spray. Agitate the aerosol thoroughly, then spray the centre of the repair area, working outwards with a circular motion. Apply the paint as several thin coats.

After a period of about two-weeks, which the paint needs to harden fully, the surface of the repaired area can be 'cut' with a mild cutting compound prior to wax polishing. When carrying out bodywork repairs, remember that the quality of the finished job is proportional to the time and effort expended

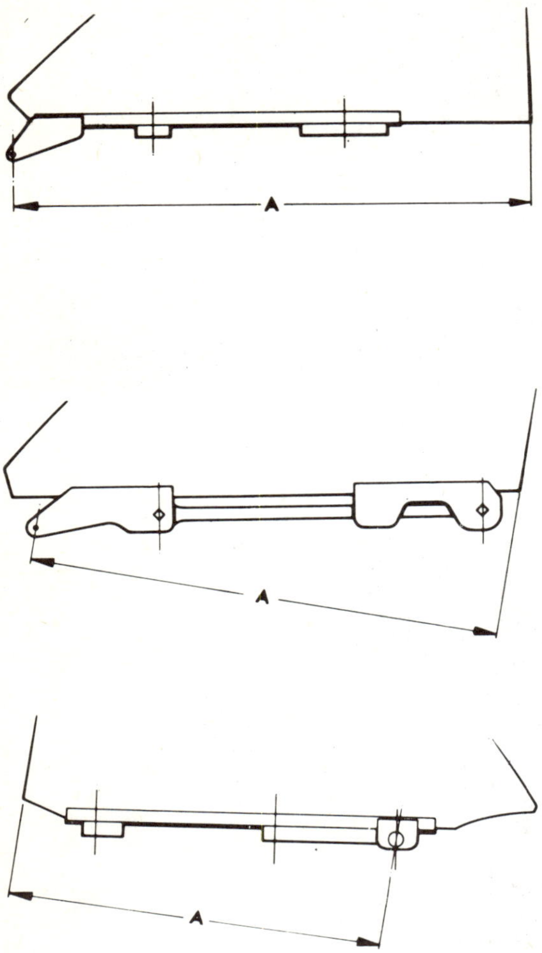

Fig 12:10 DOOR GLASS – FITTING POSITION IN SUPPORT CHANNEL
Top. 2 door models A = 37;80 inches
Centre. 4 door models - front door A = 26.80 inches
Bottom. 4 door models - rear doors A = 23.20 inches

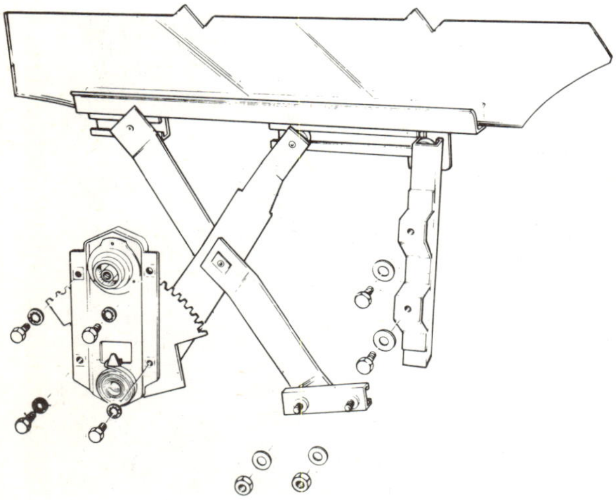

Fig 12:11 Window winding mechanism - rear doors of 4 door saloon

2 Referring to Fig 12:9, if the horizontal support is removed from the door panel the regulator arms can then be disengaged from the roller runs in the glass support channel. If the support channel fixed roller at the forward end is then disengaged from the vertical channel, the glass can be tipped and drawn out from the top of the door. On four door cars the front window horizontal guide channel is bolted to the window support channel.

3 Note that on replacement it is possible to adjust the vertical channels to keep the glass in position against the weatherstrip on the outer door window sill. Access to the adjusting bolts from the top ends of both channels is through holes in the front and rear edges of the doors. The holes are covered by press in plugs.

4 If new glass is being fitted into the channels make sure that they are positioned according to the dimensions given in Fig 12:10.

5 Rear door glasses on four door cars are removed according to the same principles.

6 When the lower horizontal support channel is replaced it should be adjusted in its slots so that the top edge of the glass is parallel with the door frame.

11 Window winder mechanism - removal and replacement

If the window mechanism operating arms are disengaged from their channels as described in the previous Section, the securing screws can be removed and the regulator mechanism taken out through the door panel aperture.

12 Door catches - removal and replacement

1 Remove the interior trim panel and close the window of the door in question.

2 Remove the glass run channel securing screws.

3 Remove the lock securing screws and the lock can then be detached from the control rods and taken out (Fig 12:12, 12:13 and 12:15). On two door models the exterior handle must be removed before the rod can be detached from it and the door lock removed.

4 On replacement smear the lock mechanism and rod ends with high melting point grease.

13 Exterior door handles and latch buttons - removal and replacement

1 All door handles are secured by nuts on studs in the handle which are accessible from inside the door once the trim pad has been removed (Fig 12:14).

2 Once the handle has been removed the push button and spring can be released by removing the three screws securing the cover plate.

3 The lock barrel may be removed from the push button by pressing down the spring loaded plunger in the end of the lock barrel and removing the extension. The barrel will then come out and another may be fitted.

14 Interior latch handles - removal and replacement

With the door trim pad removed, the latch assembly can be undone by removing the two securing nuts. If it is then turned at right angles horizontally it can be unhooked from the connecting rod (Figs 12:16 and 12:17).

15 Bonnet and boot lids and locks - elimination of rattles, removal and replacement

1 Rattles traced to the bonnet are usually due to maladjustment or wear of the catch.

2 Both front and rear ends of the bonnet are adjustable for height. Before altering the front on the catch ensure that the rear is flush with the bodywork and centrally positioned. Adjustments can be made on the hinge bracket mountings which are slotted.

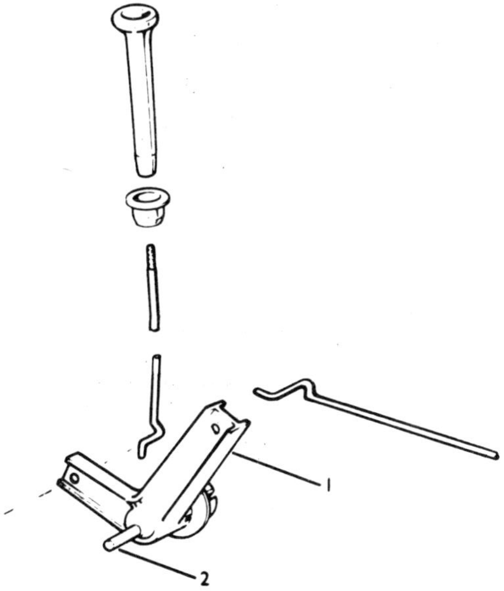

Fig 12:12 REAR DOORS – LOCKING LATCH
1 Bell crank 2 Pivot pin

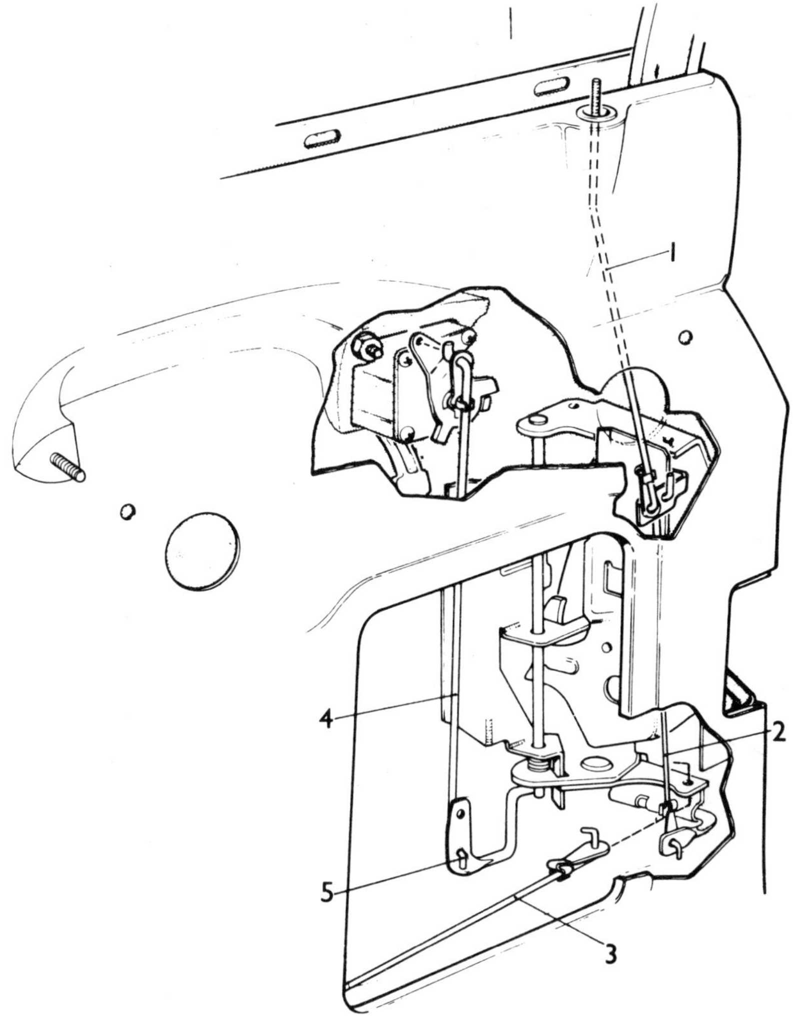

Fig 12:13 FRONT DOOR LOCK AND CATCH MECHANISM DETAIL
1 and 2 Internal lock button link rods
3 Door latch interior release handle rod
4 Door latch exterior release button rod
5 Rod link - no anti-rattle clip fitted (lower hole - 4 door models, upper hole - 2 door models)

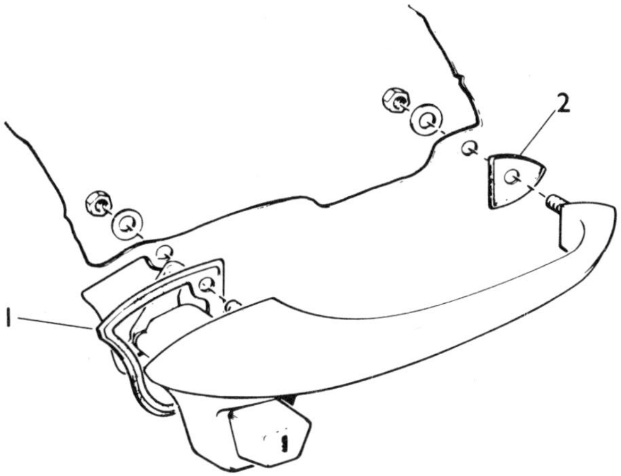

Fig 12:14 EXTERIOR DOOR HANDLES – DETAIL
1 and 2 Gaskets

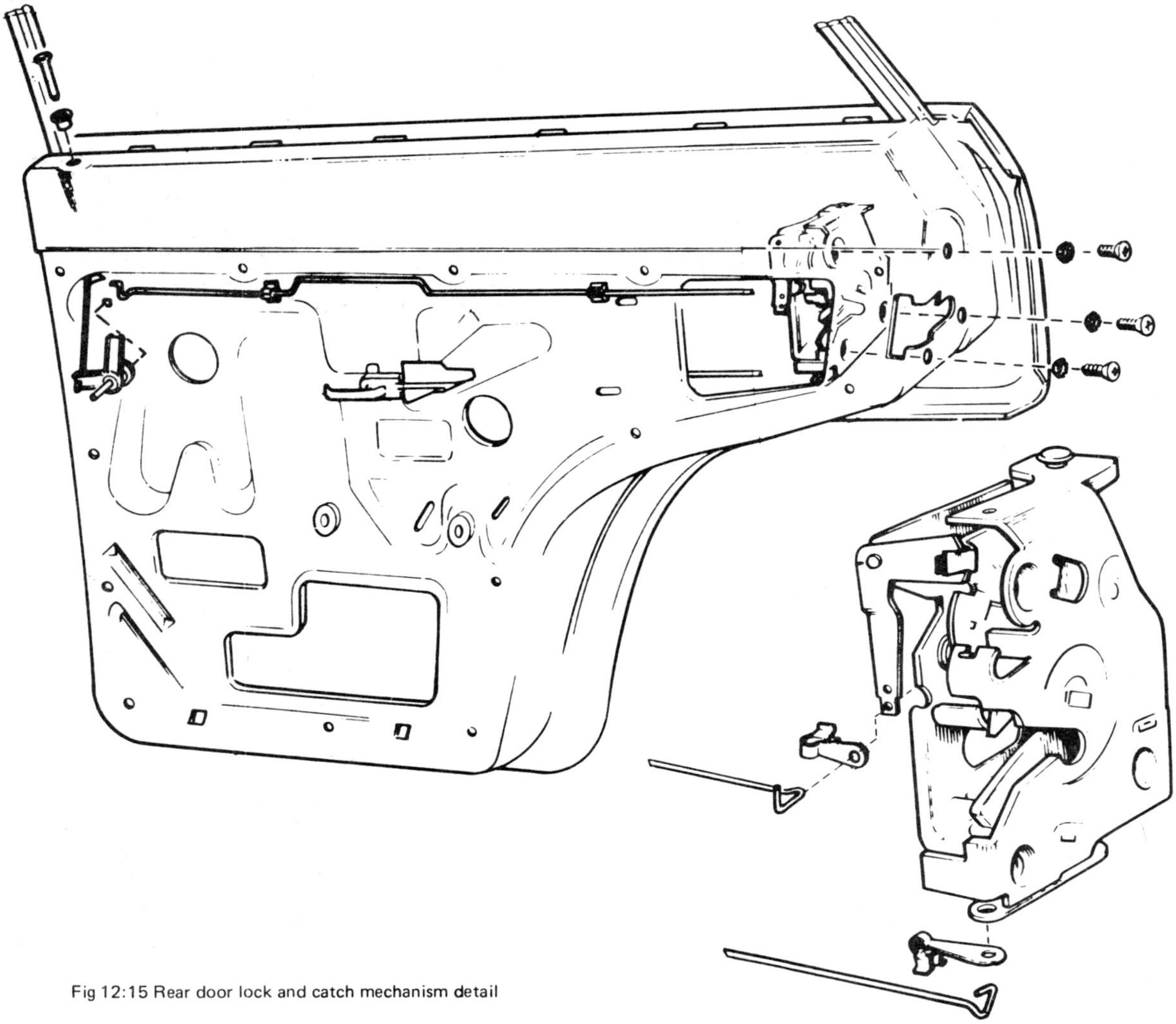

Fig 12:15 Rear door lock and catch mechanism detail

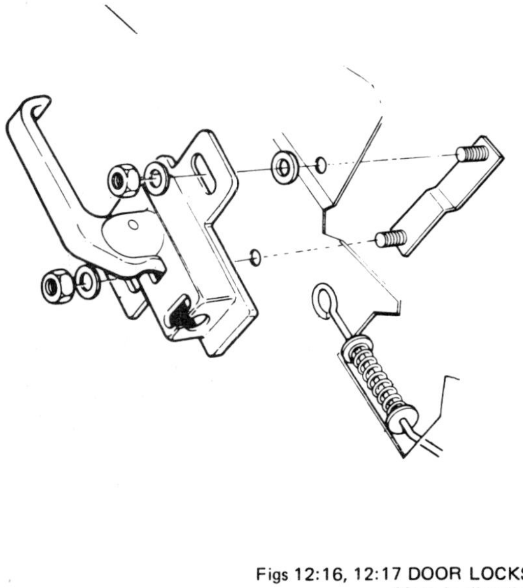

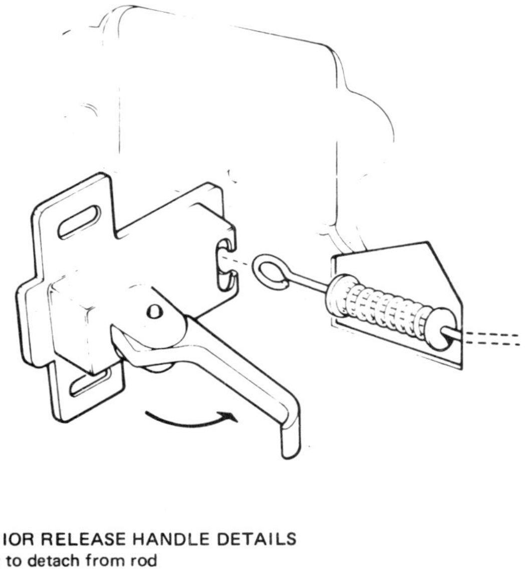

Figs 12:16, 12:17 DOOR LOCKS – INTERIOR RELEASE HANDLE DETAILS
Move in direction of arrow to detach from rod

3 If extreme pressure is required to close the bonnet so that the catch grips, or if the catch engages and the front of the bonnet can still be pressed down a noticeable amount, then the dovetail bolt which engages in the catch needs adjustment. If the locknut is slackened, the bolt may be screwed in or out as necessary (Fig 12:18).
4 If the latch mechanism is faulty it can be removed. First mark the position to assist line-up when another is fitted, remove the radiator grille and undo the three nuts holding it to the support rail (Fig 12:19).
5 On models fitted with a cable operated release mechanism, release the cable outer retaining clip and the end of the cable can then be detached from the release lever.

If the cable should break it will be necessary to detach the release lever mechanism from the side of the scuttle by undoing the two self-tapping screws and drilling out the pop rivet. The inner end of the cable may then be unhooked (Figs 12:20 and 12:21).
6 Removal and replacement of the bonnet is quite straightforward. Make sure that there is another person available to support it. Mark the position of the upper hinge brackets on the bonnet and then remove the two bolts at each side when it may be lifted off.
7 The boot lid is hinged with integral torsion bars which hold it open. The torsion bar hooks into one of three slots to give the most suitable tension. The latch engages a striker loop which is clamped to the rear panel and may be raised or lowered so that the lid may be closed with the correct amount of firm hand pressure.
8 When removing the boot lid, mark the hinges and note any packing wedges that may have been installed between hinge and lid to maintain the correct height in relation to the surrounding bodywork. If the hinges are being removed from the bodywork it will be necessary to remove the rear seat, squab and parcel shelf first in order to get at the upper mounting bolt.
9 To remove the turn button and lock assembly requires a spring retainer plate to be moved sideways. This retainer plate is accessible through an aperture on the inside of the lid.
10 The lock may be dismantled after the lock support and nut are removed and the striking lever taken off. To separate the extension from the lock barrel, depress the spring loaded plunger which fits into the extension. When reassembling, make sure that the lever recess fits over the pip on the turn button shaft (Fig 12:22).
11 The latch mechanism can be taken out after removal of the turn button assembly, by undoing the mounting screws through the same aperture. **On later models, no turn button is fitted, repeat paragraph 10, insert key and remove lock barrel from body.**

16 Estate car - rear door hinges and lock

1 The rear door turn button, lock and catch are similar to those of the saloon car boot. The only difference is that the turn button lever is connected to the latch by a link rod rather than directly engaging it. Both may be readily removed after removing the door trim pad (Fig 12:23).
2 Access to the rear door hinges is possible after removing the cover assembly. This can be lowered, complete with lamp unit, after the nine attaching screws have been removed. Detach the wires from the lamp, and the cover can be taken out.
3 The rear door hinges also incorporate torque rods to keep the door in the raised position when opened. Mark the hinges on the door panel before removing the door from the hinges. If the hinges are to be removed, take the door off first.

17 Weatherstrips

1 The doors, bonnet and boot are all fitted with rubber weatherstrips to keep water out (and in the case of the bonnet to prevent engine fumes from entering the ventilation air inlet). Damaged weatherstrips should be renewed without delay as deterioration of the bodywork interior will otherwise result.
2 The bonnet landing strip is fixed to the upper edge of the ventila-

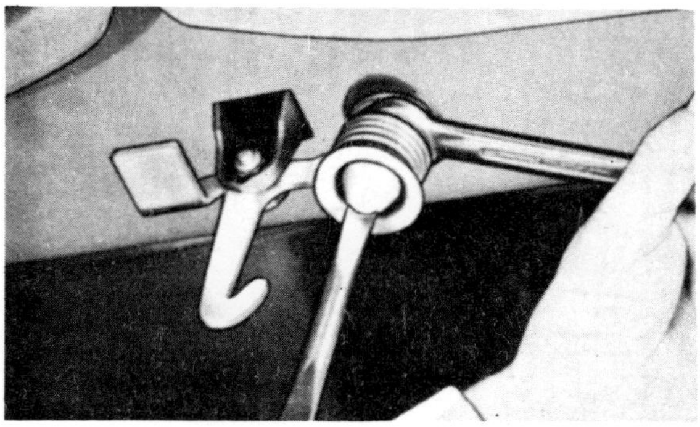
Fig 12:18 Bonnet catch dovetail bolt - adjustment

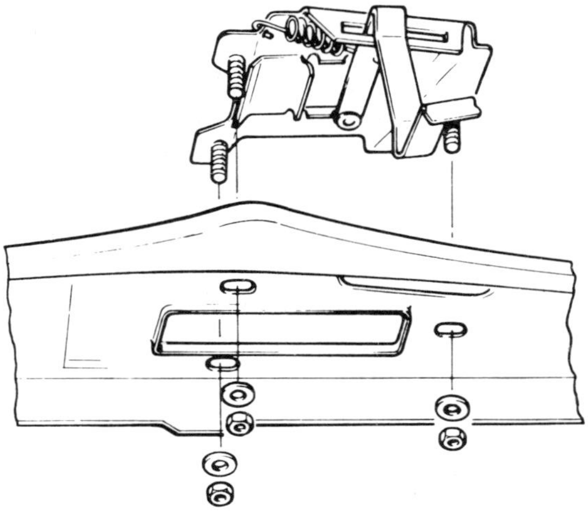
Fig 12:19 Bonnet release catch - detail of fixing

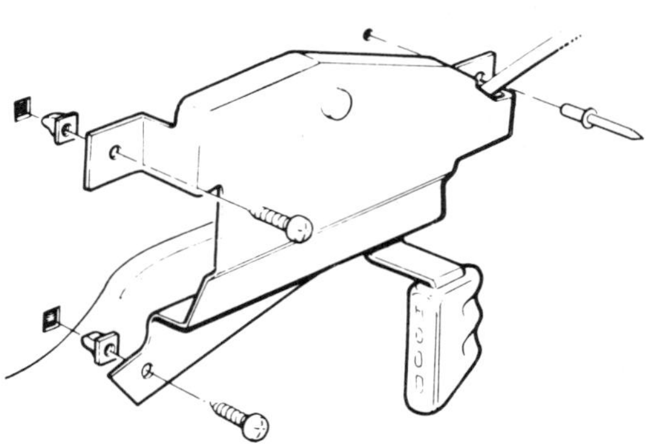
Fig 12:20 Bonnet release cable lever. Detail of fixing

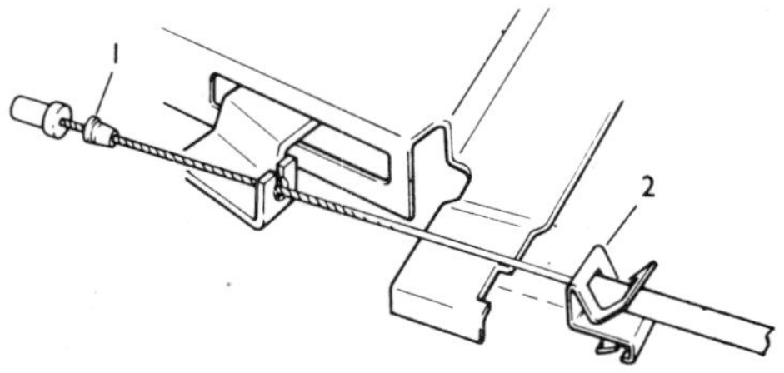

Fig 12:21 BONNET LATCH CABLE RELEASE DETAIL
1 Lever retaining sleeve 2 Cable outer clip

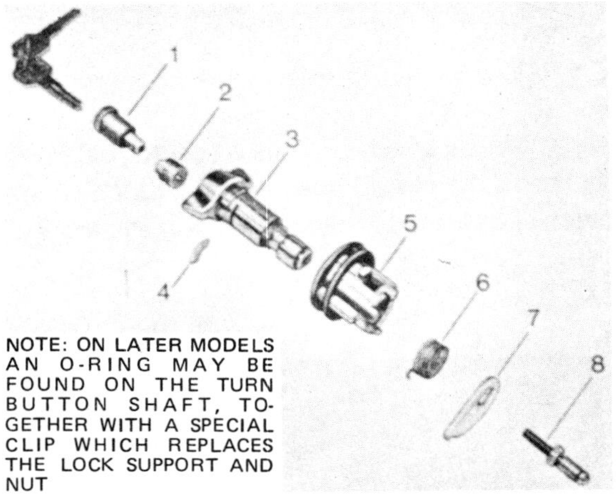

NOTE: ON LATER MODELS AN O-RING MAY BE FOUND ON THE TURN BUTTON SHAFT, TOGETHER WITH A SPECIAL CLIP WHICH REPLACES THE LOCK SUPPORT AND NUT

Fig 12:22 BOOT LATCH TURN BUTTON AND LOCK ASSEMBLY
1 Lock barrel
2 Barrel extension
3 Turn button
4 Keep segment
5 Body
6 Release spring
7 Lock lever
8 Lock support and nut

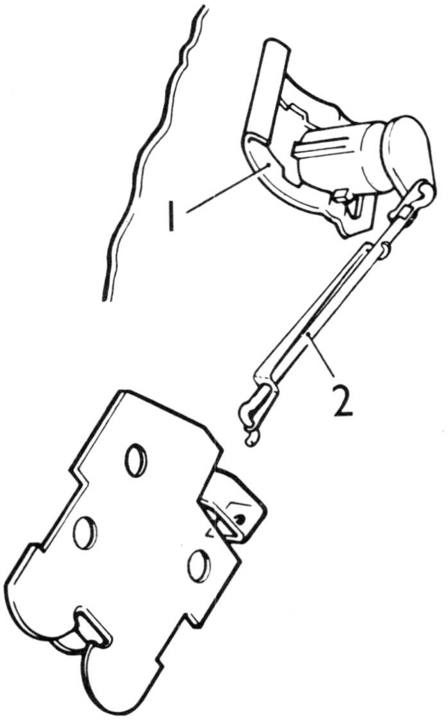

Fig 12:23 REAR DOOR – ESTATE CAR. TURN BUTTON AND LATCH ARRANGEMENT
1 Turn button retaining plate 2 Link rod

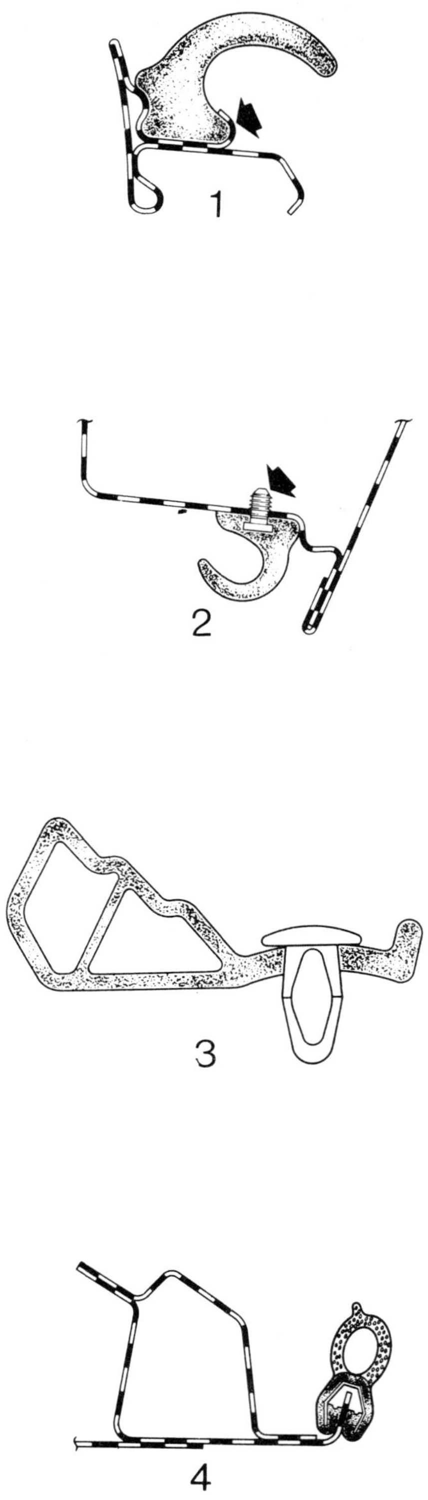

Fig 12:24 WEATHERSTRIPS – CROSS SECTIONS
1 Door upper part - retained in channel (arrowed)
2 Door lower section - held by push fit fasteners (arrowed)
3 Bonnet landing weatherstrip held by push fit fastener
4 Boot aperture and estate car rear door strip

tion cover panel by press in studs. When fitting a new strip, the raised section must face the front of the car.

3 The boot aperture weatherstrip is simply pressed over the flange, although it can help to use an adhesive to hold it in position. It is important to see that the ends of the strip are butted neatly together and sealed with adhesive. If the smooth outer skin of the strip is broken, the whole article will absorb water and encourage rust.

4 Door weatherstrips are held to the frame at the top in a channel. The lower edge is secured by press in plastic fasteners. On the estate car the rear door strip is similar to that used on a saloon car boot.

18 Grilles and exterior trim

1 The radiator grille is secured by ten screws, six along the top, two into the headlamp mounting on the lower edge and two inside from behind. On SL models, the grille sections can be separated from the frame (Fig 12:26).

2 Door sill mouldings, if fitted, are attached by seven pop rivets which must be drilled out to remove them (Fig 12:27).

3 Waist mouldings are held to the door by tabs at the front and rear which are clinched over the door panel, and along their length they are integral with the window sealing strip (Fig 12:28).

4 Door aperture tread plates are held by studs engaging in keyhole slots and a single plastic push fit stud at the rear end. To remove the tread plate pull out the stud, slide the plate forward and lift it off (Fig 12:29).

5 Fixed window reveal mouldings are retained by the glazing strip and, to remove them, the window assembly must be taken out (refer to Section 7).

19 Heater/demister and ventilating system - general description

Ventilation of all models is provided by a pair of intakes in front of the windscreen which duct air to face level outlets at each end of **the instrument panel. Air is exhausted from the car through slots under the rear screen or above the rear door on estate cars.**

In addition there is a ventilation assembly centrally mounted in the scuttle which incorporates a booster fan and heater battery where specified. The controls for this are independent of the air inlets at the ends of the instrument panel.

The ventilator assembly is a moulding with a series of ducts in it and two flaps.

Each flap is controlled by one of the two levers. The left hand lever controls the outflow of air to either the car or the screen, and the right hand lever controls the inflow of air past the heater, fully, partially, or not at all. It should be appreciated that where a heater is fitted hot water passes through it at all times. There is no hot water control valve. If the flaps are out of adjustment in hot weather, therefore, you could wonder where the heat is coming from! Note also that both levers should be in the down position to cut off air entry into the car. If the right lever is left in the upper cold position with the left lever at 'off' some **warm** air can get into the car.

On later models an illumination heater escutcheon has been used. The bulb is accessible after removing the control knobs and securing screw, and withdrawing the escutcheon.

20 Heater controls - adjustment and cable renewal

1 The controls are mounted in the centre of the instrument panel. Pull off the control lever knobs, remove the screw at the bottom of the escutcheon and lift out the escutcheon.

Remove the filler panel next to the steering column and then undo the two screws holding the lever assembly to the instrument panel. The control can be withdrawn from below (Fig 12:30).

2 Cables should be clipped to the control so that no more than 1/8 inch protrudes above the clip. Otherwise the movement of the lever will be restricted (Fig 12:31).

3 Adjustment of the cables is at the ventilation end. Both cable

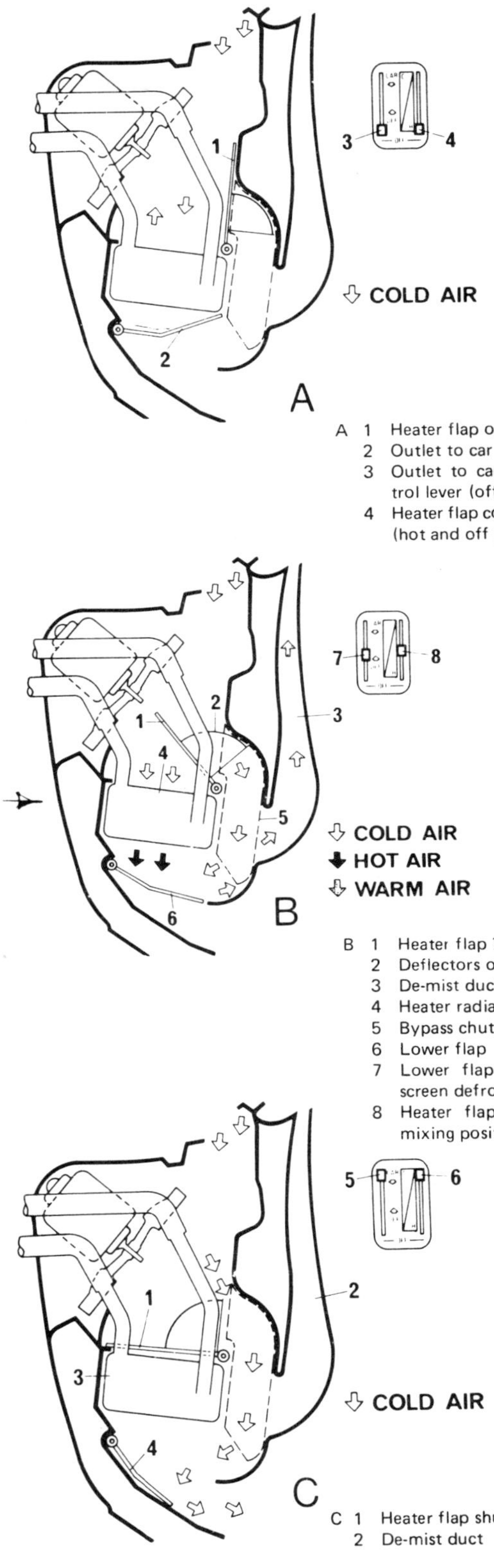

A 1 Heater flap open
2 Outlet to car flap closed
3 Outlet to car flap control lever (off)
4 Heater flap control lever (hot and off position)

B 1 Heater flap ½ open
2 Deflectors on upperflap
3 De-mist duct
4 Heater radiator
5 Bypass chute
6 Lower flap
7 Lower flap control at screen defrost position
8 Heater flap control at mixing position

C 1 Heater flap shut
2 De-mist duct
3 Heater radiator
4 Lower flap
5 Lower flap control at car in hot position
6 Heater flap control at cold

Fig 12:25 VENTILATOR UNIT – CROSS SECTION OF TYPICAL SETTINGS

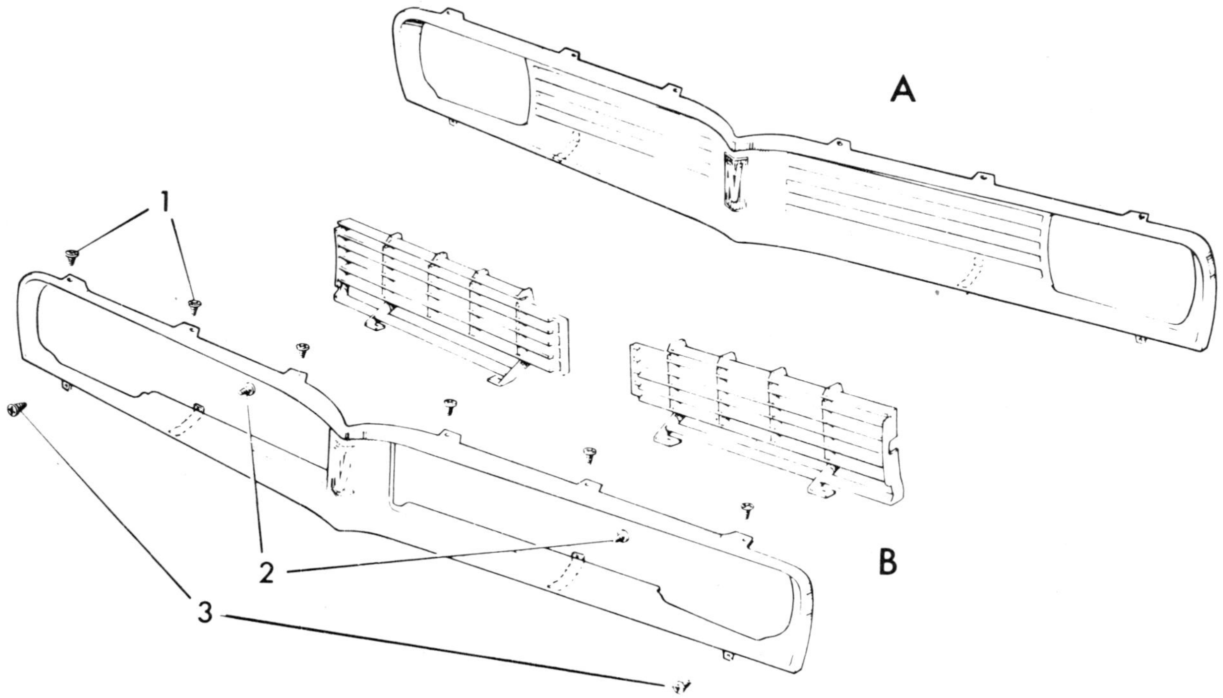

Fig 12:26 RADIATOR GRILLES
A Standard and de Luxe
B SL
1 Top fixing screws
2 Rear fixing screws
3 Lower fixing screws

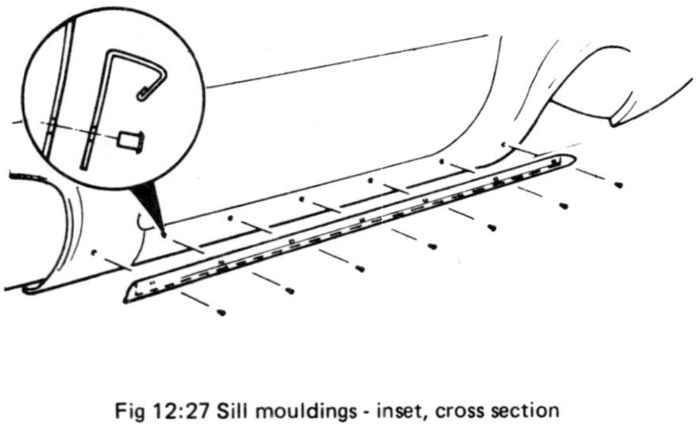

Fig 12:27 Sill mouldings - inset, cross section

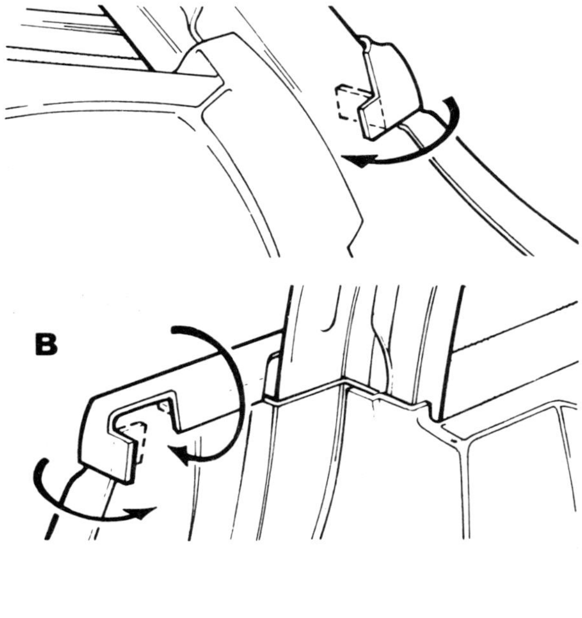

Fig 12:28 WAIST MOULDING (SL MODELS)
A Front end - one clinch tab
B Rear end - two clinch tabs

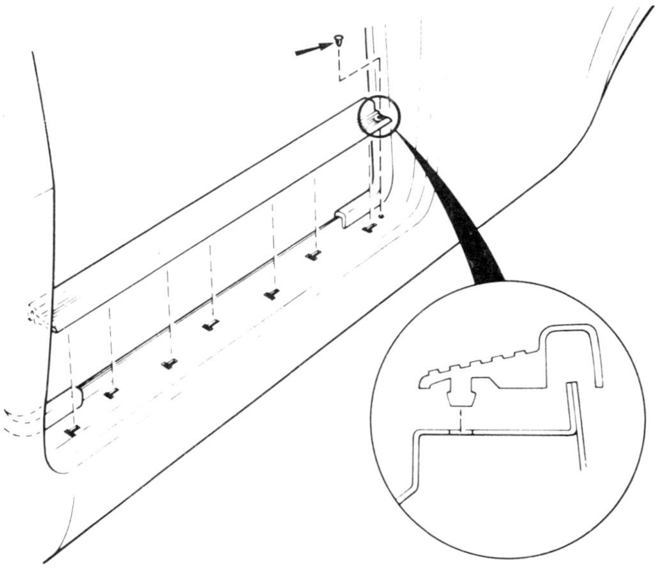

Fig 12:29 Door aperture tread plate with (inset) cross sectional view

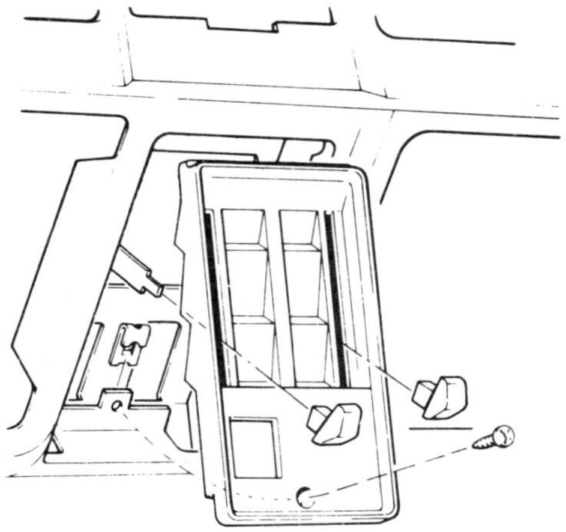

Fig 12:30 Heater control escutcheon - removal of lever knobs and escutcheon retaining screw

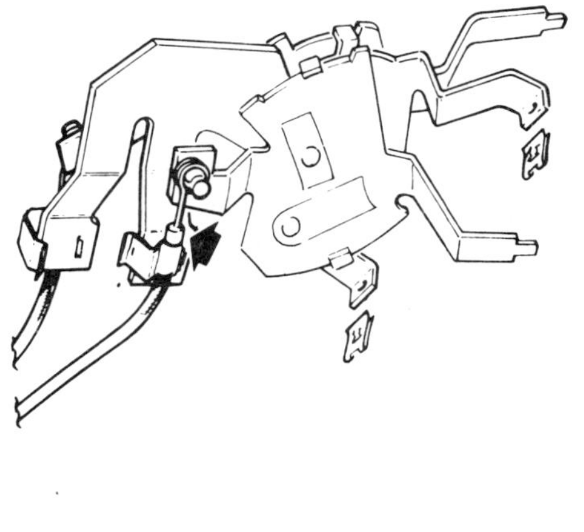

Fig 12:31 Heater control levers. Note maximum cable protrusion of $^1/_8$ inch (arrowed)

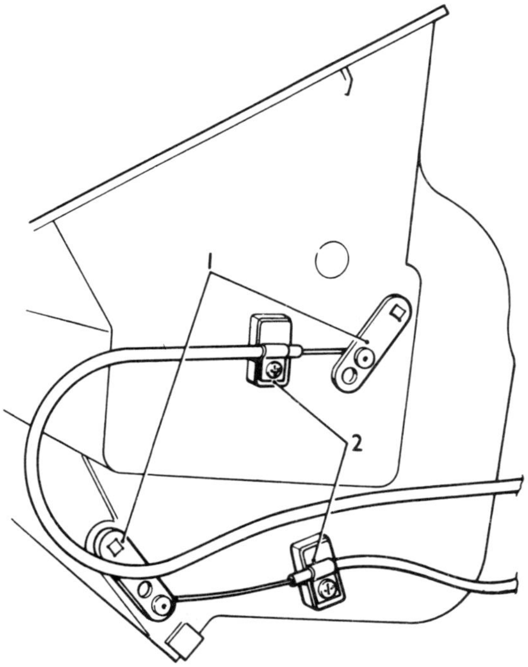

Fig 12:32 VENTILATOR FLAP CONTROL CABLE FIXINGS
1 Flap spindles 2 Cable outer clips

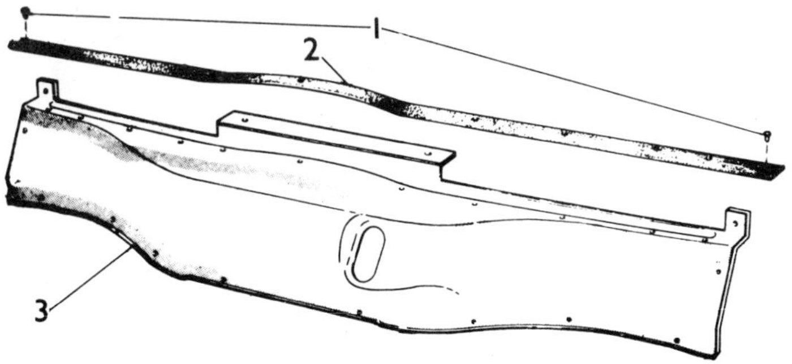

Fig 12:33 VENTILATOR COVER PANEL
1 Push fit fasteners at ends of bonnet landing weatherstrip 2 Weatherstrip 3 Cover panel

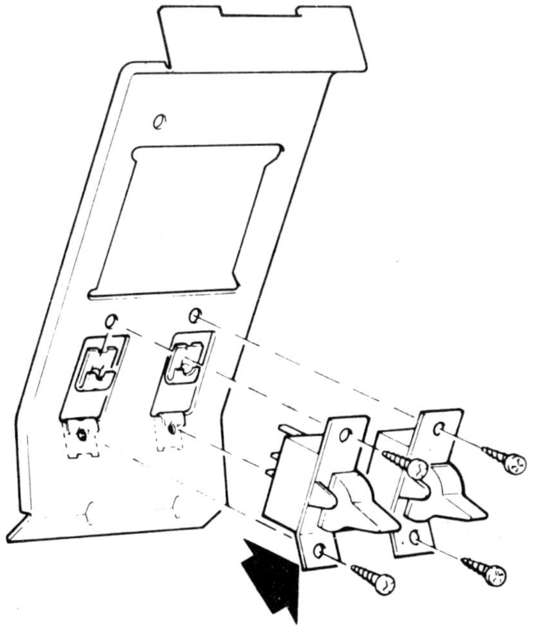

Fig 12:34 Ventilator fan switch (arrowed) - mounting details. Other switch is for optional heated rear window circuit

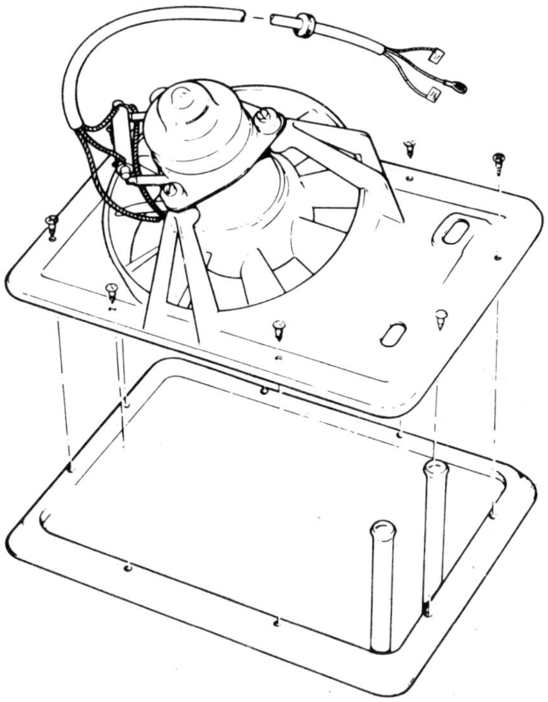

Fig 12:35 Ventilator fan assembly - detail of mounting

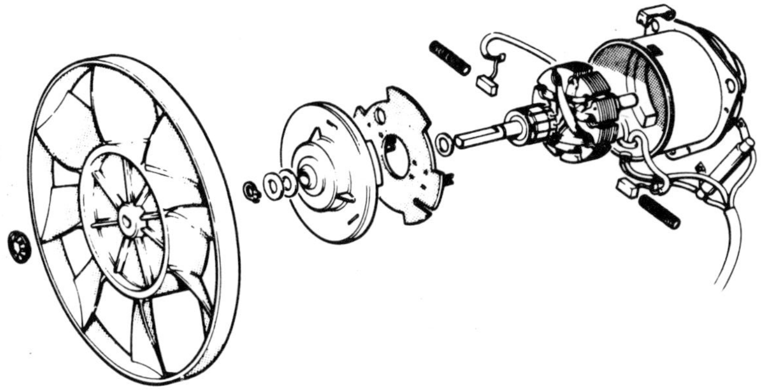

Fig 12:36 Ventilator fan motor - exploded illustration

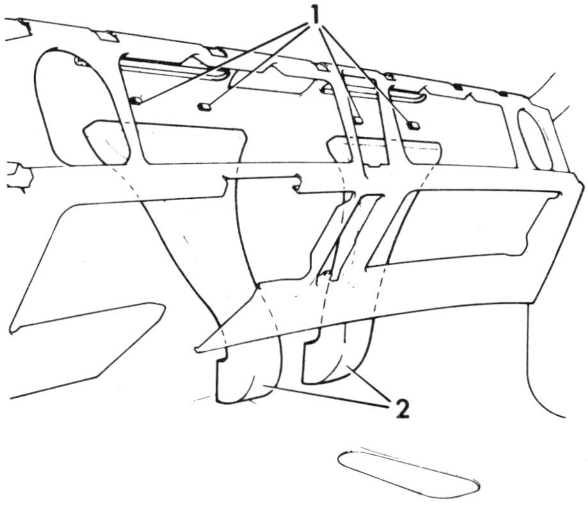

Fig 12:37 VENTILATOR DEMIST DUCTS
1 Outlet aperture clips
2 Lower ends — push fit into main ventilation casing

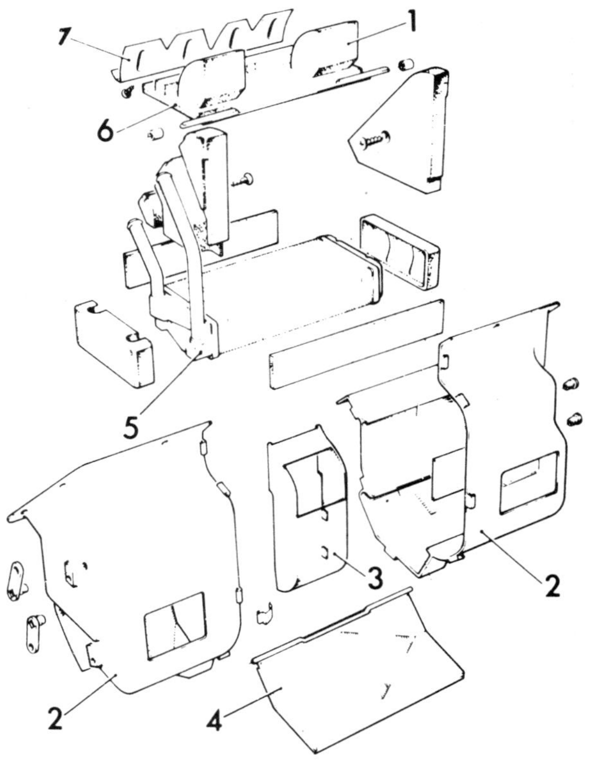

Fig 12:38 VENTILATOR/HEATER CASING ASSEMBLY — EXPLODED ILLUSTRATION
1 Deflectors
2 Casing halves
3 Bypass chute
4 Lower flap
5 Heater radiator
6 Upper flap
7 Graduation plate

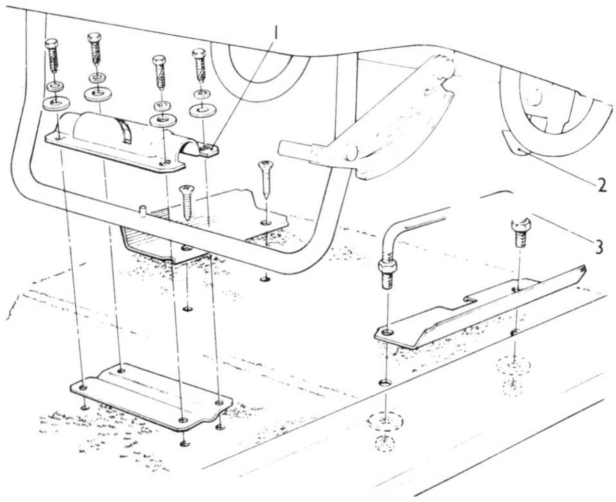

Fig 12:39 FRONT SEAT FIXING DETAILS
1 Stirrup retaining plate
2 Anti-tip safety catch (two door models)
3 Retaining loop

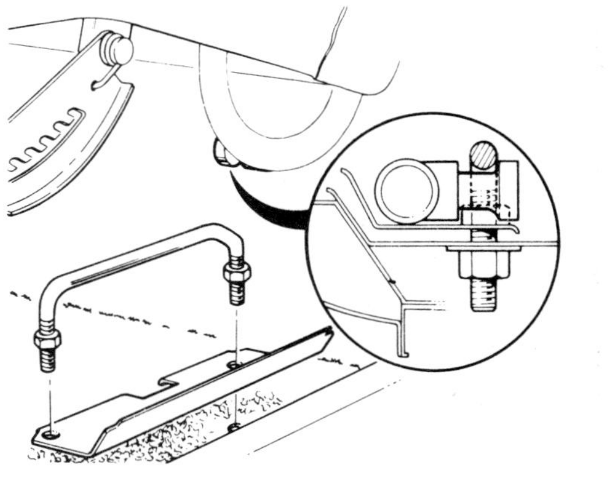

Fig 12:40 Front seat - detail of non-tip retaining lug on four door models

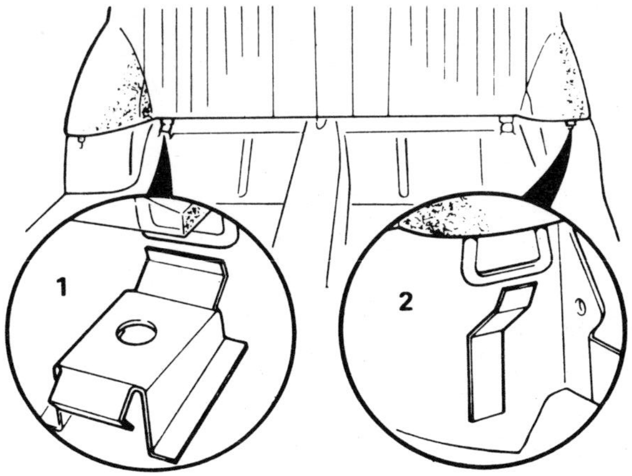

Fig 12:41 REAR SEAT SQUAB RETAINING LOOP DETAILS (INSET)

1 Floor bracket (for cushion also)
2 Wheel arch clip

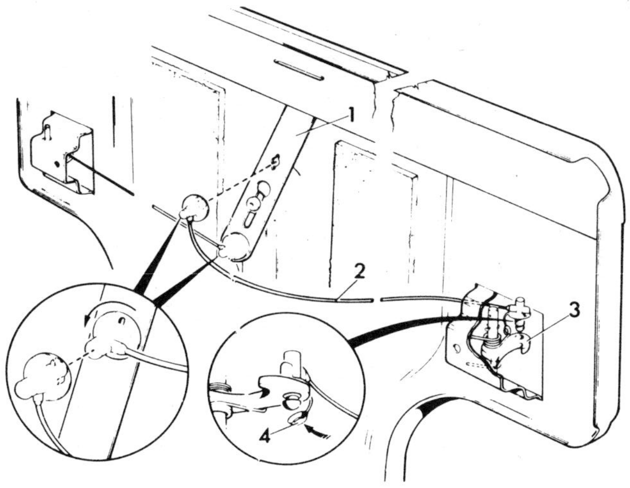

Fig 12:42 REAR SEAT SQUAB – ESTATE CAR. DETAIL OF UPRIGHT POSITION RETAINING LATCHES

1 Release lever
2 Cables
3 Release latches
4 Cable retaining washer

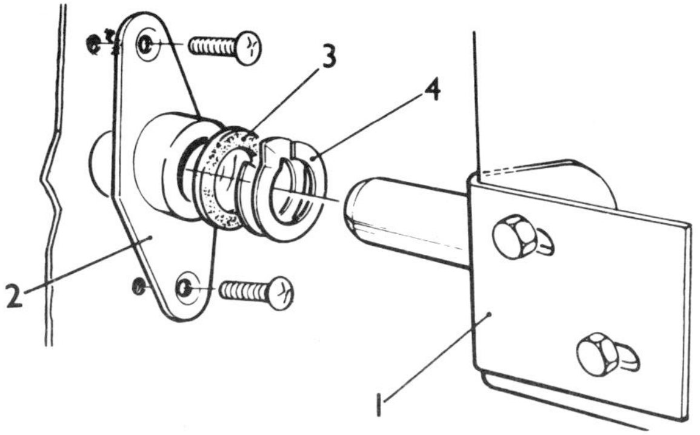

Fig 12:43 REAR SEAT CUSHION – ESTATE CAR. DETAIL OF PIVOT PIN AND BUSH

1 Pivot pin bracket
2 Bush
3 Friction washer
4 Anti-rattle spring

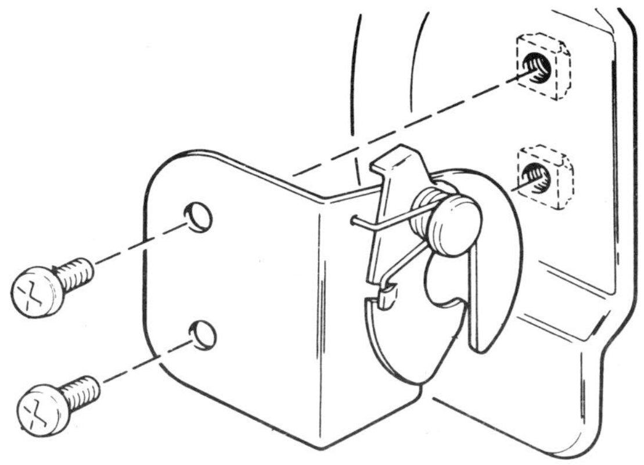

Fig 12:44 REAR SEAT – ESTATE CAR
Squab pivot catch

clips should be slackened and, with both control levers in the 'off' (down) position the upper flap should be fully forward (anticlockwise) and the lower flap fully back (clockwise) viewed from the flap lever side of the ventilation casing.

To detach the cables from the flaps, remove the clips and prise the levers off the ends of the flap spindles (Fig 12:32).

21 Fan and ventilator/heater unit - removal and replacement

1 To remove the fan and motor it is first necessary to detach the ventilator cover panel under the bonnet. Prise out the two end **fasteners holding the bonnet sealing weatherstrip, remove the fourteen screws securing the panel and disconnect the heater hoses. The panel can then be lifted off (Fig. 12:33).**

2 Remove the heater control escutcheon (Section 20) and disconnect the fan motor switch by removing the two screws and pulling the connections off the back (Fig 12:34).

3 Undo the fan motor earth wire bolt which is screwed into the dash panel behind the ashtray.

4 Remove the eight screws holding the fan motor and cover to the top of the ventilator housing and lift it out (Fig 12:35).

5 If it is wished to renew the fan motor brushes, remove the motor from the cover, detach the fan from the spindle, remove the circlip also on the spindle and undo the three spring clips holding the motor casing to the end cover (Fig 12:36).

6 When replacing the fan to the motor, fit the thicker sides of the blades towards the motor and the toothed retaining lockwasher with the concave side against the fan hub.

7 When refitting the fan assembly to the top of the ventilator, the screw holes and heater pipe apertures should be sealed to prevent water getting inside the car.

8 If it is wished to remove the ventilator and heater assembly, first remove the instrument panel cover and instrument assembly (see Chapter 10), glove box and parcel shelf.

9 The demist ducts are a push fit inot the ventilator box at the bottom and are held by clips up to the outlet slots at the top. Take these off. Disconnect the control cables at the ventilator bore. Provided the fan wires are disconnected, and ventilator shrouds are already removed, the whole unit can be lifted out (Figs 12:37 and 12:38).

10 On replacement, make sure that the demist ducts are properly secured into their clips at the outlet slots.

11 The fan wires should be correctly connected at the switch - blue/black to No 1 (top), green to No 2 (centre) and green/yellow to No 3 (lowest) terminals.

22 Seats

1 On two door models, front seats tip forward and their adjustment consists of pivoting forward and backwards on the front support frame member which is hinged at the floor. A safety catch latches the rear of the seat to the floor to prevent forward tipping in emergency (Fig 12:30).

2 On four door models, the front seat is the same except that the tipping safety catch is replaced by a peg which hooks under the retainer hoop. The retainer can be adjusted vertically to eliminate any rattle which might develop (Fig 12:40).

3 The saloon rear seat cushion is held in position by two studs and nuts under the front edge of the seat. The rear of the cushion has two wire loops which hook into brackets on the floor.

4 The saloon squab can be removed only after the cushion has been taken out. It is held by two wire loops to the same brackets holding the cushion and by two smaller loops which hook over malleable clips on the wheel arches. These clips are simply bent to release the loops (Fig 12:41).

5 The estate car rear seats fold to form a load carrying area. The cushion hinges forward into a vertical position and the squab folds flat forward to provide a platform (Fig 12:42).

The cushion pivots on a pin at each side fixed to a bracket and this engages in a bush in the body side panel. In order to remove the cushion the pivot pin brackets must be unbolted from the seat (Fig 12:43). The squab pivots on pins fixed to the side panels. Two hooks with spring catches fixed to the sides of the squab engage the pins. This is a quick release method as the spring catches can release the squab from the pivot pins so that it may be lifted out completely (Fig 12:44).

In addition there are two latches fitted to keep the squab upright. Should it be necessary to renew a latch release cable, remove the securing washer from the latch end and turn the cable through 90° at the release lever end to disengage it from the lever.

Chapter 13 Supplement

Contents

Introduction

Viva ohv 1256cc models covered by this manual underwent a number of modifications in September 1973, commencing with engine No. 1400011. These changes are described in this supplementary Chapter, which owners of later models should read in conjunction with the original relative Chapters of this manual.

Specifications

The specifications listed here are revised or additional to the main details given at the beginning of each Chapter.

Engine

Compression ratio

1256cc high compression	9.2 : 1

Camshaft and bearings

Cam dimension (minimum) peak to base	1.302 in

Connecting rods and big-end bearings

Crankpin diameter - standard	1.8302 to 1.8310 in

Cylinder head

Minimum permissible depth	3.193 in	

Pistons and rings

Ring gap in cylinder bore	0.009 to 0.020 in	

Lubrication system

Oil pressure (hot)	44 to 56 lb/in^2 @ 3000rpm	
Oil pressure relief valve:		
Spring free lengths	1.96 in	
Spring load at 1.66 in	7 lb 15oz to 8 lb 3oz	

Torque wrench settings

	lb f ft	Kg f m
Engine		
Main bearing cap bolts	82	11.3
Cooling system		
Viscous coupling centre bolt (left-hand thread)	14	1.9

Fuel system

Zenith/Stromberg carburettors 150CD-SEV

Number and metering needle:		
Early cars	3616B	B1DE
Later cars	3709B	B1DJ
Jet orifice	2.54 mm	
Air valve spring	Natural	
Fast idle cam	M2	
Needle valve	1.5 mm	
Needle valve washer thickness	1.6 mm	
Engine idling speed	800 to 850rpm	
C O Level	2.5 to 3.0%	

Ignition system

Modified distributors have been introduced at different times since 1973. For general information, refer to Chapter 4 Specifications. For details of advance curves, consult your Vauxhall dealer quoting the car engine number.

Wheels and tyres

Wheels

GLS	5 inch Rostyle
Other models	4½ inch steel

Tyres

L, SL and GLS	155SR x 13
Other models	5.64 x 13 4 ply rating
Pressures	24 lb/in^2 (front) 28 lb/in^2 (rear)

Engine

1 Pistons and Piston rings

The pistons used are similar to those installed in the 1159cc engine except, of course, for their larger diameter. The pistons used in low compression engines are shorter than the ones fitted to standard power units and are identified by an 'L' stamped on the crown. The piston rings comprise two compression and one oil control. The compression rings are marked 'TOP' on their top faces and the second compression ring is stepped on its lower face.

The oil control ring is of the three piece type. When assembling the oil control ring components to the lower ring groove, make sure that the ends of the centre expander do not overlap. Stagger the ends of the two rails and the expander at equidistant points round the piston.

2 Positive crankcase ventilation

This system is used on all 1256cc engines, regardless of the type of carburettor fitted. No maintenance is required other than occasionally checking the condition and security of the system connecting hoses.

3 Engine/transmission rear mounting

A modified rear mounting is fitted to all later 1256cc cars. When reassembling the large hexagon washer to the centre of the mounting crossmember, make sure that two of the sides of the hexagon are parallel with the flange sides of the crossmember to ensure clearance during flexing of the mounting.

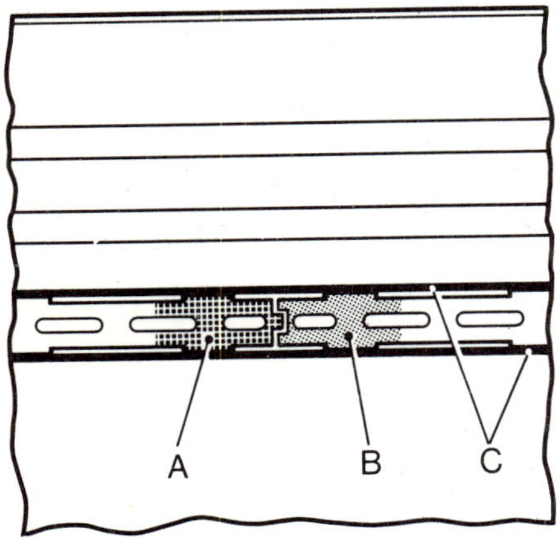

Fig. 13.1. Oil control ring (1256 cc engine)
A and B expander, C rails

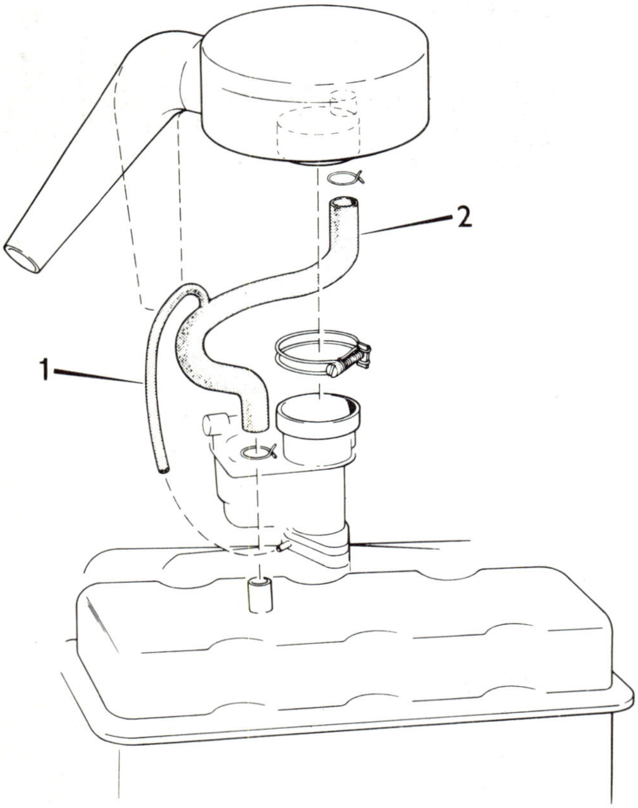

Fig. 13.2. POSITIVE CRANKCASE VENTILATION SYSTEM

1 Branch hose to inlet manifold
2 Main hose to air cleaner from rocker cover

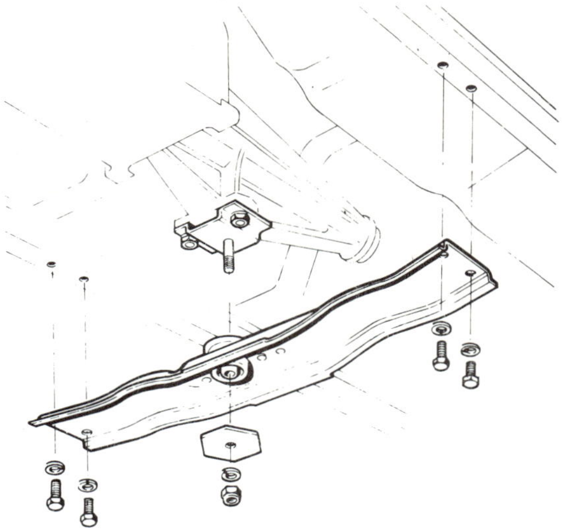

Fig. 13.3. Components of the engine/transmission rear mounting

Cooling system

4 Viscous coupling fan

On later model cars, a viscous coupling type fan is used. This is a sealed unit which is filled with oil and permits direct drive up to engine speeds of 1000rpm. At higher engine speeds, 'slip' occurs which progressively reduces the fan rotational speed and noise. Due to the higher forward speeds of the car, the cooling effect is unimpaired because of the greater ram effect of the air passing through the radiator. Fuel economy will be improved as the fan will not require so much engine power to drive it and the engine will not suffer from over cooling at high road speeds.

When removing the viscous coupling, first mark the relative positions of the fan to it so that the two components can be refitted in the same way and balance maintained.

The single centre bolt which secures the fan and pulley has a left-hand thread. When refitting it, tighten to the specified torque.

5 Water pump seal

Later model water pumps incorporate a ceramic counterface which is located between the rotor and the seal. The counterface should be installed so that its ceramic face is towards the pump seal.

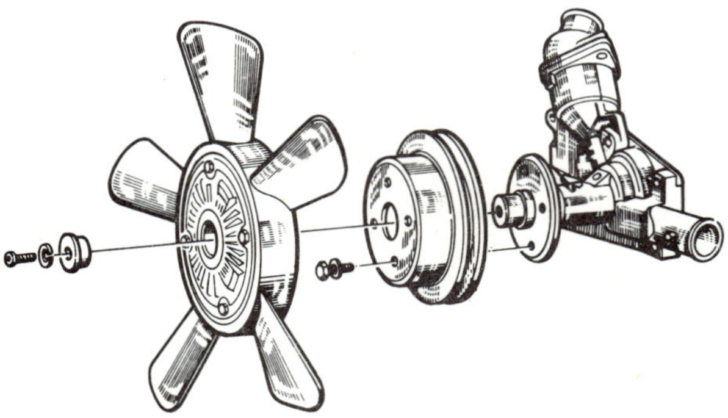

Fig 13:4 Viscous cooling fan, pulley and water pump

Fuel system

6 Air cleaner mounting

The paper element type air cleaner, used in conjunction with Stromberg carburettors, is mounted on an elbow pipe attached to the carburettor intake.

7 Fuel pump

The fuel pump is of the semi-sealed type, the only dismantling which can be carried out is removal of the pump cover for access to the filter. When reassembling, make sure that the four pegs on the gauze are towards the top cover.

8 Carburettor (Stromberg 150CD-SEV) - description

This unit is of normal constant vacuum, single variable jet type incorporating an air valve, a metering needle and a disc type cold starting device. To ensure consistent fuel flow, the metering needle is spring loaded against one side of the jet.

A temperature compensator is provided to offset minor variation in mixture strength caused by heat transfer from the engine to the carburettor body.

An idle trim screw is provided which regulates an air bleed into the mixing chamber. After initial running-in, should the idling quality become poor, turn the screw clockwise until smooth running is restored.

Note: From October 1976 all carburettors have a modified float chamber housing which necessitates the use of a special tool VR 2242 for jet adjustment. With this type of carburettor, adjustment is as given in the following Section, except that a CO meter *must* be used for mixture adjustment.

9 Carburettor (Stromberg 150CD-SEV) - maintenance and adjustment

1 Maintain the oil level in the hydraulic damper at 0.30 in (8 mm) below the top of the air valve guide. When refitting the damper, raise the air valve by inserting a finger into the carburettor air intake (air cleaner removed) and make sure that the cap is fully seated before screwing down the damper cap.

2 *Setting of the cold start setting pin* is as described in Chapter 3, Section 10.

3 *Idling adjustment* should be restricted normally to varying the speed, if necessary, by turning the throttle stop screw. Where the mixture strength is incorrect, then the jet adjuster must be rotated. To do this a special tool (VR 2156) must be obtained or made up to engage in the holes in the jet-adjuster.

4 The correct idling mixture may be ascertained in one of three ways:

i) Using a C.O meter

ii) Using a Colortune or similar device

iii) Lifting the air valve by inserting a small screwdriver into the air intake and raising the valve 0.04 in (1 mm). If the engine speed rises momentarily and then resumes its normal idling speed, the mixture is correct. If the speed continues to rise, the mixture is rich and if the engine stalls then it is too weak.

5 Before adjusting the jet, have the engine at normal operating temperature and idling at the specified revs. When viewed from below, turning the jet adjuster anti-clockwise enrichens the mixture and turning it clockwise weakens it. If the jet has been dismantled and reassembled, set the jet adjuster initially two turns from the fully screwed up position.

6 *To set the float chamber ventilation valve,* adjust the stop screw(1) (Fig. 13.14) until a 2 mm dia. drill can be inserted between the valve lever and the stop screw post. Adjustment of

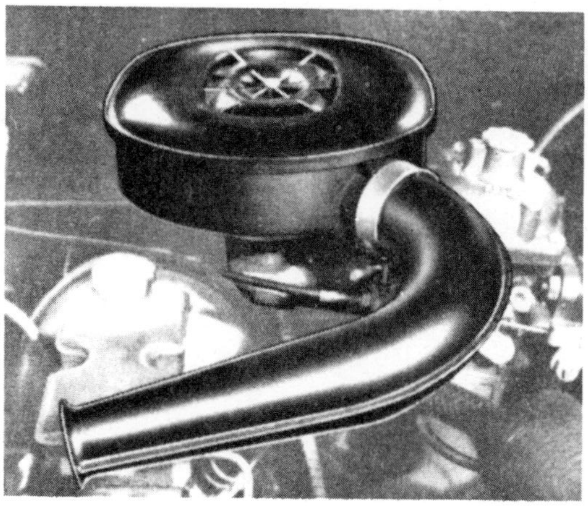

Fig. 13.5. Air cleaner fitted to Stromberg carburettor

Fig. 13.6. Fuel pump location

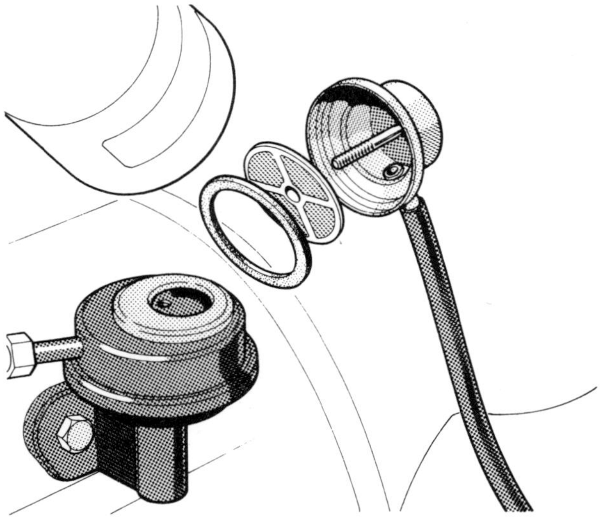

Fig. 13.7. Access to filter on semi-sealed type fuel pump

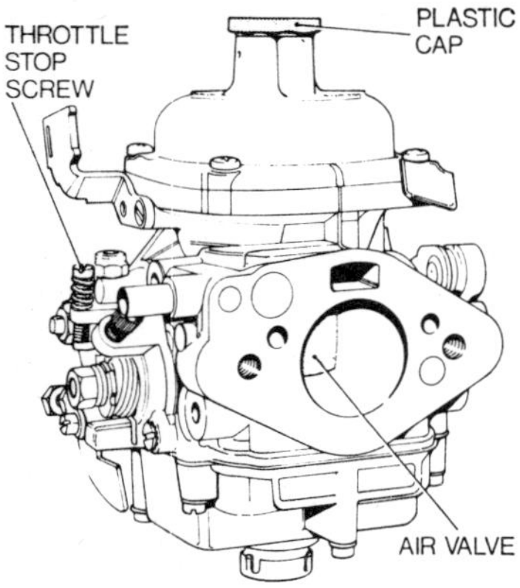

Fig. 13.8. Stromberg 150CD-SEV carburettor

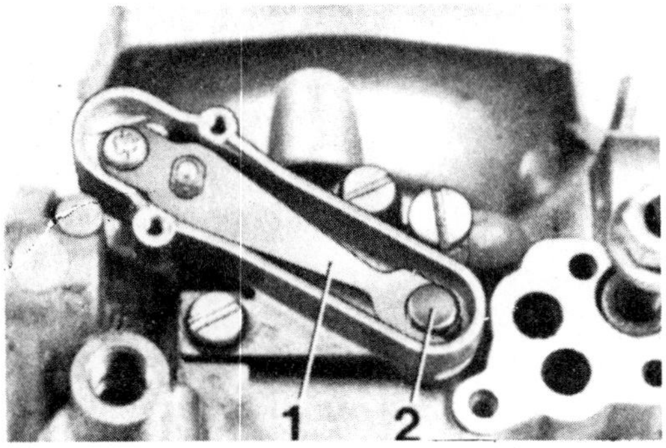

Fig. 13.9. TEMPERATURE COMPENSATOR ON STROMBERG 150CD-SEV CARBURETTOR

1 Bi-metal strip 2 Plug

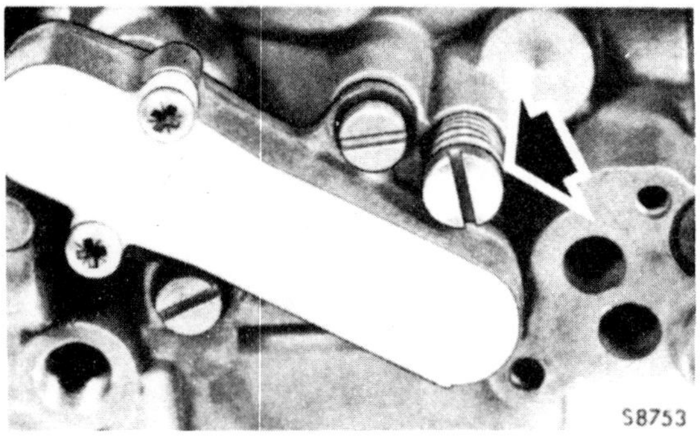

Fig. 13.10. Idle trim screw on Stromberg 150CD-SEV carburettor

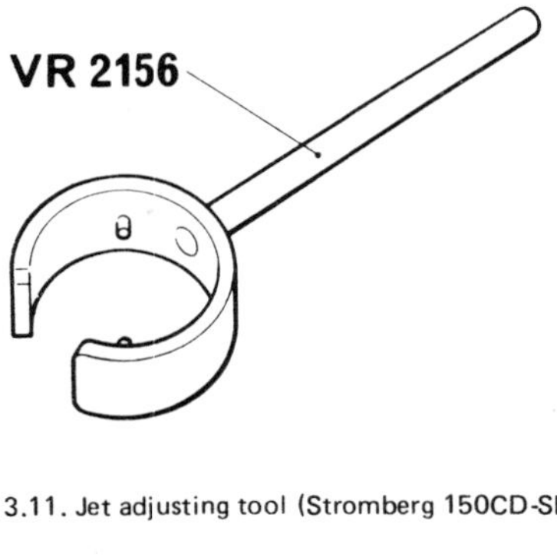

Fig. 13.11. Jet adjusting tool (Stromberg 150CD-SEV)

Fig. 13.12. Stromberg 150CD-SEV damper and cap

Fig. 13.13. Jet adjusting tool engagment holes in Stromberg 150CD-SEV carburettor

Fig. 13.14. SETTING THE FLOAT CHAMBER VENTILATION VALVE

1 Stop screw 2 Throttle stop screw

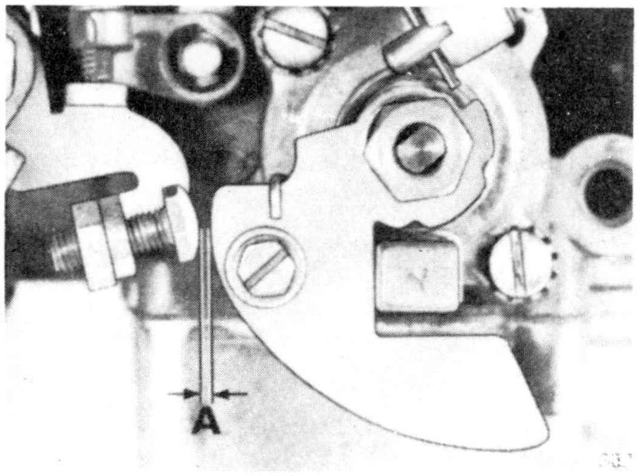

Fig. 13.15. Fast idle screw and cam (Stromberg 150CD-SEV)
A = 0.10 in (2.50 mm)

the throttle stop screw may be required if the engine idling speed has altered.

7 *To set the fast idle screw,* have the choke pushed fully in and, using feeler gauges, adjust the screw until a gap exists between the screw and the cam of 0.10 in (2.50 mm).

10 Carburettor (Stromberg 150CD-SEV) - removal, overhaul and refitting

1 Remove the air cleaner.

2 Disconnect the choke and accelerator controls from the carburettor.

3 Disconnect the fuel inlet hose from the carburettor and plug it.

4 Disconnect the distributor vacuum pipe.

5 Unscrew and remove the carburettor flange securing nuts and lift the carburettor from the inlet manifold.

6 Clean the external surfaces of the carburettor and then remove the hydraulic damper.

7 Extract the four screws and lift off the top cover.

8 Lift out the air valve, metering needle and diaphragm assembly.

9 Extract the float chamber securing screws and pull off the float chamber. If it is stuck tight on its 'O' ring seal, rotate the jet adjuster in both directions to break the seal.

10 The cold start device can be removed from the carburettor body (two screws) and the jet assembly unscrewed.

11 Remove the float by prising its pivot from the spring supports.

12 Unscrew the fuel inlet needle valve, retaining the washer for refitting. Note that the filter gauze on the valve is non-detachable.

13 It is not recommended that the throttle valve plate or spindle are removed unless absolutely essential. If wear has taken place in these components then it will probably be a wise course to purchase a new carburettor.

14 To remove the needle from the air valve piston, extract the small securing grub screw.

15 The flexible diaphragm can be removed after withdrawing its retaining ring (four screws).

16 Examine all components for wear and renew as necessary. To check that the needle is to the correct specification, pull the needle partially from its casing to expose the identification number.

17 Reassembly is a reversal of dismantling, but observe the following:

- a) *When installing the jet-assembly, ensure that the nylon washer is located on the jet stem between the flange and the plain washer.*
- b) *Having refitted the float, invert the carburettor and check that the highest points of the floats are between 16 and 17 mm above the body flange (gasket removed).*
- c) *Before fitting the float chamber, screw the jet adjuster fully into the carburettor body to prevent the float chamber fouling the shoulder on the jet adjuster.*
- d) *Install the metering needle so that the nylon washer is flush with the face of the air valve. Make sure that the needle is so positioned that the securing grub screw tightens against the flat which is ground onto the side of the needle casing.*
- e) *Install the diaphragm to the valve so that the small tongue engages in the valve slot.*
- f) *Install the valve/diaphragm assembly so that the tongue, on the outer edge of the diaphragm, engages in the slot in the carburettor body.*
- g) *Make sure that, when refitting the temperature compensator, the rubber seals are installed in the carburettor body and over the post of the compensator.*

18 Carry out the adjustment described in Section 9 of this Chapter.

19 Refitting the carburettor is a reversal of removal, but use a new flange gasket and check that the choke control cable is correctly set to ensure complete return of the cold start device when the choke control is pushed in.

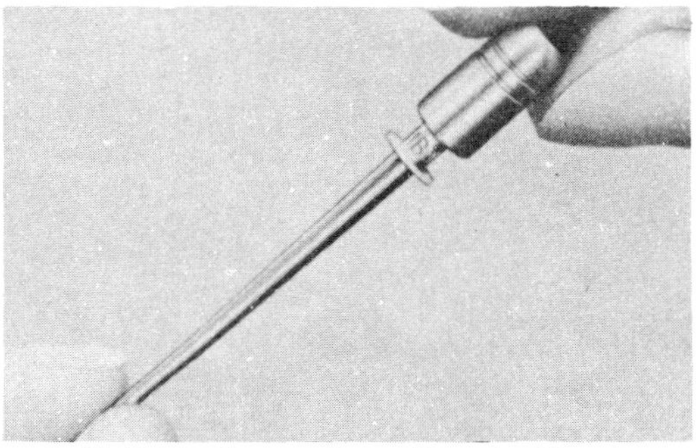

Fig. 13.16. Checking metering needle number (Stromberg 150CD-SEV)

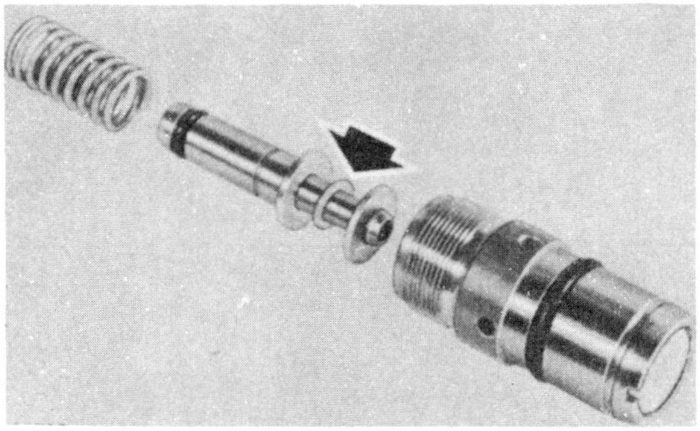

Fig. 13.17. Jet assembly (Stromberg 150CD-SEV)
Nylon washer arrowed

Fig. 13.18. Float setting diagram (Stromberg 150CD-SEV)
A = 16.0 to 17.0 mm

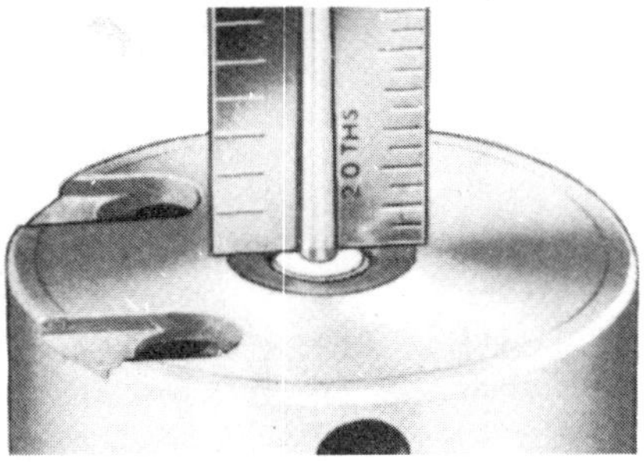

Fig. 13.19. Installing metering needle (Stromberg 150CD-SEV)

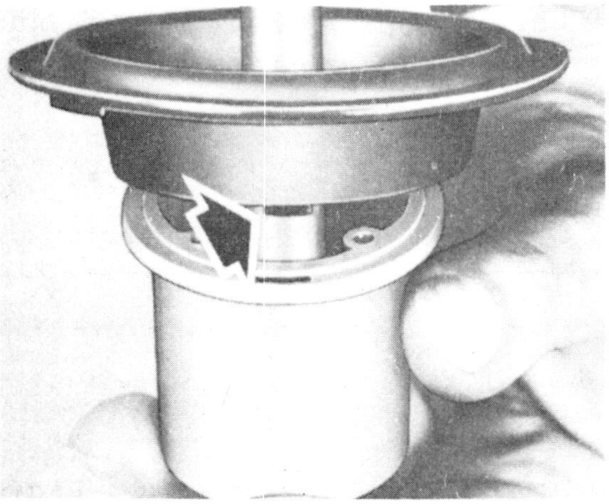

Fig. 13.20. Diaphragm to valve alignment (Stromberg 150CD- SEV)

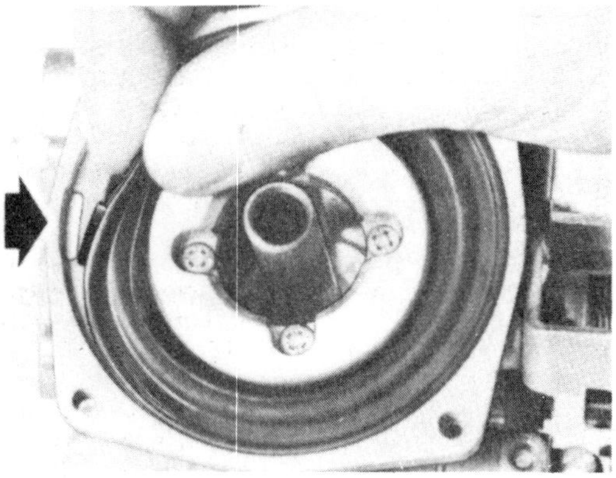

Fig. 13.21. Valve/diaphragm to body alignment (Stromberg 150CD-SEV)

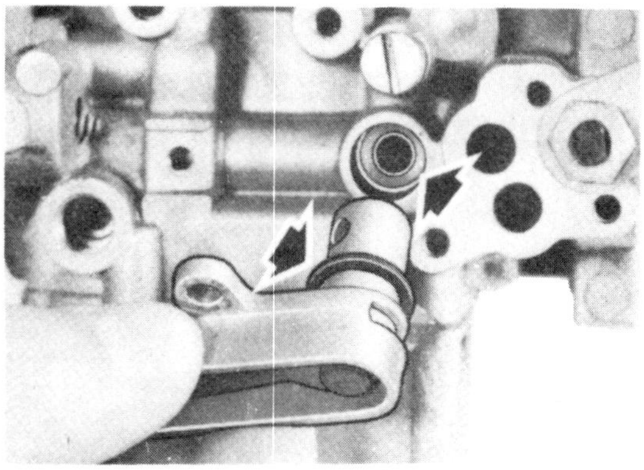

Fig. 13.22. Temperature compensator to body seals (Stromberg 150CD-SEV)

Ignition system

11 Ignition timing

For 1974 models, the static ignition setting for both high and low compression engines is 9°BTDC.

Clutch

12 Clutch modifications

1 The sizes of the pressure plate assembly and driven plate were increased to 7.25 in (184 mm) diameter commencing with engine No. 1153840. The necessary modifications to the flywheel and clutch bellhousing were carried out at the same time.
2 From chassis number FX104461, the clutch cable is correctly adjusted when the pedal is level with the brake pedal. There should be no free play.

Gearbox

13 Synchromesh units

Some splines on the synchro sleeves fitted to later gearboxes have the keystone contour on one side only.

Before dismantling this type of unit, mark the components to ensure exact refitting. The sliding keys must engage with the sleeve splines which have the keystone contours on both sides.

It will be observed that with this type of synchro unit, all the sliding keys are recessed allowing the turned out end of the retaining springs to be located in any key. Make sure that the springs are assembled in an anti-clockwise fashion when viewed from either side of the hub. The turned out end of each spring must engage the same key.

Fig. 13.23. Later type synchro unit

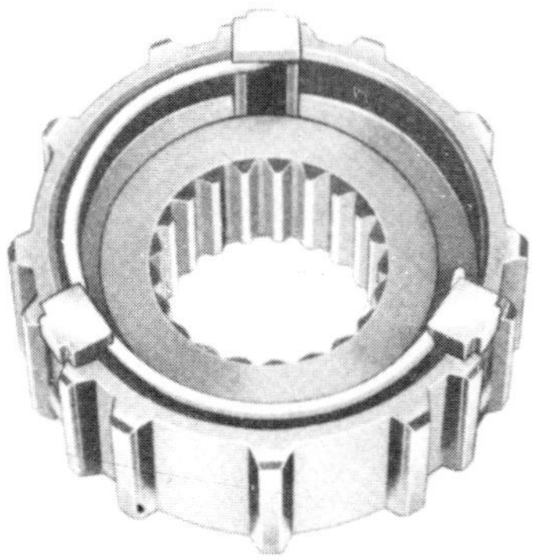

Fig. 13.24. Later type synchro unit showing engagement of spring in anticlockwise direction

Rear axle

14 Differential

Later differential units have a strengthened casing and pinion shaft. The clearance between this type of shaft and the differential pinions should be between 0.0019 and 0.0056 in (0.48 and 0.14 mm). The side bearing spacer thickness has also been increased from 0.179 to 0.243 in (4.5 to 6 mm).

Braking system

15 Self-adjusting rear brakes

1 Brakes of this type (Lockheed) are fitted to later cars. The mechanism incorporates toothed ratchet plates attached to the brake shoes, which are actuated each time the foot pedal or handbrake lever is applied and ensure that the minimum clearance between the shoe friction lining and drum is always maintained.
2 To renew the shoes, jack up the rear of the car, support securely on axle stands and remove the roadwheels.
3 Fully release the handbrake and remove the brake drum after unscrewing its single retaining bolt.
4 Brush away any accumulated dust from the shoes. If the friction material has worn down to the level of the rivet heads, all four rear shoes must be renewed. With bonded linings, the minimum thickness of the friction material, before renewal is necessary, is 1/16 in (1.6 mm).
5 To remove the shoes, turn each of the dished washers on the shoe steady posts through 90° and remove the washers, springs and posts.
6 Pull the shoes from the slots of the lower anchorage and then pull the shoe which is nearer the front of the car from the slot in the wheel cylinder piston. Allow this brake shoe to move nearer the centre of the brake assembly under the action of the shoe return springs and then withdraw both shoes complete with return springs.
7 With the shoes removed, on no account depress the brake pedal and if necessary plaçe a rubber band or a piece of wire round the wheel cylinder to prevent the piston dropping out.
8 The automatic adjuster ratchet plates should be removed from the shoe by extracting the circlips, washers and clevis pins.
9 Reassemble the new shoes on the bench in the same relative position as the original ones in respect of leading and trailing edges, and then attach the automatic adjuster ratchet plates and the shoe return springs, making sure that the ends of the springs engage in the correct holes in the shoe webs. Apply a smear of brake component grease to the moving surfaces of the adjuster mechanism (not the ratchet teeth) and to the shoe contact areas of the brake backplate.
10 Install the shoes complete with return springs by reversing the removal operations and ensuring that the handbrake operating lever is fully engaged in the slot in the trailing shoe.
11 Depress the lower ratchet plate against the pressure of its spring and move the plates to the position of least adjustment (shoes to point of maximum retraction).
12 Fit the shoe steady posts, springs and washers and install the brake drum.
13 Apply the footbrake or handbrake several times to actuate the shoe automatic adjuster and then refit the roadwheel.
14 Repeat all the foregoing operations on the opposing rear brake and then lower the car to the ground.

16 Disc caliper anti-squeal shims

On later model Lockheed type calipers, a re-designed anti-squeal shim is used. With this type of shim, the larger 'D' shaped cut-out must be at the bottom.

17 Pressure regulating valve

On some export models, a pressure regulating valve is incorporated in the rear braking hydraulic circuit. Its purpose is to prevent the rear wheels locking in advance of the front ones during heavy brake applications. Testing for correct operation of the valve should be carried out on a section of dry concrete roadway and the tyre marks examined. The front tyres should leave burn marks on the road, the rear ones should not. Incorrect operation of the valve can only be remedied by renewal of the valve, which is located just in front of the rear axle on the right-hand bodyframe member.

18 Load sensitive pressure reducing valve

On some models, this type of valve is incorporated in the braking system to ensure that the rear wheels do not lock before the front ones under all vehicle loading conditions. The valve mounting bracket is attached to the body floor pan and the valve is connected to the rear hydraulic circuit. The metering plunger in the valve is operated by a spring arm which is anchored to the valve and connected to the rear axle by a link.

Should the valve not appear to be operating correctly, check the clearance between the end of the metering plunger and the end of the adjusting screw. This should be 0.020 in (0.50 mm) with the car containing one person and a full fuel tank and having the valve connecting link *temporarily* assembled in the axle bracket lower hole.

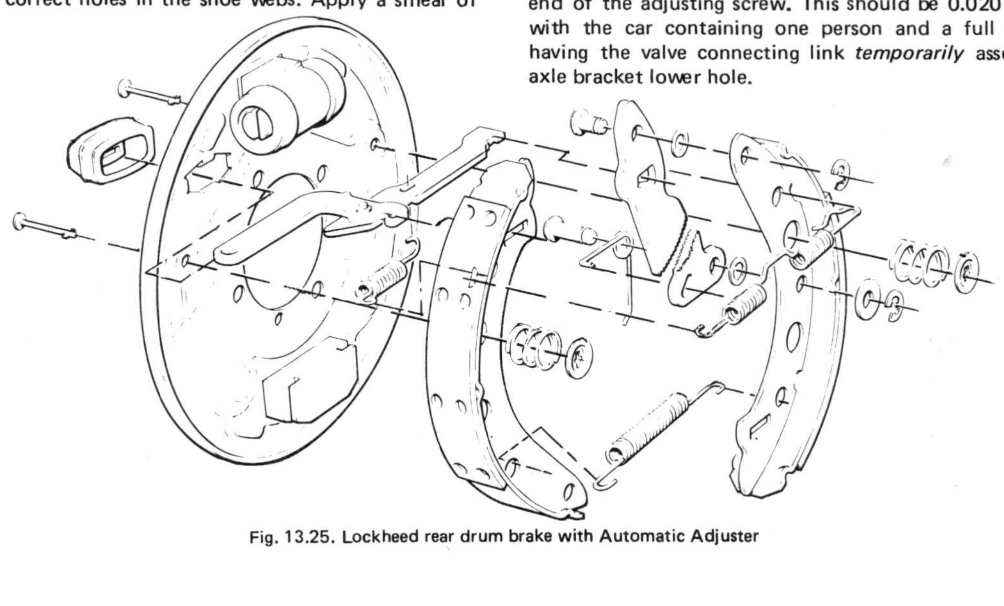

Fig. 13.25. Lockheed rear drum brake with Automatic Adjuster

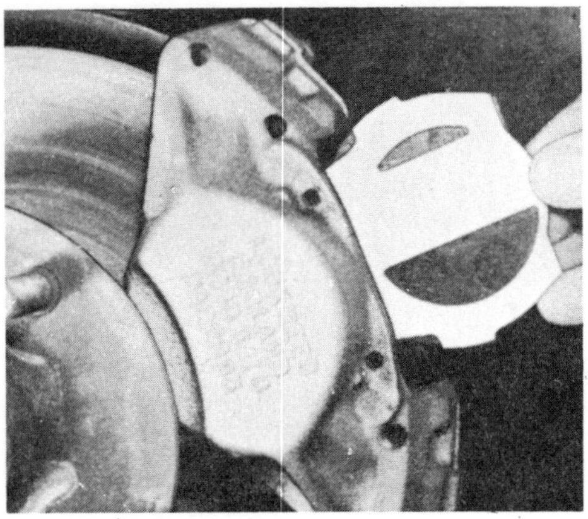

Fig. 13.26. Anti-squeal shim on later type Lockheed caliper

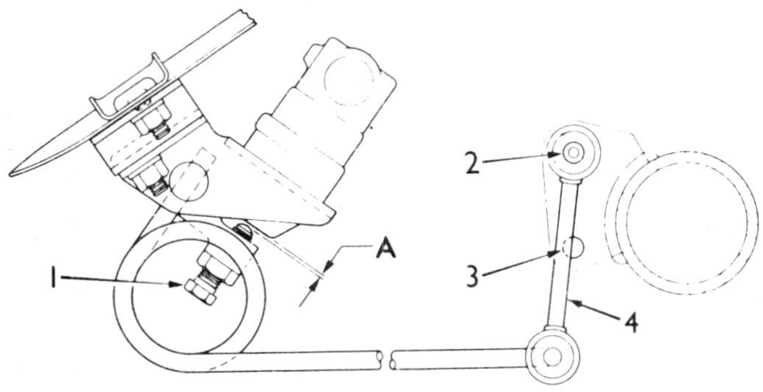

Fig. 13.27. LOAD SENSITIVE PRESSURE REDUCING VALVE

1 Adjusting screw
2 Bracket upper hole
3 Bracket lower hole
4 Link
A=Clearance

Electrical system

19 Pre-engaged starter motors - additional types

Two additional types of pre-engaged starter motors may be encountered on models built for certain overseas territories. The Lucas M35J/PE and the 3M100/PE are both designed with face type moulded commutators.

20 Fuses and fusible links

1 The fusebox on later models includes two spare fuses. Where a heated rear window is fitted, the rating of No. 2 fuse is now specified at 50amp instead of the original rating of 35amp.
2 Two fusible links are used from 1974 onwards, and these are connected into the main battery feed instead of the single link previously used. One of the new links protects the lighting circuits while the other one protects the rest of the electrical system with the exception of the starter circuit.

21 Twin dial type instrumentation

1 To gain access to the warning lamps at the side of this type of panel and to the speedometer bulb, reach behind the panel by inserting the hand above the parcel shelf on the driver's side.
2 Access to the remaining bulbs and switches can only be gained after removing the instrument panel as described in the following paragraphs. The instrument panel warning and illumination lamps have bayonet type holders containing capless wedge-base bulbs.
3 To remove the instrument panel, first detach the panel aperture filler (four screws). Release the speedometer trip control and the drive cable.
4 Unscrew and remove the four securing screws from the instrument panel and withdraw the panel far enough to allow disconnection of the wiring harness plugs.
5 Switches can be removed after squeezing their securing spring clips inwards.
6 The instrument voltage stabiliser is secured to the rear of the instrument assembly by one screw.
7 The gauges can be removed from the panel after withdrawing their securing screws and the gauge glasses and bezels can be detached if necessary by prising back the bent-over tags.

22 Combined wiper/washer/direction indicator switch

1 These switches are mounted on a single plate which is clamped to the steering column and correctly located by engagement of the plate spigot in a slot in the steering column.
2 To remove the switch assembly, first withdraw the steering wheel and column shrouds. Disconnect the switch multi-socket connectors and then release the clamp screw and remove the switch.
3 The direction indicator, wiper and washer switches, complete with mounting plate and striker bush, are supplied as an assembly and will have to be renewed complete, should a fault develop in the direction indicator components. However, should the wiper or washer switches become faulty, these can be renewed separately after drilling out the original rivets and removing the crosshead screw. Install the new switch using the nuts, bolts and crosshead screw supplied with the switch.

23 Headlamp units - Cibie type

1 This make of headlamp may be encountered as an alternative to the Lucas design. The headlamp mounting plate is secured by plastic inserts riveted to the lamp unit which clip over the tapered ends of the adjustment screws.
2 Bulb renewal is carried out in a similar way to that described in Chapter 10, Section 26 for Lucas type units.

24 Heated rear window

1 A heated rear window is a standard feature on later models.
2 Connections are to an illuminated switch and to earth under a screw within the luggage compartment.
3 The printed circuit type heating element on the rear window glass must be treated carefully, cleaning it only with a soft cloth and water and taking care not to scratch it. If the printed circuit should become damaged, it can be repaired using an electrically conductive paint.

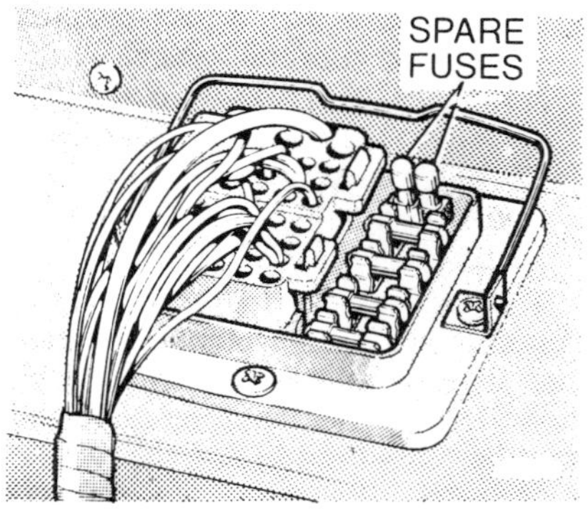

Fig. 13.28. Fuse box

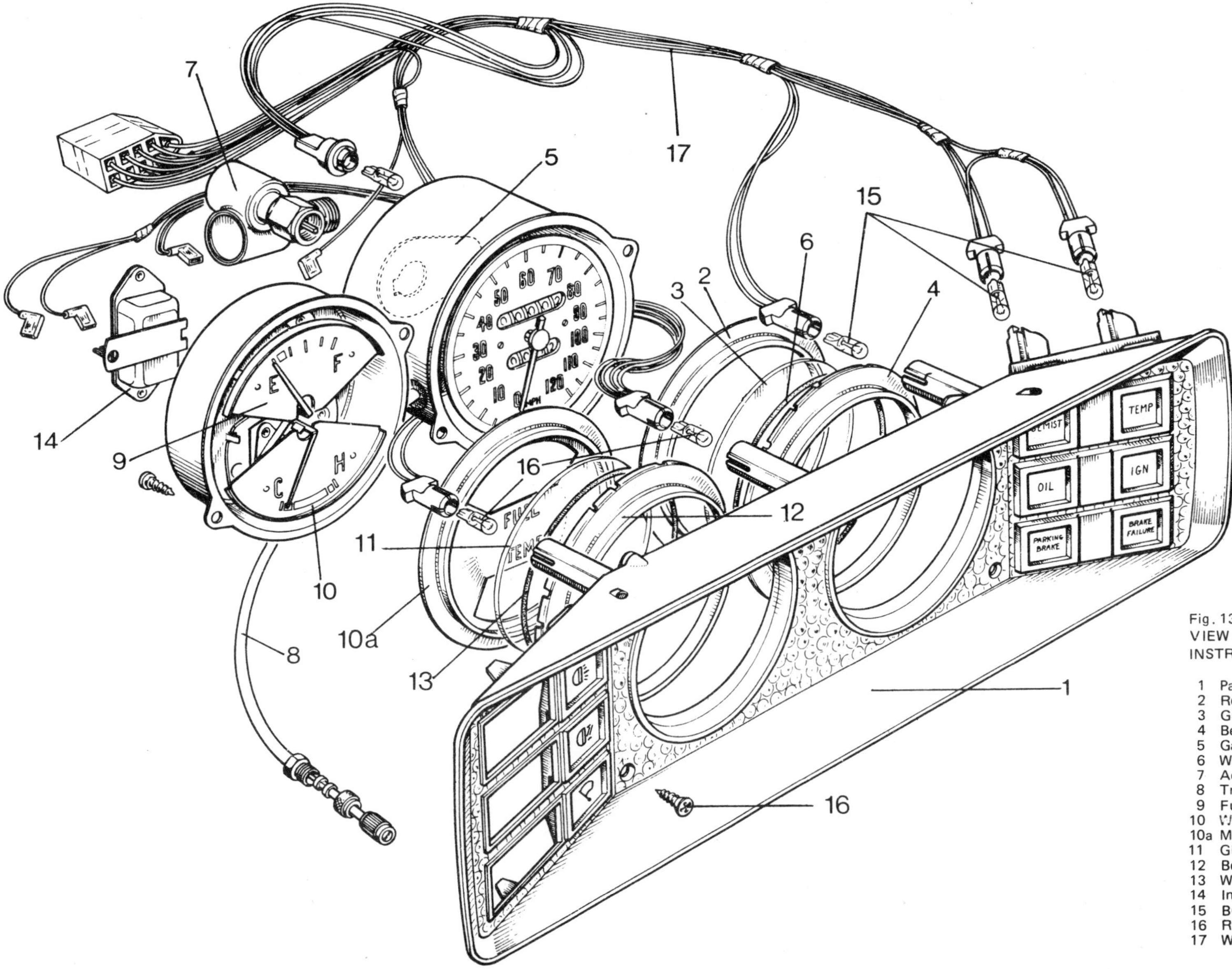

Fig. 13.29. EXPLODED VIEW OF TWIN DIAL INSTRUMENTATION

1 Panel
2 Retainer
3 Glass
4 Bezel
5 Gasket
6 Washer
7 Adaptor
8 Trip recorder cable
9 Fuel gauge
10 Water temperature gauge
10a Mask and retainer
11 Glass
12 Bezel
13 Washer
14 Instrument voltage regulator
15 Bulbs
16 Retaining screw
17 Wiring harness

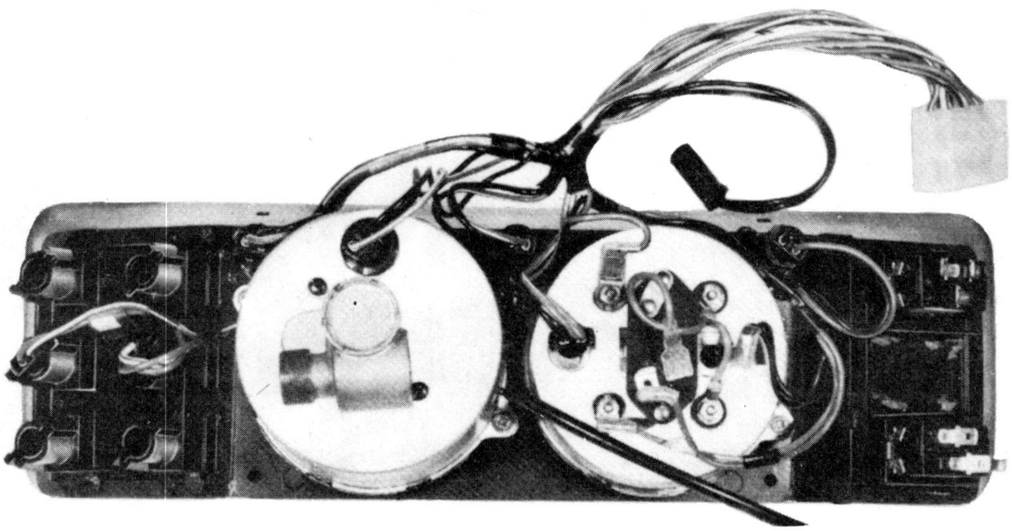

Fig. 13.30. Rear view of twin dial instrument panel

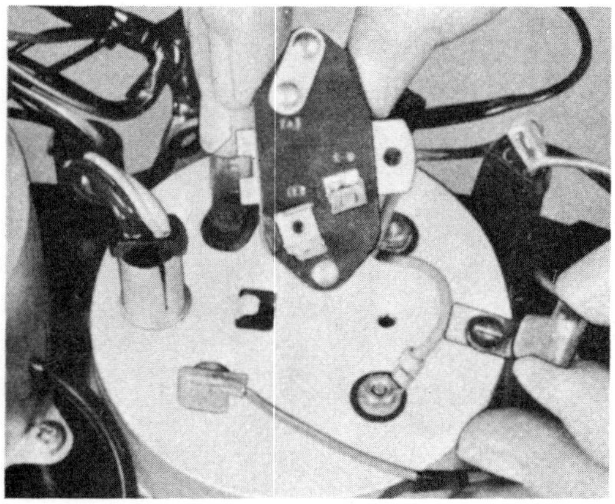

Fig. 13.31. Removing voltage stabiliser (twin dial instrumentation)

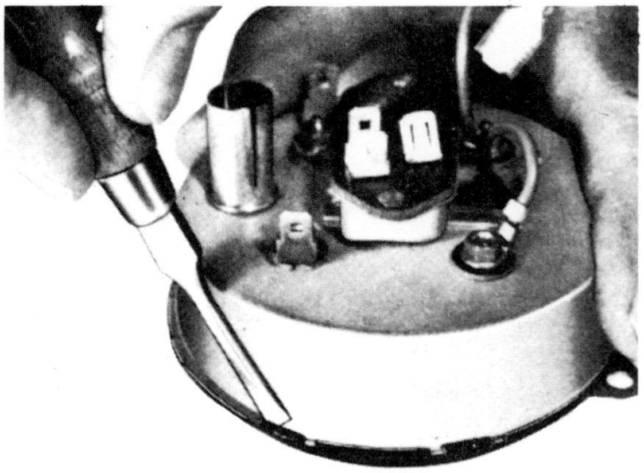

Fig. 13.32. Prising off bezel tag (twin dial instrument)

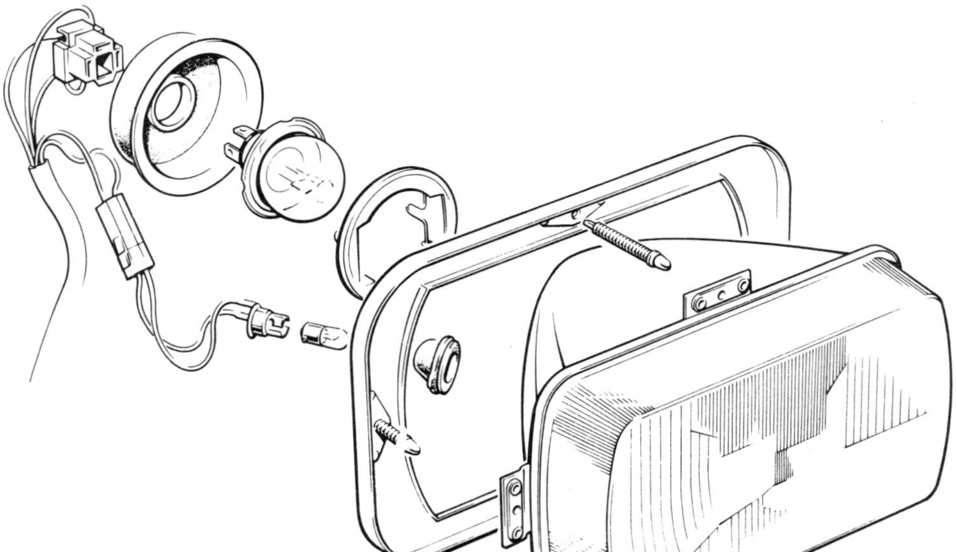

Fig. 13.33. Exploded view of Cibie type headlamp

Suspension and steering

25 Steering column lock

In order to prevent the starter/ignition switch being inadvertently switched off when the car is in motion, a safety button is incorporated which must be depressed before the key can be turned to the 'LOCK' position. The operation therefore calls for the use of two hands and is a valuable safety feature.

Body and underframe

26 Heater water valve

1 In order to ensure a more positive shut off of hot coolant from the engine to the heater when the controls are set to 'COLD', a heater water valve is mounted on the ventilator shroud panel and is cable operated.
2 The connecting hoses run from the cylinder head to the valve and from the valve to the heater matrix inlet pipe.
3 Where a heater water valve is installed, a control panel of modified design is used in which the heat control lever moves upwards to the 'HOT' position.

27 Centre console

1 GLS models have a centre console, removal of which is straightforward after taking out the self-tapping screws.
2 Access to the illumination lamps is either by removing four screws and detaching the insert or by removing the screws and detaching the whole assembly.
3 To gain access to the switches, remove the insert, then remove the appropriate switch(es) by removing two retaining screws.

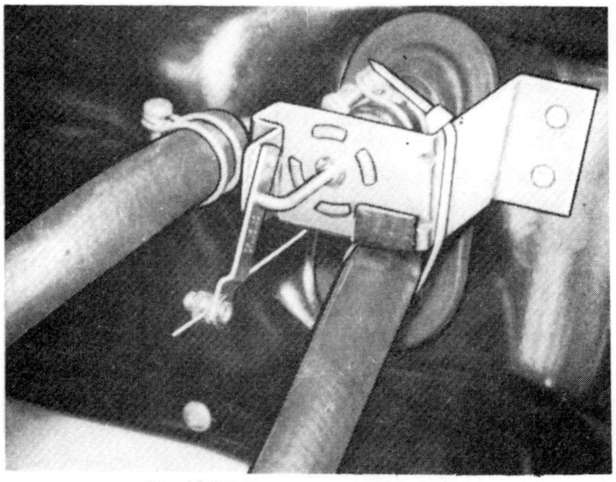

Fig. 13.35. Heater water control valve

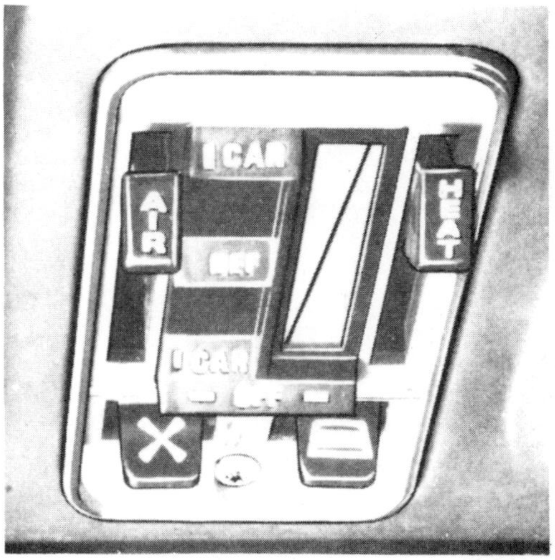

Fig. 13.36. Heater control panel used in conjunction with water valve

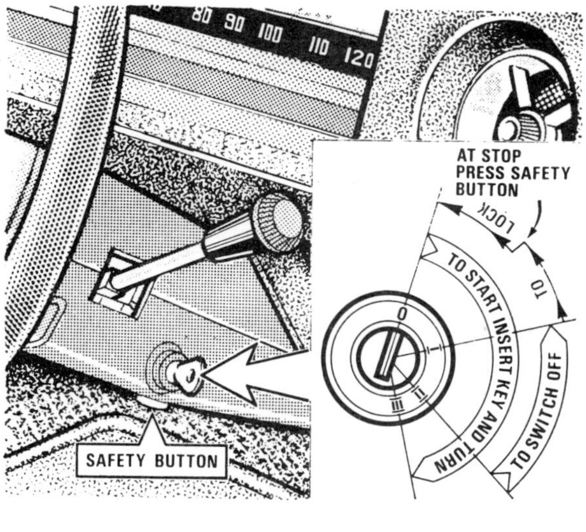

Fig. 13.34. Steering lock incorporating safety button

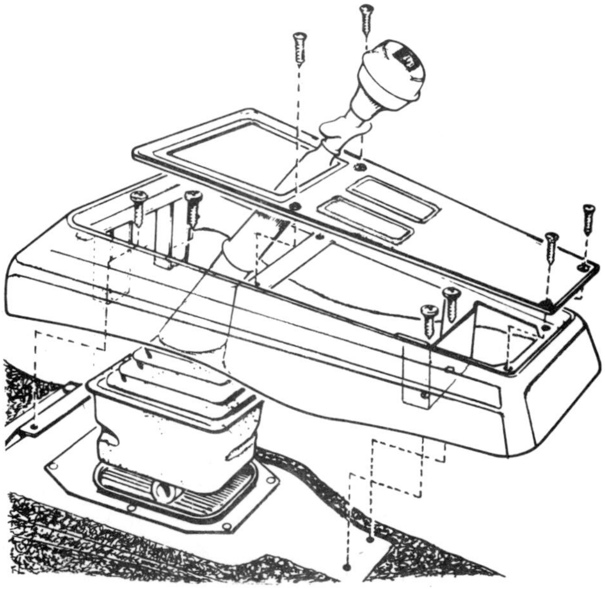

Fig. 13.38. Gear change lever console assembly (typical)

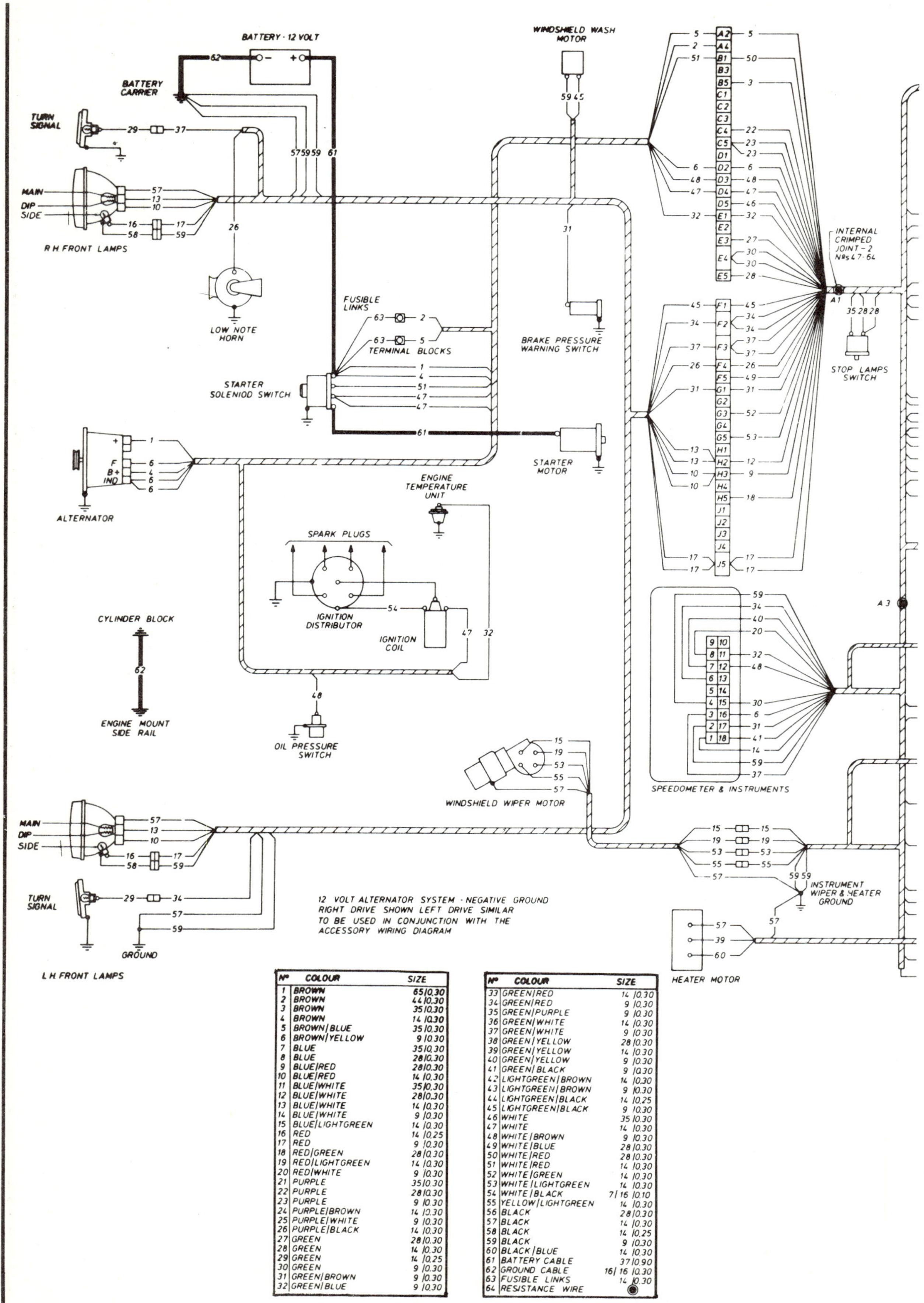

12 VOLT ALTERNATOR SYSTEM - NEGATIVE GROUND
RIGHT DRIVE SHOWN LEFT DRIVE SIMILAR
TO BE USED IN CONJUNCTION WITH THE
ACCESSORY WIRING DIAGRAM

Nº	COLOUR	SIZE
1	BROWN	65/0.30
2	BROWN	44/0.30
3	BROWN	35/0.30
4	BROWN	14/0.30
5	BROWN/BLUE	35/0.30
6	BROWN/YELLOW	9/0.30
7	BLUE	35/0.30
8	BLUE	28/0.30
9	BLUE/RED	28/0.30
10	BLUE/RED	14/0.30
11	BLUE/WHITE	35/0.30
12	BLUE/WHITE	28/0.30
13	BLUE/WHITE	14/0.30
14	BLUE/WHITE	9/0.30
15	BLUE/LIGHTGREEN	14/0.30
16	RED	14/0.25
17	RED	9/0.30
18	RED/GREEN	28/0.30
19	RED/LIGHTGREEN	14/0.30
20	RED/WHITE	9/0.30
21	PURPLE	35/0.30
22	PURPLE	28/0.30
23	PURPLE	9/0.30
24	PURPLE/BROWN	14/0.30
25	PURPLE/WHITE	9/0.30
26	PURPLE/BLACK	14/0.30
27	GREEN	28/0.30
28	GREEN	14/0.30
29	GREEN	14/0.25
30	GREEN	9/0.30
31	GREEN/BROWN	9/0.30
32	GREEN/BLUE	9/0.30

Nº	COLOUR	SIZE
33	GREEN/RED	14/0.30
34	GREEN/RED	9/0.30
35	GREEN/PURPLE	9/0.30
36	GREEN/WHITE	14/0.30
37	GREEN/WHITE	9/0.30
38	GREEN/YELLOW	28/0.30
39	GREEN/YELLOW	14/0.30
40	GREEN/YELLOW	9/0.30
41	GREEN/BLACK	9/0.30
42	LIGHTGREEN/BROWN	14/0.30
43	LIGHTGREEN/BROWN	9/0.30
44	LIGHTGREEN/BLACK	14/0.25
45	LIGHTGREEN/BLACK	9/0.30
46	WHITE	35/0.30
47	WHITE	14/0.30
48	WHITE/BROWN	9/0.30
49	WHITE/BLUE	28/0.30
50	WHITE/RED	28/0.30
51	WHITE/RED	14/0.30
52	WHITE/GREEN	14/0.30
53	WHITE/LIGHTGREEN	14/0.30
54	WHITE/BLACK	7/16/0.10
55	YELLOW/LIGHTGREEN	14/0.30
56	BLACK	28/0.30
57	BLACK	14/0.30
58	BLACK	14/0.25
59	BLACK	9/0.30
60	BLACK/BLUE	14/0.30
61	BATTERY CABLE	37/0.90
62	GROUND CABLE	16/16/0.30
63	FUSIBLE LINKS	14/0.30
64	RESISTANCE WIRE	

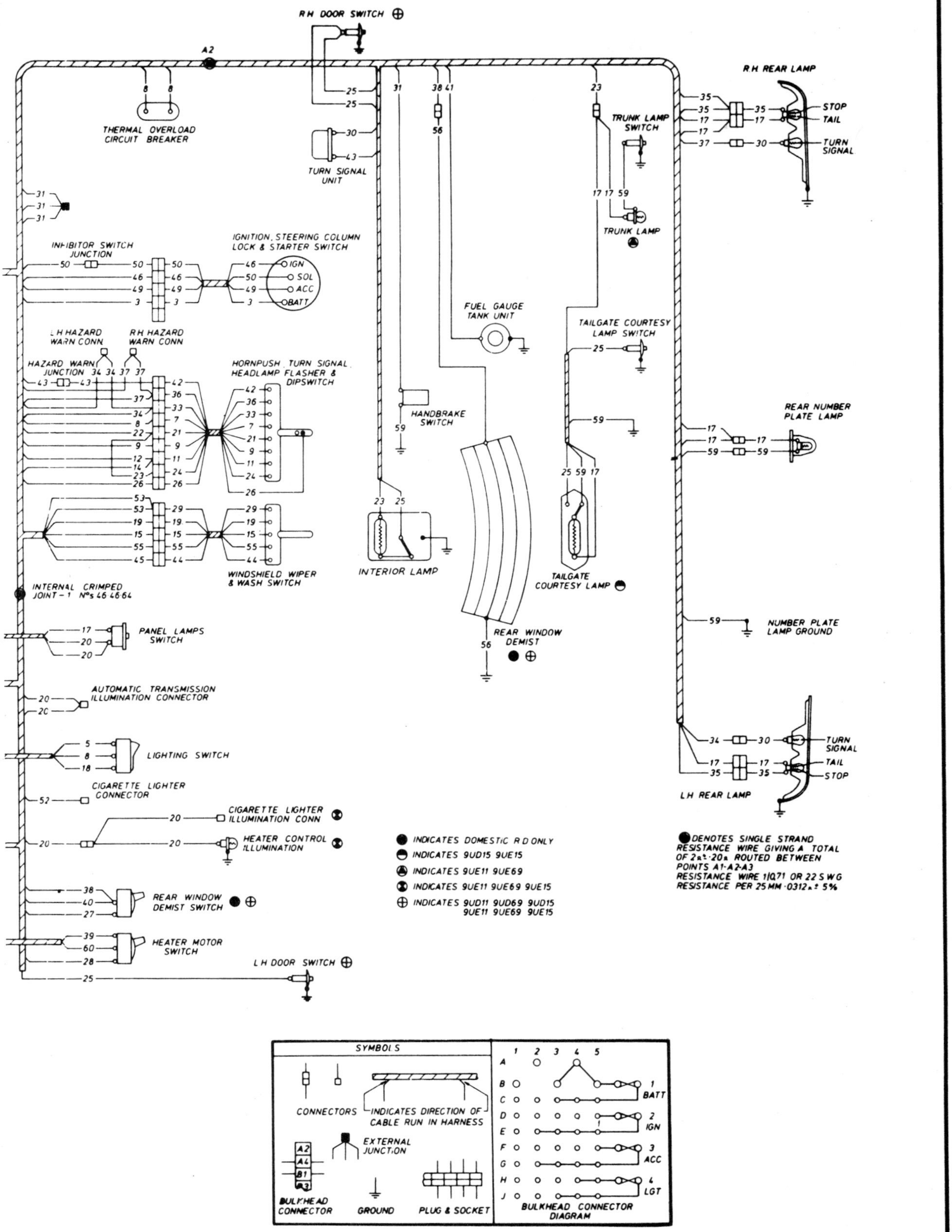

Fig. 13.37. Wiring diagram Viva with standard instrumentation (1974 models)

Metric conversion tables

			Millimetres to Inches		Inches to Millimetres	
Inches	Decimals	Millimetres	mm	Inches	Inches	mm
1/64	0.015625	0.3969	0.01	0.00039	0.001	0.0254
1/32	0.03125	0.7937	0.02	0.00079	0.002	0.0508
3/64	0.046875	1.1906	0.03	0.00118	0.003	0.0762
1/16	0.0625	1.5875	0.04	0.00157	0.004	0.1016
5/64	0.078125	1.9844	0.05	0.00197	0.005	0.1270
3/32	0.09375	2.3812	0.06	0.00236	0.006	0.1524
7/64	0.109375	2.7781	0.07	0.00276	0.007	0.1778
1/8	0.125	3.1750	0.08	0.00315	0.008	0.2032
9/64	0.140625	3.5719	0.09	0.00354	0.009	0.2286
5/32	0.15625	3.9687	0.1	0.00394	0.01	0.254
11/64	0.171875	4.3656	0.2	0.00787	0.02	0.508
3/16	0.1875	4.7625	0.3	0.1181	0.03	0.762
13/64	0.203125	5.1594	0.4	0.01575	0.04	1.016
7/32	0.21875	5.5562	0.5	0.01969	0.05	1.270
15/64	0.234275	5.9531	0.6	0.02362	0.06	1.524
1/4	0.25	6.3500	0.7	0.02756	0.07	1.778
17/64	0.265625	6.7469	0.8	0.3150	0.08	2.032
9/32	0.28125	7.1437	0.9	0.03543	0.09	2.286
19/64	0.296875	7.5406	1	0.03937	0.1	2.54
5/16	0.3125	7.9375	2	0.07874	0.2	5.08
21/64	0.328125	8.3344	3	0.11811	0.3	7.62
11/32	0.34375	8.7312	4	0.15748	0.4	10.16
23/64	0.359375	9.1281	5	0.19685	0.5	12.70
3/8	0.375	9.5250	6	0.23622	0.6	15.24
25/64	0.390625	9.9219	7	0.27559	0.7	17.78
13/32	0.40625	10.3187	8	0.31496	0.8	20.32
27/64	0.421875	10.7156	9	0.35433	0.9	22.86
7/16	0.4375	11.1125	10	0.39270	1	25.4
29/64	0.453125	11.5094	11	0.43307	2	50.8
15/32	0.46875	11.9062	12	0.47244	3	76.2
31/64	0.484375	12.3031	13	0.51181	4	101.6
1/2	0.5	12.7000	14	0.55118	5	127.0
33/64	0.515625	13.0969	15	0.59055	6	152.4
17/32	0.53125	13.4937	16	0.62992	7	177.8
35/64	0.546875	13.8906	17	0.66929	8	203.2
9/16	0.5625	14.2875	18	0.70866	9	228.6
37/64	0.578125	14.6844	19	0.74803	10	254.0
19/32	0.59375	15.0812	20	0.78740	11	279.4
39/64	0.609375	15.4781	21	0.82677	12	304.8
5/8	0.625	15.8750	22	0.86614	13	330.2
41/64	0.640625	16.2719	23	0.90551	14	355.6
21/32	0.65625	16.6687	24	0.94488	15	381.0
43/64	0.671875	17.0656	25	0.98425	16	406.4
11/16	0.6875	17.4625	26	1.02362	17	431.8
45/64	0.703125	17.8594	27	1.06299	18	457.2
23/32	0.71875	18.2562	28	1.10236	19	482.6
47/64	0.734375	18.6531	29	1.14173	20	508.0
3/4	0.75	19.0500	30	1.18110	21	533.4
49/64	0.765625	19.4469	31	1.22047	22	558.8
25/32	0.78125	19.8437	32	1.25984	23	584.2
51/64	0.796875	20.2406	33	1.29921	24	609.6
13/16	0.8125	20.6375	34	1.33858	25	635.0
53/64	0.828125	21.0344	35	1.37795	26	660.4
27/32	0.84375	21.4312	36	1.41732	27	685.8
55/64	0.859375	21.8281	37	1.4567	28	711.2
7/8	0.875	22.2250	38	1.4961	29	736.6
57/64	0.890625	22.6219	39	1.5354	30	762.0
29/32	0.90625	23.0187	40	1.5748	31	787.4
59/64	0.921875	23.4156	41	1.6142	32	812.8
15/16	0.9375	23.8125	42	1.6535	33	838.2
61/64	0.953125	24.2094	43	1.6929	34	863.6
31/32	0.96875	24.6062	44	1.7323	35	889.0
63/64	0.984375	25.0031	45	1.7717	36	914.4

1 Imperial gallon = 8 Imp pints = 1.16 US gallons = 277.42 cu in = 4.5459 litres

1 US gallon = 4 US quarts = 0.862 Imp gallon = 231 cu in = 3.785 litres

1 Litre = 0.2199 Imp gallon = 0.2642 US gallon = 61.0253 cu in = 1000 cc

Miles to Kilometres		Kilometres to Miles	
1	1.61	1	0.62
2	3.22	2	1.24
3	4.83	3	1.86
4	6.44	4	2.49
5	8.05	5	3.11
6	9.66	6	3.73
7	11.27	7	4.35
8	12.88	8	4.97
9	14.48	9	5.59
10	16.09	10	6.21
20	32.19	20	12.43
30	48.28	30	18.64
40	64.37	40	24.85
50	80.47	50	31.07
60	96.56	60	37.28
70	112.65	70	43.50
80	128.75	80	49.71
90	144.84	90	55.92
100	160.93	100	62.14

lb f ft to Kg f m		Kg f m to lb f ft		lb f/in^2 : Kg f/cm^2		Kg f/cm^2 : lb f/in^2	
1	0.138	1	7.233	1	0.07	1	14.22
2	0.276	2	14.466	2	0.14	2	28.50
3	0.414	3	21.699	3	0.21	3	42.67
4	0.553	4	28.932	4	0.28	4	56.89
5	0.691	5	36.165	5	0.35	5	71.12
6	0.829	6	43.398	6	0.42	6	85.34
7	0.967	7	50.631	7	0.49	7	99.56
8	1.106	8	57.864	8	0.56	8	113.79
9	1.244	9	65.097	9	0.63	9	128.00
10	1.382	10	62.330	10	0.70	10	142.23
20	2.765	20	144.660	20	1.41	20	284.47
30	4.147	30	216.990	30	2.11	30	426.70

Index

Printed by
Haynes Publishing Group
Sparkford Yeovil Somerset
England